Implementation of Sensors and Artificial Intelligence for Environmental Hazards Assessment in Urban, Agriculture and Forestry Systems

Implementation of Sensors and Artificial Intelligence for Environmental Hazards Assessment in Urban, Agriculture and Forestry Systems

Editors

Sigfredo Fuentes
Ranjith R Unnithan
Eden Jane Tongson
Nir Lipovetzky

MDPI • Basel • Beijing • Wuhan • Barcelona • Belgrade • Manchester • Tokyo • Cluj • Tianjin

Editors

Sigfredo Fuentes
The University of Melbourne
Australia

Ranjith R Unnithan
The University of Melbourne
Australia

Eden Jane Tongson
The University of Melbourne
Australia

Nir Lipovetzky
The University of Melbourne
Australia

Editorial Office
MDPI
St. Alban-Anlage 66
4052 Basel, Switzerland

This is a reprint of articles from the Special Issue published online in the open access journal *Sensors* (ISSN 1424-8220) (available at: http://www.mdpi.com).

For citation purposes, cite each article independently as indicated on the article page online and as indicated below:

LastName, A.A.; LastName, B.B.; LastName, C.C. Article Title. *Journal Name* **Year**, *Volume Number*, Page Range.

ISBN 978-3-0365-2904-2 (Hbk)
ISBN 978-3-0365-2905-9 (PDF)

Cover image courtesy of Eden Tongson
Modified from Nearmap image from the city of Melbourne, Australia. Downloaded on the 22nd of Nov 2021.

© 2021 by the authors. Articles in this book are Open Access and distributed under the Creative Commons Attribution (CC BY) license, which allows users to download, copy and build upon published articles, as long as the author and publisher are properly credited, which ensures maximum dissemination and a wider impact of our publications.

The book as a whole is distributed by MDPI under the terms and conditions of the Creative Commons license CC BY-NC-ND.

Contents

About the Editors vii

Sigfredo Fuentes and Eden Jane Tongson
Editorial: Special Issue "Implementation of Sensors and Artificial Intelligence for Environmental Hazards Assessment in Urban, Agriculture and Forestry Systems"
Reprinted from: *Sensors* **2021**, *21*, 6383, doi:10.3390/s21196383 1

Adegbite Adesipo, Oluwaseun Fadeyi, Kamil Kuca, Ondrej Krejcar, Petra Maresova, Ali Selamat and Mayowa Adenola
Smart and Climate-Smart Agricultural Trends as Core Aspects of Smart Village Functions
Reprinted from: *Sensors* **2020**, *20*, 5977, doi:10.3390/s20215977 5

Sigfredo Fuentes, Eden Tongson and Claudia Gonzalez Viejo
Urban Green Infrastructure Monitoring Using Remote Sensing from Integrated Visible and Thermal Infrared Cameras Mounted on a Moving Vehicle
Reprinted from: *Sensors* **2021**, *21*, 295, doi:10.3390/s21010295 27

Ivana Lučin, Luka Grbčić, Zoran Čarija and Lado Kranjčević
Machine-Learning Classification of a Number of Contaminant Sources in an Urban Water Network
Reprinted from: *Sensors* **2021**, *21*, 245, doi:10.3390/s21010245 43

Luka Grbčić, Lado Kranjčević, Siniša Družeta and Ivana Lučin
A Machine Learning-Based Algorithm for Water Network Contamination Source Localization
Reprinted from: *Sensors* **2020**, *20*, 2613, doi:10.3390/s20092613 59

Ayub Mohammadi, Sadra Karimzadeh, Khalil Valizadeh Kamran and Masashi Matsuoka
Extraction of Land Information, Future Landscape Changes and Seismic Hazard Assessment: A Case Study of Tabriz, Iran
Reprinted from: *Sensors* **2020**, *20*, 7010, doi:10.3390/s20247010 77

Sigfredo Fuentes, Claudia Gonzalez Viejo, Brendan Cullen, Eden Tongson, Surinder S. Chauhan and Frank R. Dunshea
Artificial Intelligence Applied to a Robotic Dairy Farm to Model Milk Productivity and Quality based on Cow Data and Daily Environmental Parameters
Reprinted from: *Sensors* **2020**, *20*, 2975, doi:10.3390/s20102975 105

Sigfredo Fuentes, Claudia Gonzalez Viejo, Surinder S. Chauhan, Aleena Joy, Eden Tongson and Frank R. Dunshea
Non-Invasive Sheep Biometrics Obtained by Computer Vision Algorithms and Machine Learning Modeling Using Integrated Visible/Infrared Thermal Cameras
Reprinted from: *Sensors* **2020**, *20*, 6334, doi:10.3390/s20216334 117

Sigfredo Fuentes, Vasiliki Summerson, Claudia Gonzalez Viejo, Eden Tongson, Nir Lipovetzky, Kerry L. Wilkinson, Colleen Szeto and Ranjith R. Unnithan
Assessment of Smoke Contamination in Grapevine Berries and Taint in Wines Due to Bushfires Using a Low-Cost E-Nose and an Artificial Intelligence Approach
Reprinted from: *Sensors* **2020**, *20*, 5108, doi:10.3390/s20185108 135

Vasiliki Summerson, Claudia Gonzalez Viejo, Colleen Szeto, Kerry L. Wilkinson, Damir D. Torrico, Alexis Pang, Roberta De Bei and Sigfredo Fuentes
Classification of Smoke Contaminated Cabernet Sauvignon Berries and Leaves Based on Chemical Fingerprinting and Machine Learning Algorithms
Reprinted from: *Sensors* **2020**, *20*, 5099, doi:10.3390/s20185099 . . . . **151**

Sigfredo Fuentes, Eden Tongson, Ranjith R. Unnithan and Claudia Gonzalez Viejo
Early Detection of Aphid Infestation and Insect-Plant Interaction Assessment in Wheat Using a Low-Cost Electronic Nose (E-Nose), Near-Infrared Spectroscopy and Machine Learning Modeling
Reprinted from: *Sensors* **2021**, *21*, 5948, doi:10.3390/s21175948 . . . . **175**

About the Editors

Sigfredo Fuentes is an Associate Professor in Digital Agriculture, Food and Wine Sciences. Professor Fuentes' scientific interests range from climate change impacts on agriculture, development of new computational tools for plant physiology, food, and wine science, new and emerging sensor technology, proximal, short and long-range remote sensing using robots and UAVs, machine learning, and artificial intelligence.

Ranjith R Unnithan is a Research Group Leader and Senior Lecturer at the Department of Electrical and Electronic Engineering at the University of Melbourne. He is also the Director of Sensor Research at Hort-Eye Pty Ltd. His research areas include CMOS image sensors, spectral image sensors, augmented reality (AR) displays, drone-based sensors and applications, electronic sensors for biomedical applications, thermal image cameras, and nanophotonic engineering.

Eden Jane Tongson is a Postdoctoral Fellow in Digital Agriculture, Food and Wine Sciences. Dr. Tongson's research interests are in the areas of genetics and high throughput phenotyping of crops and the implementation of digital tools and machine learning in agriculture and food. She is also a professional scientific illustrator.

Nir Lipovetzky's research interests include AI planning, searching, learning, verification, and intention recognition, with a special focus on how to introduce different approaches to the problem of inference in sequential decision problems, and applications in autonomous systems. He is involved in the development of the Lightweight Automated Planning ToolKiT (LAPKT), which aims to make your life easier if you are involved in the creation, use or extension of basic to advanced automated planners.

Editorial

Editorial: Special Issue "Implementation of Sensors and Artificial Intelligence for Environmental Hazards Assessment in Urban, Agriculture and Forestry Systems"

Sigfredo Fuentes * and Eden Jane Tongson

Digital Agriculture, Food and Wine Research Group, Faculty of Veterinary and Agricultural Sciences, School of Agriculture and Food, The University of Melbourne, Parkville, VIC 3010, Australia; eden.tongson@unimelb.edu.au
* Correspondence: sfuentes@unimelb.edu.au

Citation: Fuentes, S.; Tongson, E.J. Editorial: Special Issue "Implementation of Sensors and Artificial Intelligence for Environmental Hazards Assessment in Urban, Agriculture and Forestry Systems". *Sensors* **2021**, *21*, 6383. https://doi.org/10.3390/s21196383

Received: 14 September 2021
Accepted: 18 September 2021
Published: 24 September 2021

Publisher's Note: MDPI stays neutral with regard to jurisdictional claims in published maps and institutional affiliations.

Copyright: © 2021 by the authors. Licensee MDPI, Basel, Switzerland. This article is an open access article distributed under the terms and conditions of the Creative Commons Attribution (CC BY) license (https://creativecommons.org/licenses/by/4.0/).

Artificial intelligence (AI), together with robotics, sensors, sensor networks, internet of things (IoT) and machine/deep learning modeling, has reached the forefront towards the goal of increased efficiency in a multitude of application and purpose. The development and application of AI requires specific considerations, approaches, and methodologies. This special issue focused on the applications of AI to environmental systems related to hazard assessment in Urban, Agriculture and Forestry. A total of ten papers were published in this special issue, with topics ranging from reviewing the current climate-smart agriculture approaches for smart village development [1] to the integration of visible and infrared thermal cameras for automated urban green infrastructure monitoring on top of moving vehicles [2]; the implementation of machine learning to classify contaminant sources for urban water networks [3]; water network contamination assessment using machine learning in the UK [4]; future landscape changes, seismic and hazard assessment tested in Tabriz, Iran assessed using satellite remote sensing [5]; AI applied to a robotic dairy farm to assess milk productivity and quality traits using meteorological and cow data [6]; AI and computer vision from visible and infrared thermal images to obtain non-invasive biometrics from sheep to assess welfare [7]; the assessment of smoke contamination and smoke taint in wines due to bushfires using a low-cost electronic nose and AI [8]; the classification of smoke contaminated grapevine berries and leaves using chemical fingerprinting and machine learning [9]; and the detection of aphid infestation in wheat plants and insect-plant physiological interactions using low-cost electronic noses, chemical fingerprinting and machine learing [10].

The development of smart villages in Europe requires a framework to secure sustainability based on climate-smart agriculture. As argued by Adesipo et al. [1], these considerations need to be based on advances in technology to increase yield and minimize the farming losses associated with biotic and abiotic stresses. This approach will help for the efficient planning and management of smart villages with smart agriculture. The proposed frameworks will secure the success of these smart-agriculture practices under current and future climate change scenarios, making the system flexible and reactive based on recent smart technological advances related to sensor technologies for automated monitoring, data processing and reporting. Digital technological advances were reported for an automated urban green infrastructure monitoring using integrated visible and infrared thermal cameras in Fuentes et al. [2]. Studied in Melbourne, Australia, this system is a novel assessment method which utilizes moving vehicles as monitoring robots to assess tree by tree growth and water status using computer vision algorithms. It was suggested that this system could be used on public transport to support the city council's management, maintenance and improvement of green infrastructure and as a potential tool to increase urban resiliency to climate change, specifically against the urban heat island effect.

One of the detrimental effects of reduced green infrastructure is the contamination of waterways and water networks. The study by Lučin et al. [3] proposed a classification system based on machine learning models (Neural Network and Random Forest) to predict the number of contaminant injections in the Richmond water supply, UK. This study also proposed that the implementation of these algorithms can be used to run simulations to detect potential contamination risks and nodes with a high probability of contamination, making a management system more predictive than reactive with such vital urban resources. Similar work was conducted by Grbčić et al. [4] to locate contamination sources in water networks with a combination of Artificial Neural Network (ANN) to classify pollution sources. Other types of hazard assessments in urban systems were based on a case study in Tabriz, Iran, by Mohammadi et al. [5]. Using satellite remote sensing to extract land information made it possible to predict landscape changes due to seismic activity with high accuracy ranging from 94 to 96%. These technological advances can be extrapolated to other cities with similar risks.

For agricultural systems, novel digital technologies were applied for farm animal welfare assessment based on weather information and cow data to predict milk productivity and quality through supervised machine learning [6]. The models developed presented high accuracies for correlation models in the range of $R = 0.86$ and $R = 0.87$, respectively. The proposed AI system's automation can be implemented in robotic and conventional dairy farms to respond more efficiently to climatic anomalies, such as cold stress or heat waves, to maintain animal welfare. Heat stress in animal transport has recently been a focus of public concern due to the high mortality of animals transported by sea passing; for example, the Persian Gulf with 50 °C. A high level of heat stress can result in serious health issues to animals and ultimately death. Digital and AI technologies based on integrated visible and infrared thermal cameras were proposed by Fuentes et al. [7] to assess the physiological parameters of sheep in heat stress environments. The proposed models showed high accuracies to monitor the heart rate, respiration rate and skin temperature of animals. These digital technologies could help farmers manage their livestock more efficiently through objective assessments of animal welfare.

Climate change effects include the increased incidence, number and severity of climatic anomalies such as heatwaves and bushfires. These climatic anomalies have a specific impact on viticulture and winemaking, specifically with bushfires producing smoke contamination on leaves and berries, which are later passed to the wine through the fermentation process. These have been investigated in two studies focused on implementing digital technologies and machine learning modeling using low-cost electronic noses [8] and near-infrared spectroscopy to assess the levels of smoke contamination in berries and smoke taint in wines [9]. The models demonstrated high accuracy, showing the good potential of these approaches as practical options for grape-growers. The application of these tools offers an accurate, cost-effective and objective assessment of smoke contamination and taint in wines for efficient management purposes.

Finally, low-cost electronic noses and near-infrared spectroscopy were also implemented to assess the infestation of insects in plants and the insect-plant interaction [10]. This study presented a novel way to sniff aphid infestation in wheat plants and estimate plant physiological parameters using machine learning modeling. Models developed resulted in the high accuracy of monitoring insect numbers, early infestation and physiological parameters such as photosynthesis, transpiration and stomatal conductance, which usually require expensive instrumentation for single leaf measurements. This research also proposed a deployment system using unmanned aerial vehicles (UAV) to increase the spatial and temporal monitoring scales for more efficient assessments.

Funding: This research received no external funding.

Institutional Review Board Statement: Not applicable.

Informed Consent Statement: Not applicable.

Data Availability Statement: Data availability is specified for every published paper.

Conflicts of Interest: The authors declare no conflict of interest.

References

1. Adesipo, A.; Fadeyi, O.; Kuca, K.; Krejcar, O.; Maresova, P.; Selamat, A.; Adenola, M. Smart and Climate-Smart Agricultural Trends as Core Aspects of Smart Village Functions. *Sensors* **2020**, *20*, 5977. [CrossRef] [PubMed]
2. Fuentes, S.; Tongson, E.; Gonzalez Viejo, C. Urban Green Infrastructure Monitoring Using Remote Sensing from Integrated Visible and Thermal Infrared Cameras Mounted on a Moving Vehicle. *Sensors* **2021**, *21*, 295. [CrossRef] [PubMed]
3. Lučin, I.; Grbčić, L.; Čarija, Z.; Kranjčević, L. Machine-learning classification of a number of contaminant sources in an urban water network. *Sensors* **2021**, *21*, 245. [CrossRef] [PubMed]
4. Grbčić, L.; Lučin, I.; Kranjčević, L.; Družeta, S. A machine learning-based algorithm for water network contamination source localization. *Sensors* **2020**, *20*, 2613. [CrossRef] [PubMed]
5. Mohammadi, A.; Karimzadeh, S.; Valizadeh Kamran, K.; Matsuoka, M. Extraction of land information, future landscape changes and seismic hazard assessment: A case study of Tabriz, Iran. *Sensors* **2020**, *20*, 7010. [CrossRef] [PubMed]
6. Fuentes, S.; Gonzalez Viejo, C.; Cullen, B.; Tongson, E.; Chauhan, S.S.; Dunshea, F.R. Artificial Intelligence Applied to a Robotic Dairy Farm to Model Milk Productivity and Quality based on Cow Data and Daily Environmental Parameters. *Sensors* **2020**, *20*, 2975. [CrossRef] [PubMed]
7. Fuentes, S.; Gonzalez Viejo, C.; Chauhan, S.S.; Joy, A.; Tongson, E.; Dunshea, F.R. Non-Invasive Sheep Biometrics Obtained by Computer Vision Algorithms and Machine Learning Modeling Using Integrated Visible/Infrared Thermal Cameras. *Sensors* **2020**, *20*, 6334. [CrossRef] [PubMed]
8. Fuentes, S.; Summerson, V.; Gonzalez Viejo, C.; Tongson, E.; Lipovetzky, N.; Wilkinson, K.L.; Szeto, C.; Unnithan, R.R. Assessment of Smoke Contamination in Grapevine Berries and Taint in Wines Due to Bushfires Using a Low-Cost E-Nose and an Artificial Intelligence Approach. *Sensors* **2020**, *20*, 5108. [CrossRef] [PubMed]
9. Summerson, V.; Gonzalez Viejo, C.; Szeto, C.; Wilkinson, K.L.; Torrico, D.D.; Pang, A.; De Bei, R.; Fuentes, S. Classification of smoke contaminated Cabernet Sauvignon berries and leaves based on chemical fingerprinting and machine learning algorithms. *Sensors* **2020**, *20*, 5099. [CrossRef] [PubMed]
10. Fuentes, S.; Tongson, E.; Unnithan, R.R.; Gonzalez Viejo, C. Early Detection of Aphid Infestation and Insect-Plant Interaction Assessment in Wheat Using a Low-Cost Electronic Nose (E-Nose), Near-Infrared Spectroscopy and Machine Learning Modeling. *Sensors* **2021**, *21*, 5948. [CrossRef] [PubMed]

Review

Smart and Climate-Smart Agricultural Trends as Core Aspects of Smart Village Functions

Adegbite Adesipo [1], Oluwaseun Fadeyi [2,3], Kamil Kuca [3], Ondrej Krejcar [3,4,*], Petra Maresova [5], Ali Selamat [3,4] and Mayowa Adenola [6]

[1] Department of Soil Protection and Recultivation, Brandenburg University of Technology, Konrad-Wachsmann-Alle 6, 03046 Cottbus, Germany; adesiade@b-tu.de
[2] Department of Geology, Faculty of Geography and Geoscience, University of Trier, Universitätsring 15, 54296 Trier, Germany; phar2kind@gmail.com
[3] Center for Basic and Applied Research, Faculty of Informatics and Management, University of Hradec Kralove, Rokitanskeho 62, 50003 Hradec Kralove, Czech Republic; kamil.kuca@uhk.cz (K.K.); aselamat@utm.my (A.S.)
[4] Malaysia Japan International Institute of Technology (MJIIT), Universiti Teknologi Malaysia, Jalan Sultan Yahya Petra, Kuala Lumpur 54100, Malaysia
[5] Department of Economy, Faculty of Informatics and Management, University of Hradec Kralove, Rokitanskeho 62, 500 03 Hradec Kralove, Czech Republic; petra.maresova@uhk.cz
[6] Department of Urban and Regional Planning, School of Environmental Technology, Federal University of Technology, PMB 704, Akure 340252, Nigeria; adenolamt@gmail.com
* Correspondence: ondrej.krejcar@uhk.cz

Received: 1 September 2020; Accepted: 17 October 2020; Published: 22 October 2020

Abstract: Attention has shifted to the development of villages in Europe and other parts of the world with the goal of combating rural–urban migration, and moving toward self-sufficiency in rural areas. This situation has birthed the smart village idea. Smart village initiatives such as those of the European Union is motivating global efforts aimed at improving the live and livelihood of rural dwellers. These initiatives are focused on improving agricultural productivity, among other things, since most of the food we eat are grown in rural areas around the world. Nevertheless, a major challenge faced by proponents of the smart village concept is how to provide a framework for the development of the term, so that this development is tailored towards sustainability. The current work examines the level of progress of climate smart agriculture, and tries to borrow from its ideals, to develop a framework for smart village development. Given the advances in technology, agricultural development that encompasses reduction of farming losses, optimization of agricultural processes for increased yield, as well as prevention, monitoring, and early detection of plant and animal diseases, has now embraced varieties of smart sensor technologies. The implication is that the studies and results generated around the concept of climate smart agriculture can be adopted in planning of villages, and transforming them into smart villages. Hence, we argue that for effective development of the smart village framework, smart agricultural techniques must be prioritized, viz-a-viz other developmental practicalities.

Keywords: smart village; smart agriculture; climate-smart agriculture; technology; sustainability

1. Introduction

The need to develop rural communities in terms of productivity and convenience, so as to curb urban migration has received much attention in the last decade. First, the Institute of Electrical and Electronics Engineers (IEEE), as part of its mission, commenced the installation of solar-powered bulbs in many rural communities worldwide [1]. This was followed in 2016 by the Cork Declaration, agreed amongst 340 representatives of European states towards ensuring that rural communities enjoy better lives. These efforts culminated into the coining of the word "smart village", defined as a community that tries to develop current strength and resources, while making futuristic developmental plans on the basis of technology [2,3]. While there are several thematic areas of priority within the smart village development framework, agriculture is seen as the most important of them all [3]. Furthermore, the need to bridge the digitization gap between cities and villages, is also an important aspect, so that lives and livelihood can be improved. Since a smart village is one that seemingly accepts new technologies, precision agriculture uses ultra-modern techniques for animal and crop production, which saves time and reduces wastage, and meets the requirements of smart villages. This is crucial for the sustainability of smart villages [4]. This is because improved food production and efficient animal management systems must be at par with village development, and must be continually transformed to influence the different aspects of smart villages, in terms of policy and practice [5].

To effectively play its role in smart villages, precision agriculture covers smart and climate smart agriculture (CSA) techniques, and other aspects that are capable of ensuring higher agricultural production output in an environment-friendly manner, provides optimum income for the farmer, and is able to feed a growing population. Many studies showed that these processes can be realized through the adoption of ultra-modern agricultural techniques such as bio and nano technologies [6], IoT and blockchain-based methods [7], and drone technologies [8], among other climate smart ideas. On the basis of this argument, efforts that tend to reduce farming losses, increase yield, as well as monitor, detect, and potentially prevent plant and animal diseases are now being automated, finding growing applications, and offering optimal solutions. Based on the forgone explanations, the current study attempts to establish smart and CSA trends in smart village research, in order to see how much they are useful for smart village development.

The rest of this study is arranged as follows. Section 1.1 draws a foundation for this study, by focusing on the research question. Section 2 briefly builds a background for smart village research by listing existing projects, and describes a few state-of-the-art smart agricultural solutions. In Section 3, attention is drawn to climate-smart agriculture, with specific reference to what makes up the concept, a few challenges in its framework, as well as the latest progress in its development. Section 4 describes the challenges created by the interplay of adopting CSA in smart villages, and also tries to answer the research question. The section also conceptualizes climate-smartness, as it influences sustainable development of smart villages. Finally, Section 5 describes future research directions in smart-village and smart agricultural research, and draws relevant policy recommendations and conclusions

1.1. Research Question

Based on the vast importance of agriculture in smart village development, this study adapts its research question from the editorial note presented by the editors of MDPI's special issue within the Sustainability journal published in August 2018. Within the report, Visvizi and Lytras [9] gave a revealing background of future directions for smart village research. The editors pointed to a few research questions that future smart village research should strive to answer. One of these is: "How will smart and CSA research give account of, and conceptualize transformation and change in the smart village context?"(p. 8) [9]. This question is what the current study modifies and seeks to answer.

2. Related Literature

2.1. Current Smart-Village Projects Around the World

Before delving completely into smart agricultural systems in smart villages, it is important to consider existing smart village initiatives in order to have an updated knowledge of smart village trends, and why smart agriculture might need to be prioritized. Zavratnik et al. [10] described the IEEE smart village project, and the EU smart village initiative, which are further elaborated in subsequent paragraphs.

The IEEE smart village program is one of the most popular today. It has a goal of advancing education in off-grid societies, and fostering sustainability in the entire value chain of the smart village energy sector. Initially taking off as an initiative that seeks to provide community solutions in 2009, the current name was coined 5 years later. The IEEE smart village plan is a global initiative, touching lives in Asia, some parts of North America, and mostly in Africa [1], through the promotion of smart energy production in rural areas, and is mostly financed through fundraising. Major efforts that were developed from the initiative include the so-called SunBlazer II—a movable power base solar station [11]; "Learning beyond the Light Bulb" [12]—a program aimed at training locals on the development and design of off-grid solar electricity panels and fostering its sustainability and scalability. As reported by Zavratnik [10], the program also comes with a remote study event that runs for about nine months, and allows practice exchange amongst involved communities, for knowledge sharing and skill enhancement.

Within the framework of the Consultative Group on International Agricultural Research (CGIAR), several smart-village projects took off around the world. Many of which were funded through international research organizations with clear impacts in areas that were worse hit by climate change [10]. These projects mostly focus on training smallholder farmers on agricultural resilience, through the adoption of practices that support food security [13], so that persons within these affected communities are able to maintain a livelihood through agricultural methods that help decrease GHG emissions. For instance, farmers in the Lower Nyando valleys of Kenya are benefitting from improved agroforestry systems that adopts knowledge of Information and Communication Technology (ICT) [13]. Thanks to the CGAIR initiative, they are able to cultivate cash crops in-between rows of multi-purpose trees, thereby improving soil stability and enrichment. Given the increased demand for trees, several nurseries were developed, adding farmers' incomes, and with women as the highest beneficiaries. The state of Bihar in India also benefitted from the CGIAR's smart village initiative. It previously had soils that were greatly affected by water-logging, but new drainage construction changed the channel of rapidly flowing flood waters, out of the farming areas [14]. This improved system also ensured that underground aquifers were steadily recharged. Improved technological ideas also saw better rainwater harvesting in areas benefitting from the Climate Change Agriculture and Food Security (CCAFS) program. Overall, weather and planting can now be monitored from smartphone applications by the farmers, in order to avoid unwanted losses [15].

The European Union's smart village initiative is by far the most organized and detailed system. Having undergone several fine-tuning, the initiative has improved tremendously since the Cork Declaration 2016 [10]. Notable amongst the goals of the EU smart village drive is agricultural boost, mainly because the rural areas are where European foods are mostly produced [16]. There is also the goal of reduced youth exodus to urban centers [17]. In the first assembly of the newly adopted "Intergroup SMART Villages for Rural Communities", György Mudri, a former Members of European Parliament stressed that smart villages are not only for the development of new infrastructures, but also for building capacity of locals [18]. In response to this statement, The Austrian Chamber of Agriculture commenced online training for about 10,000 farmers, who now have remote access to latest agricultural researches and can subsequently implement such ideas on their farms [19]. There is also the so-called COWOCAT rural initiative, which currently trains youth to commence working in villages [19].

Description of smart village drives of the above initiatives show that agriculture is one of the most prominent aspects of the smart village plans. As a result, this study delves into smart agricultural practices that can the build capacity of smart villages, if adopted.

2.2. Ultra-Modern Smart Agricultural Solutions

While there are varieties of smart technologies adopted in agriculture nowadays, this section focuses only on bio-sensors, agricultural drones, IoT and Blockchain-based sensors, and a number of combined technologies that adopted animal husbandry, as well as in crop, soil, and pest management.

2.3. Nanostructured Biological Sensors

Biological sensing devices are some of the new technological interventions reshaping agricultural systems today, which might be adopted in smart villages. As reported by Antonacci [6], bio-sensors with extremely small structures were found to possess the ability to help in crop maturity evaluation, management of amount of pesticides and fertilizers, as well as detection of humidity levels in soils for effective irrigation. To carry out these functions, bio-sensors rely on the characteristics of nano-materials, such as immobilizing bio-receptors on transducers, integrating and miniaturizing some biological components of plants, transducer systems, and micro-fluids, into very complex plants make-up [20,21]. Although the use of harmful pesticides is gradually being phased out in agricultural systems, less harmful pesticides are still very much in use [22]. In areas known for prior pesticides usage, modern agricultural techniques often aim at detecting pesticide presence, as well as their levels within the soil, before cultivation. To do this, cutting-edge bio-sensors with very high sensitivity (because of their surface-volume ratio), extremely rapid response time, and quick electron-transfer kinetic are utilized. The sensors possess stable strength to map pesticide quantities within soils, and longer lifespan, when compared to the earliest bio-based sensors [23]. Newly improved bio-sensors with extremely small structures are also able to surpass soil pre-treatment, due to the presence of pesticides, herbicides, and fungicides, without losing their potency [6].

Yu et al. [24] developed tyrosinase/TiO2 biosensor to determine the presence of atrazine pesticides. This was done by fabricating a structure through the allowance of vertical growth of TiO_2 nanotubes. This meant that well-arranged nanotubes would provide large surface areas for immobilizing the tyrosinase enzyme. The structure gave room for excellent loading of enzymes, as well as transfer of electrons, which yielded improved system robustness and sensitivity. The system was tested in well-grinded, air-dried paddy soils, gathered at varying depths. The soil also passed through a sieving process using a 1.0 mm filter, and a 35 °C re-drying process that lasted for 48 hrs. It was subsequently mixed with acetone, prior to undergoing shaking at a temperature of 25 °C for 60 min. Results given by [24] showed that after carrying out analysis of supernatants, atrazine was observed to be present in 0.2 ppt to 2 part-per-billion. Standard deviation was subsequently found to be below 0.05 ppt when compared to high performance liquid chromatography (HPLC).

Dong et al. [25] introduced a novel nano-structured bio-sensor technique for detecting very low pesticide traces in soil. The technique works by electrochemically reducing Ellman's reagent via the inhibition of acetylcholinesterase. This bio-sensor adopts amperometric, designed to immobilize acetylcholinesterase on multiple walls of carbon-type nanotubes-chitosan nanocomposites modified glassy carbon electrode. High sensitivity of the system is offered by the very good conductivity and biological compatibility of multiple walls of carbon-type nanotubes-chitosan [25]. This can be additionally improved by electrochemically reducing 5,5-dithiobis (2-nitrobenzoic) acid. In testing the system, methyl parathion pesticide was observed to exhibit an inhibitive effect on acetylcholinesterase. An electrochemical change in the reduction response of 5,5-dithiobis (2-nitrobenzoic) acid was also observed. Overall, the system was found to possess a pesticide detection precision of 7.5×10^{-13} M when tested on spiked soil.

In another nano structured bio-sensor study, Shi et al. [26] observed the presence of soil acetamiprid using SELEX; a new 20 mer bio-sensing unit that is able to bind acetamiprid, using aptamer made of nanoparticles of gold. The unit works to detect the pesticide optically at values ranging from 75 nM to 7.5 µM. It bears the combined characteristics of a nanomaterial, and those of artificial molecules. Tested soils were collected around Tongji University, China, with initial air-drying carried out before the sample was grinded, to allow 1.0 mm sieving. A second drying was also done using an oven at 35 °C for 2 days, prior to acetone mixing and shaking at 25 °C for 60 min. Dichloromethane was subsequently added to the mixture, and then removed ultrasonically before the sample was filtered.

Beyond sensing pesticides within soils, bio-sensors with very tiny structures were also employed in monitoring diseases of crop plants. A very important aspect of any smart village is the effective management of farm economy, achievable through the protection of crops against diseases. Quantum dots offer classical examples of materials that are useful for monitoring plant diseases, as they possess broad excitation spectra. Safarpour et al. [27] identified the vector responsible for sugar beet's yellow vein and Rhizomania disease like Polymyxa betae. This was detected using quantum dot techniques that subjected the plant root sap samples to several pre-treatment in order to extract the virus. The quantum dots unit utilizes Förster Resonance Energy Transfer (FRET) modeling in its detection operation [27]. By using a similar technology, Bakhori et al. [28] detected synthetic oligonucleotide of Ganoderma boninense. However, this work employed adjusted quantum dots with carboxylic groups that are then conjugated using a DNA probe. This gave rise to an improved sensitivity of the system, yielding 3.55×10^{-9} M as the detection limit [28].

By adopting bio-sensors in the detection of soil nutrients and fertilizers, Ali et al. [29] revealed that soil nitrates can be detected using a system that relies on microfluidic impedimetric sensing. The unit works by adopting nano-sheets of graphene oxide and the nanofibers of the so-called poly (3,4-ethylenedioxythiophene). The researchers showed that poly (3,4-ethylenedioxythiophene) composite can bear the enzyme; nitrate reductase, and also measure the amount of nitrate ions in soil samples on which sweet corn was cultivated. This is done at 0.44-442 mg/L concentration, so the detection limit was 0.135 mg/L [29]. Carrying out the procedure, however, involves sample drying at 105 °C, and subsequent nitrate extraction through the addition of 2 M KCl solution. The mixture was shaken for 60 min, and filtered using Whatman filter paper. Finally, sample extraction was kept in a syringe for infusion into the experimental device.

Several other research examples exist for bio-sensor utilization in agricultural work. Nevertheless, a summary of some state-of-the art techniques, some of which are already described elsewhere, is presented in Table 1.

Table 1. Summary of some biological sensing techniques for soils and plants (Adapted from Antonacci [6]).

Group	Analyte	Bio-Sensing Method	Conversion	Nanomaterial Media	Detection Limit/Time	Reference
Herbicide	Soil glyphosate; soil glufosinate	Specified dual polymers with imprinted template	Anodic stripping voltammetry done with differential pulse that makes use of nanoparticles of gold adjusted pencil graphite electrode	Nanotubes with multi-walled carbon	0.35 ng mL^{-1}; 0.19 ng mL^{-1}	[30]
	Soil atrazine	Tyrosinase inhibition	Utilizing an amperometric analysis adopts a conventional 3 electrode cell	Nanotubes of titanium dioxide	0.1 ppt (approx. 600 s)	[24]
Fungus/ Fungicide	Trichoderma harzianum present within the soil	DNA probe in a single strand	Electrochemical analysis that utilizes an electrode made from gold	Nanoparticles of zinc oxide- chitosan nanocomposite membrane	1 × 10^{-19} mol/L (600 s)	[31]
Fertilizer & Nutrient	Soil nitrates	Polypyrrole electrode that is in solid state, and easily selects ions	Experimenting a potentiometric analysis through the use of adjusted glass carbon	Oxide of graphene	0.00001 M (≤15 s)	[32]
	Soil nitrates	Reduction of nitrate	Carrying out an impedimetric analysis via the use of a gold electrode	Nano-fibers of poly(3,4-ethylenedioxythiophene) polystyrene sulfonate - nanosheets composite derived from graphene oxide	0.68 mg/L (few hundreds of seconds)	[29]
	Soil urease; Soil urea	Nanoparticles of gold is adopted as catalyst, acting like horseradish peroxidase	pH indicator; Colorimetric;	Nanoparticles of gold	1.8 U/L (600 s); 5 µM (600 s)	[33]
Disease	Ganoderma boninense (synthetic DNA)	DNA probe	Transfer of energy through fluorescence resonance	Quantum dots	3.55 × 10^{-9} M (600 s)	[28]
	Sweet corn seed: Pantoea stewartii sbusp. Stewartii NCPPB 449	Immuno-sensor	Immunosorbent assay linked to enzyme	Nanoparticles of gold	7.8 × 10^3 cfu/mL (below 1800 s)	[34]
Virus	For orchid plant: Odontogloss um ringspot virus; Cymbidium mosaic virus;	plasmon resonance of particle; Fiber optic	Utilizing nano-rods made of gold as sensing device (Immuno-sensor)	Nano-rods made of gold	42 pg/mL (600 s) 48 pg/mL (600 s)	[35]
Pesticide	Soil acetamiprid	Affinity with 20mer specific aptamer	Carrying out an colorimetric analysis	Nanoparticles of gold	5 nM (300 s)	[26]
	Soil methyl parathion	Acetylcholinesterase inhibition	Adopting adjusted glassy electrode of carbon to cause voltametric differential pulse	Nanotubes with multi-walled carbon -chitosan nanocomposites	7.5 × 10^{-13} M (2 s)	[25]

2.4. Drone Technologies (Unmanned Aerial Vehicles)

Unmanned Aerial Vehicles (UAV), also known as drones, have become popular in agricultural production work. In a review study by Mogili [8], the researchers reported that drones can be used in pesticide and fertilizer application, so that humans do not come in contact with the some of these pesticides, which are harmful, and are gradually being phased out. Drones can also function as water sprinkling systems [8,36].

Primicerio et al. [37] adopted VIPtero, a UAV for managing a vineyard in an experimental set-up in Italy. The system, which is made up of an aerial platform with six rotors and a camera, can fly in a self-governed manner to a particular point in the air, in order to take measurement of the vegetation canopy reflectance. Prior to flight take-off, accuracy of the camera is evaluated in relation to ground-based measurements with high resolution, which were gathered using field spectrometer. Subsequently, VIPtero gets air-bound in the vineyard, and gathers as many as 63 multi-spectral images in a 600 seconds time period. The recorded images are analyzed and classified, prior to the production of vigor maps on normalized difference vegetation index. Results showed the heterogeneity conditions of the crops, implying that they were in line with those gathered using the ground-based spectrometer [37]. This smart system appears to be promising as an effective and detailed data gathering system in agriculture, and can be adopted over larger areas in smart villages.

In another UAV based research, Burgos et al. [38] used a 4 cm Sensefly Swinglet UAV to differentiate green cover from grape canopy. A digital surface model (DSM) with 3 dimensions was adopted to create an exact digital terrain models (DTM), acquired via the use of processing libraries of python, and subsequently subtracted from DSM, so as to arrive at a differential digital model (DDM) for the measured terrain (a vineyard). Vine pixels within the DDM were obtained by selection of pixels >50 cm elevation from the ground. The results indicated that there is a possibility of separating vine row pixels from green cover pixels, as a differential digital model pointed to values ranging from –0.1 m to +1.5 m. Furthermore, manual polygon delineation, which depended on an RGB image of the vine rows and green cover, revealed huge differences averaging 1.23 m and 0.08 m for vine and ground, respectively. Elevation of the vine rows was good and tallied with its topping height of 1.35 m from the field [38]. The authors noted that vine pixels extraction would aid future analyses, such as pixels' supervised classification.

Berni et al. [39] also demonstrated the possibility of generating remotely sensed data over an agricultural field, using a UAV that had a relatively cheap narrowband and thermal multispectral imaging sensors of 20 cm and 40 cm resolutions, respectively. The system gave rise to surface reflectance and temperature data, after adapting MODTRAN-based atmospheric correction. Biophysical parameter estimation was carried out using a number of vegetation indices, leading to the production and validation of chlorophyll content, detection of water stress from PRI index, as well as the temperature of the canopy. These results showed that the system yielded the same results as the conventional, expensive, and risky manned airborne sensors.

2.5. IoT-based Sensors with Complimentary Blockchain Technology

Many villages face severe agricultural challenges and that require upgrading to smart agriculture, which offers a wide range of state-of-the-art solutions. For example, in villages where access to water is a challenge, Khoa [7] maintained that IoT-based sensors can be useful in water-management irrigation systems on large rural farms. In their research, the authors developed a novel system that is able to monitor soil water level and schedule sprinkling/spraying times in well-calculated amounts. This relatively cheap technique, functions by receiving real-time data from sensors fixed within strategically arranged tunnels, in and around the farm. Based on the information supplied by sensors, which can be received through a mobile phone application, the user might decide to water the farm. Subsequently, when soil water level increases to an optimal level, the system notifies the user, who can remotely or manually switch-off the water-pumps. A unique feature of this system is its usability in up to two farms [7]. In a similar study, Nagpure et al. [40] described another IoT-based system that works by using a similar routine as [7]. However, two differences include; scaring animals away using current pulses, and wireless sensor monitoring of the ecological conditions (e.g., altitude and humidity) to ascertain the amount of irrigation water needed each time [40], which the latter unit possesses.

Mat et al. [41] presented an IoT-based mushroom cultivation, which produced a better yield when compared to the conventional system. This tool is based on an automated sensors for fertilizer application and water sprinkling on the farm, which can be controlled from the farmer's mobile phone or manually, from a centralized point within the farm. The system ensures that the timing for wetting the crop is strictly adhered to, so that the farming operation can progress even without the farmer. Overall, it was observed that average mushroom size in thickness and weight exceeded conventional cultivation by 0.3 cm and 5 gram, respectively.

As reported by Prathibha [42], it is important to curb the effect of environmental conditions on crop yield output. To do this, an efficient measurement method of the elements of weather might be required. Prathibha's research, therefore, proposed a CC3200 combined sensor unit, which comprises a processor for network, a micro-controller, a Wi-Fi unit, a camera, as well as temperature and humidity sensors. This weather utility device comes as a portable unit with low power consumption for longer battery-life. The system monitors temperature (using a thermopile sensor that uses infrared technology) and humidity across the agricultural field, which are subsequently processed as camera images and sent via Wi-Fi to the farmer's mobile phone as multi-media messages. Information of this nature helps the farmer to know how good the soil water is to support the grown crop. A similar study [43] designed another unit that can also send immediate signals to a farmer, after recording real-time data on weather, in and around the farm. This unit was made up of a breadboard, a combination of sensors that can monitor UV Index/IR/Visible light (SI1145 Digital Sensor), soil moisture content, humidity, temperature (DHT11), an ESP32s Node MCU, all of which are connected to a monitor, which is in turn linked remotely to the famer's mobile phone, using an LED visual alert and Blynk mobile phone application. Two very special characteristics of this unit are; its ability to save power in sleep mode, for a battery life that averages 10 days, and the speed of sending signal (180 s) [43].

By using a pre-coded algorithm, also known as the "*Cuckoo Search Algorithm*" [44], a framework for automated watering of a piece of farmland was designed. Based on pre-analysis of different kinds of soils, the researchers found that a soil moisture value of 700 meant dry, and would require immediate watering. The IoT sensor was, therefore, designed on the basis of this information, comprising the so-called ThingSpeak, which also gives direction on the most suitable soil type for a specific crop. A temperature-based sensor was initially used on the soil, and the result was sent to a converter called Arduino. Depending on the measured value, the Arduino was connected to an automated watering system, which could be controlled from the mobile device handled by the farmer. When soil wetting gets to optimal levels, the soil sensor sends a feedback signal to the farmer who then stops the watering [44].

While IoT-based sensing techniques are available for improving precision agriculture, optimizing these ideas with blockchain technologies might offer even more robust results. Patil et al. [45] noted that IoT-based sensing technologies might sometimes be flawed on the grounds of; extremely large scale, a lack of homogeneity of different IoT-based sensing operations, as well as standardization. This meant that the data gathered using IoT-based sensing technologies comes with privacy concerns for the farmer [45]. Hence, the researchers developed a blockchain greenhouse farming tool to cater to security and privacy. The model is made up of a smart greenhouse (a covered piece of farmland protected from environmental conditions) that comprises a series of sensors and actuators, smart hub (a local blockchain which manages the connectivity of all sensors and equipment in the smart greenhouse); an overlay network connection that manages the nodes; a cloud storage platform and an end-user platform. A system of this nature addresses security challenges across all fronts within the farm [45]. In another Blockchain-based smart agricultural study [46], a traceability platform for food safety was designed. In collaboration with the Internet of Things, the system involved "Enterprise Resource

Planning" where farmers, processing plants, and organizations involved in the logistics of agricultural and food products, and the consumers, can assess on their mobile phones, a blockchain node that gives detailed description of how the products were cultivated, harvested, stored, processed, and sold [46]. The essence of this technique was to build virtual trust in food processing, using the so-called *Trusted Trade Blockchain Network Cloud Platform (TTBNCP)*

2.6. Smart Animal Production, Management, and Monitoring

The use of machine vision in body condition scoring of dairy received extensive research attention in the last few decades. Fox et al. [47] listed animal nutrition, insemination, and health as core reasons for body condition monitoring. When this process is carried out by the farmer or veterinary doctor (human monitoring), there is a possibility for biased data gathering, due to the individual's mental state, level of experience, and residual knowledge [48]. Furthermore, the process might also be time-consuming. Improvement in body condition scoring in the 1970s employed ultrasounds [49], which was flawed on the ground that mastering the collection of reliable ultrasonic body condition scoring required more time in comparison to data gathering by humans [50]. Additionally, the cost of purchasing the ultrasound device, and hiring an expert, made the process too expensive. This led to camera-recording of animals applied to body condition monitoring, based on the belief that this would yield better results [48]. Fourier descriptor cameras [51], thermal [52], and RGB cameras [53] were adopted. Nevertheless, images could not be processed automatically until seven years ago [54]. As a step in the right direction for body condition scoring, Spoliansky et al. [55] used a 3D camera that was well-equipped to carry out automatic image processing, leading to the development of effective and unbiased collection of body condition scoring, in 2017 [55]. The system provides real-time data, useful for commercial milking purposes, and genetic evaluation (based on lactation). In addition, automated image gathering of body condition scoring might provide ease of monitoring when there are more than one animal. This is pivotal to early warning signs for morphological changes in animal body size [48].

Yanmaz et al. [56] suggested adopting thermography in the early detection of lameness in horses. Two types of thermography are—contact and contactless types. Contactless thermography has higher precisions via infrared radiation. Internal temperature of the affected animal part can be viewed on a medical thermogram, so that treatment can be planned early. Similarly, temperature around a sick cattle's gluteal region differs significantly from those of other parts when studied using thermal infrared scanning [57]. As reported by Steensels et al. [58,59], temperature management in poultry, as well as early mastitis identification are some other areas where thermal scanning was reported to be useful.

Accelerometers exist nowadays for remote measurement of animal gait [60]. This can determine when the animal is lying or walking [48]. The method is also useful in the determination of lameness, animal sickness, or how much the animal feeds [61], based on the distance covered by the animal against time. The device can be mounted on the leg, ear or other parts of the animal, so that it continuously sends signals to the farmers' mobile phone. When the animal is lying down, the instrument automatically changes its processing speed. The device was also utilized in monitoring the health of fishes, by attaching it to their fins [62]. A demerit of the system, however, is the fact that continuous processing of mobility data tends to rapidly reduce battery life [63].

Wearable belts that are able to tap animal sweat and measure the amount of sodium it contains are some of the latest smart technologies in animal husbandry. In the work of Glennon et al. [64], the authors developed a smart technique for quick detection of the sodium content of sweat. The unit appears in two ways, one that resembles a vertically placed watch and can be worn round an animal leg, and the other that looks like a horizontal pod. Both come with a Velcro strap that can be used to attach it to the animal skin. The systems, through capillary action, receive sweat via its orifice, and send it through an electrode that is sensitive to sodium. The electrode in turn sends it into a storage section containing an adsorbent substance. Sweat flow rate can be improved by varying the width of the sweat flow channel in-between electrode and the storage section of the system. Sweat flow rate generally decreases with decreased width of the flow channel. This also determines the length of time the system will be used before the electrode and the adsorbent material are changed. Stored sweat is available for measurement as total harvested sweat volume as well as its sodium concentration per time. Electrode signals are moved to an electronic board that possesses high input impedance for voltage capturing. Results of this analysis are sent to a remote base station, which is either a laptop, or mobile phone, using bluetooth technology for onward visualization and possible storage.

In livestock management [65] a pregnancy detection method was developed, which makes use of Xbee transmitters linked to LM35 temperature sensors. Two animals were experimented, with temperatures recorded at five days and twelve days from insemination. The sensing system was attached to the tail of the cows. Temperature records were found to be high in pregnant cow, especially in the evenings. The sensing unit works effectively within a distance of 40 m, and serves as a low-cost technique, when compared to some invasive pregnancy detection methods in livestock.

In an ongoing study aimed at finding a sensor that is able to detect the level of progesterone in milk fed to cattle [66], interdigital sensors are used, which are able to yield single side access to the substance being tested. Within the study, progesterone hormone of about 20 mg was allowed to dissolve in 0.5 mL of approximately 100 per cent ethanol. The solution was subsequently poured into about 1000 mL of pure Milli-Q water, so that a stock solution was achieved. Successive mixture dilution led to 0.02 ng/mL progesterone concentration. The sensor offered different results at varying progesterone concentration with sensitivity in the pitch range of 50 µm [66].

Infectious coughs in piggery need rapid detection and treatment. To detect this kind of cough, Ferrari et al. [67] fixed 1 m multi-directional microphones of 50 to 16,000 Hz around a farm. The microphone was connected to a laptop, and animal cough sound patterns were recorded, digitized, and analyzed using Matlab 7. Having earlier injected healthy animals with citric acid, acoustic parameters such as time difference between coughs, peak frequency, and root mean square were used to differentiate coughs from a healthy pig from those of infected ones. While healthy pigs relaxed for about 52 s after each cough attack, the infected pigs coughed after every 37 s. Peak frequency for infected and non-infected pigs was observed to be 1600 Hz and 600 Hz, respectively.

3. Moving towards Climate-Smart Agriculture

Having established in Section 2.1 that agriculture is one of the most important factors to be considered in smart village development, it is crucial to stress that climate change is a major stressor for agricultural development of rural communities [68]. The implication is that developing an agriculturally-smart village entails accepting the concept of climate-smart agriculture. Agricultural risk posed by climate is a threat to food security. As a result, there is an urgent need to effectively manage agricultural production, while fighting climate change through adaptation, resilience and mitigation [69]. This is what climate-smart agriculture offers.

There is currently no unified definition for climate-smart agriculture (CSA). In fact, almost every new study within the framework of smart-agriculture, views CSA in a slightly unique way. Nevertheless, to build a strong foundation for climate-smart agricultural framework in smart village development, the current study adopts existing knowledge and definitions, to coin a new and more robust definition for the term. Table 2 presents some definitions put forward by climate change and agricultural scholars and research organizations. Keywords derived from the definitions show that each has one or more shortcomings. As a result, it might be difficult to build the concept of smart village on a definition that lacks one or more fundamental aspects.

Given the definitions in Table 2, considerable aspects of climate-smart agriculture include; capacity building, sustainability, emission reduction, vulnerability reduction, profit, food security, transformation, new knowledge, new technology, and productivity. By linking the above keywords together, we define climate-smart agriculture as a "transformative and sustainable kind of agriculture that tries to increase efficiency (productivity) in food security and production systems, using a combination of the pillars of climate change (adaptation, resilience, and mitigation) as well as smart and new technological knowledge, that do not only build capacity of farmers' in terms of farming techniques, but also increase profit, reduces vulnerability of the systems as well as their results (farm products/animals), through the reduction of GHG emissions."

Table 2. Definitions of CSA.

Definition	Keywords	Reference
The combination of activities that helps to: build adaptive measures that increase productivity, increase resilience to stresses posed by climatic change, and reduce GHG emissions.	Capacity building; emission reduction	[70]
A sustainable method through which improved productivity and income is achieved in agricultural production via the adoption of adaptation, resilience and GHG emissions mitigation	Sustainability; Emission reduction; productivity; profit; capacity building	[71]
Processes that transform agricultural systems to boost food security, given current changes in climate	Productivity; transformation; food security	[68]
A system of agriculture that supports emission reduction while creating improved productivity profits, nonetheless reducing vulnerability	Vulnerability reduction; emission reduction; profit growth	[72]
A system of agriculture that improves production in a sustainable manner, while building capacity to ward-off agricultural and climate change challenges	Sustainability; capacity building; productivity	[73]
Strategies that are able to curb agricultural challenges through the increment of resilience activities to extreme weather conditions, building adaptive capacities to climate change and mitigating agriculture-based GHG emission increase.	Capacity building; emission reduction.	[74]
Practices that add to improved food security globally, and further enable farmers to effectively adapt to the incidence of climate change and global emission levels	Capacity building; emission reduction; food security	[75]
Combined use of ultramodern technologies and processes that work together to boost farming productivity and incomes, while increasing the farm's and farmers' ability to manage climate change through GHG emission reduction.	New technology adoption; productivity, profit; capacity development; emission reduction	[76]
A technique that combines a number of sustainable techniques to fight particular climate challenges within a specified farming area	Sustainability; GHG emission reduction	[77]
An agricultural framework that tries to develop and adopt technique that will improve rural livelihoods, food security, and facilitate adaptation to climate change, while also providing mitigation benefits	New knowledge; food security; capacity building.	[78]

While it can be argued that the list of keywords suggested within the current study is not exhaustive, many other definitions tend to be built around at least one of these keywords. Figure 1 is a diagrammatical representation of the main aspects of climate-smart agriculture for which it stands as a significant part of a smart village. The implication of the above expository listing of the fundamental parts of climate-smart agriculture means that for a smart village to be so called, it must strive to maintain within its agricultural systems all different aspects of CSA. Furthermore, other aspects of the smart culture within the smart village setting; smart energy management, smart living and smart healthcare, etc., must tap from these fundamental attributes of CSA, in order to provide robust services in their smart village functions.

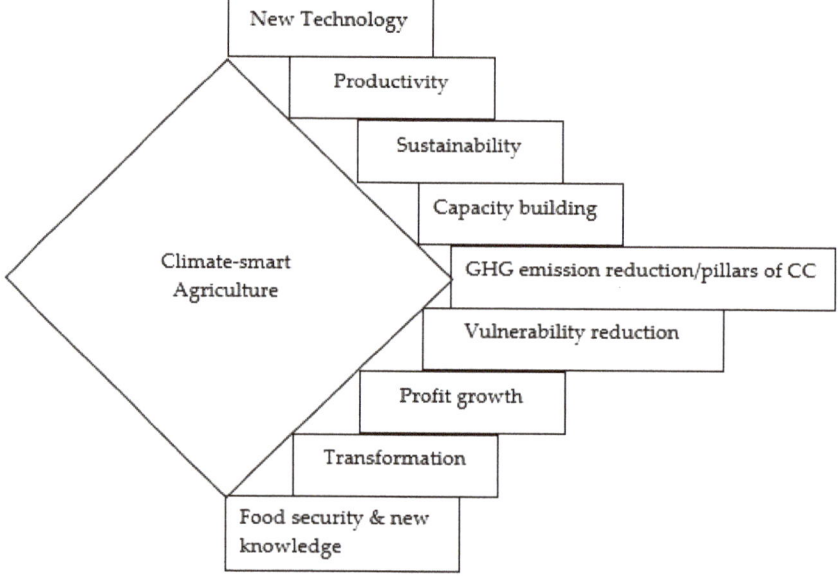

Figure 1. Key aspects of climate-smart agriculture (CSA).

In demonstrating whether CSA could increase rice yield in China, Xiong et al. [79] used crop simulation models; version 0810 of the Environmental Policy Integrated Climate (EPIC) model [80], and version 4.0 of the so-called DSSAT, an acronym for Decision Support System for Agro-technology Transfer [81], respectively. It was observed that these software simulations that gave ideas on cultivar improvement and optimization of management practices for rice due to climate change, led to increased rice production. The EPIC models specifically yielded over 2000 kgha^{-1} during the 30-year period under review [79].

Rural African farmers tend to suffer a lot from adverse weather conditions. This further creates a need for cheap and reliable weather forecast system. To attend to such needs in Nigeria [82], a cheap automatic weather station that functions on solar energy was designed. By linking meteorological sensors to microcontrollers, the farmer could gain access to processed information related to weather, through a television screen. A thermometer collects temperature information, while the anemometer and LDR measures wind speed and sunlight, respectively. Embedded temperature sensors within the microcontroller receives analog information gathered by the thermometer and converts it to digital signals [82]. In some cases, unprocessed data can also be sent to farmer's mobile phones. The cheap rate of the unit shows that it can serve as a very good system for crop management and food security, in the least developed nations.

In a research carried out by Tenzin et al. [83], to ensure effective weather monitoring around a farm, the authors designed a very cheap cloud-based weather measurement unit, using an integration of different unique weather sensors. The system, which is made up of a base and a weather station, as well as a display unit, is capable of effectively gathering humidity, temperature, wind direction, wind speed, and many other weather data types. By experimenting its usage and statistically analyzing gathered data, it was observed that the unit provided similar results as the Davis Vantage Pro2 weather monitor, which was pre-installed on the same farm, thus, offering a cheaper option [83].

In a bid to design an integrated farm that efficiently manages water and reduces climate-demanding inputs, Doyle et al. [84] designed an aquaponics unit for vegetables and fish. The design consists of a 12V DC pump that delivers water from the fish tank to the flood tank, which then supplies the area where the crops are planted at a constant rate. As soon as water is removed from the fish pond, it is carried by gravity through the grow bed area, where it is stored until it is needed for watering the vegetable bed. The pump is powered using a solar panel of 150-Watt with a 120 Ah battery.

Having described some smart agricultural and climate-smart agricultural studies, it is important to note that while smart agriculture is mostly developed, research on CSA is relatively new and still at the level of policy and framework description [85]. In a systematic review study by Chandra et al. [85], the authors observed that research on CSA is mainly divided into three parts; global policy and plans around the world concerning further development of the concept, scientific research directions, and integration of pillars of the concept (which includes; adaptation, resilience, mitigation, and food security). With respect to CSA policy framework developed by the World Bank, Taylor [86] faulted the fundamental make-up of the concept on the following grounds.

- There are no explicit conditions that can be referred to as success of CSA, which makes certain fundamental aspects like productivity, completely implicit.
- Being an important part of sustainability, resilience as pointed out within World Bank's CSA framework is not defined, thus, leaving the term implicit.
- Given an absence of conceptual framework for CSA, literature relating to the topic are merely based on success stories of some normative research on agricultural improvement.
- CSA tries not to be involved with how consumer sovereignty influences food production around the world, towards the consumption demands of the elite.

Given these fundamental shortcomings of CSA [86] 'climate-wise food system' is suggested as a more direct term that should be used to refer to sustainable food production systems, rather than CSA. Another criticism on the policy and framework of CSA comes with the injustice meted to smallholder farmers, as a result of the implementation of the concept [87]. By administering interview to some CSA experts, analysis based on a number of ethical positions showed that implementation of climate-smart agricultural approaches is not fair, especially with respect to allocation of income benefits and challenges of cost associated with emission reduction [87], among smallholders farmers and small agricultural processing industries. Budiman [87] further argued that based on how climate justice works, sharing of income benefits should depend on the financial capability of farmers.

In a comparative study of Philippines and Timor-Leste, five important features of climate-smart agricultural practices were observed by Chandra and McNamara [88]; strategies at country-specific institutional levels; delegated financial procedures; the state of the market; technology; and knowledge. In the two countries, CSA was used to resolve climate vulnerability challenges more than it was associated with emission reduction goals [88]. Overall, the researchers observed that advancing the course of CSA in these countries might involve multi-stakeholder approaches that cuts across different levels of participation, both within and outside the farm, rather than mere technical CSA developmental inputs [88]. From the above arguments for and against CSA, it is clear that while there are still fundamental challenges revolving round the CSA concept, the terms might likely continue to be utilized for agricultural problem solving, until it attains uniformity and intersection of ideologies, amongst researchers and policy makers.

What does Smart- and CSA Offer Smart Villages?

Having described in previous sections how the concept of CSA has evolved amidst the challenges faced within its developmental framework, an examination of the utility of climate- and technology-driven agriculture to smart villages is important. According to Azevedo [5], there is a big chance that CSA will empower and strengthen the conceptualization and execution of smart village in different ways. Safdar and Heap [89] noted that development of small grids to power certain climate-smart technologies has so far spurred a re-imagination of the possibility of home solar powering in many Indian villages. Items such as solar lanterns, and street solar lighting systems have become very popular. Nevertheless, a new concern is the way to enhance local productions and repairs of these materials, in order to cater for higher tariffs of importing them to interior villages, and shipping them back for repairs, when the items develop technical faults. The report also stressed how CSA has so far upheld gender equality, for instance, the CCAFS project in Kenya's Nyando valley has mostly favored women whose incomes have improved due to new technology for growing their vegetables [13].

In documenting how CSA could provide smart village farmers with possible economic benefits, Khatri-Chhetri et al. [90] carried out a research using farmers of India's Indo-Gangetic Plains. Major CSA practices by the farmers include diversifying crops, land levelling using laser, nutrient management in a site-specific mannerism, management of residue, and zero tillage, among others. The researchers started by calculating how much the farmers spent to adopt three most prominent CSA systems (variety of crops, land levelling using laser, and zero tillage). These values were estimated as +1402, +3037 and −1577 INR ha-1, respectively, for rice-wheat cultivation system. By improving their varieties in terms of crop production, the study results showed that the farmers of the Indo-Gangetic Plains can have their net return increase to up to INR 15,712 per-hectare, per-year. Similarly, when cultivating wheat and rice with no tillage, farmers could make up to INR 6951 per-hectare, per-year, and INR 8119 per-hectare, per-year with laser-based land levelling. Given the analyses of this results, it implies that integrating individual systems together would result in an even higher yield as well as income for the farmers. In econometric terms, adoption and execution of CSA practices for crop production in the north Indian River plain would significantly influence the cost of production, which decreased, but produced an increased yield of rice and wheat.

Scherr et al. [78] reported that CSA offers to rebrand villages by providing them with embrace 'climate-smart landscapes'. This means that integrated landscape management principles that adopts the pillars of climate change must be in place prior to agricultural land allocation. The development of CSA objectives also requires strong institutional mechanism. When such systems are in place, its effects transcends to other parts of the village. Steenwerth [74] noted that while smart village residents might consider migrating to big cities, climate-smart agriculture could cause a rethink, as it gives room for entrepreneurial development in the agricultural sector, as seen in the case of youth training embarked upon in rural areas across Europe. Additionally, CSA also caters for increased demand for food due to the world's growing population. This is achieved through methods that do not jeopardize environmental health [74]. With respect to animal husbandry, some zoonotic diseases can be detected early, so that treatment plans are set underway to prevent the farmer from infection. CSA also motivates the achievement of sustainable development goals through agricultural practices that use techniques that can drive food security, improve resilience, and effectively manage emissions [70]. CSA practices are also able to curb environmental challenges related to water pollution through the use of agrochemicals [91]. A notable aspect where smart agriculture surpasses expectations is the possibility of using it as a tool for enterprise resource planning, through which the safety of agricultural products/foods can be monitored [46].

4. Discussion

Revisiting the Research Question

How will smart- and climate-smart agricultural research give account of, and conceptualize transformation and change in the smart village context?

In responding to the modified research question above, it is important to draw important ideas from the definitions of smart- and CSA. Albeit, CSA bears all characteristics of smart agriculture, with a step further in lowering GHG emissions. Consequently, accounts of conceptualizing transformation and change in smart village context might tend towards the adoption of key aspects of climate-smart agriculture (see Figure 1), which are somewhat multi-disciplinary in nature [92]. What this implies is that for smart villages to reach desired level in terms of development through research and policy frameworks, ideas of climate-smartness must be fully embedded across the facets of smart village agenda. According to Katara et al. [93], continuous adoption of new technologies is the first way to conceptualize transformation of smart villages. Since technology is bound to continually change, it becomes easy to bring evolving and smarter changes to smart village progress. This means that rural population must fully embrace ICT, especially since smart village idea is based on the fact that technology is adopted to hasten the growth of sustainable development [93]. Secondly, efficiency and productivity are not completely new words in smart village research. Nevertheless, it might be useful for smart village policy analysts to learn from prevention of losses for which CSA is known [94]. Another important aspect through which transformation of smart villages can be conceptualized is through capacity building of rural dwellers. As in the case of climate-smart agriculture, building capacities would bring about self-sufficiency for persons within these communities, thus reducing urban migration [15,19]. This is part of the current efforts within the different smart village initiatives. Of all the initiatives that smart village research can draw from climate-smart agricultural practices, the idea of seeking and promoting "new knowledge" [95] might be technically referred to as the most significant. Given that the world has now embraced a knowledge-based economy for which smart village development has to be a part.

On the basis of towing a part of steady development in its processes, future smart village developmental projects need to adopt successful projects of the past as a yardstick for planning. For instance, tremendous success was recorded by the IEEE smart village initiative; the EU smart village-drive, as well as the CCACFS projects, to mention a few. By adopting the recipe for success within these projects, more smart-village projects would be actualized in many parts of the world. Furthermore, it is noteworthy to state that existing smart village projects also have unique challenges. Notable amongst the challenges faced by smart villages within the IEEE project is the issue of maintenance and repairs [96]. Although as part of the project framework, two individuals are often selected and trained within the villages to fix damaged smart inputs, when demands for these inputs become high, the number of technicians might no longer be sufficient to cater for repair and maintenance needs. This is one aspect where smart village development must learn from climate-smart ideas where capacity development is well-planned and readily available.

Another aspect for which smart village development can gain from climate-smart agriculture is in its sustainability approach. While CSA strives for the cheapest routes to progress in agriculture, smart village development mostly depends on donations and funding, which slows down the pace of making progress and achieving sustained growth. As a result, for any smart villages project to achieve lasting success, such a project would have to plan self-funding strategies [10], where inputs within the village is used to generate income that would fund new projects for growth, rather than unduly wait for funding before progress is made. In building its growth, smart village planners might need to prioritize new knowledge and link it to new technology for early warning measures against potential environmental disasters. Furthermore, proponents also need to ensure that pillars of climate change are largely considered in building infrastructures [13]. This is because the impact of climate change might continue to be felt for a long time.

While ideas drawn from smart and climate-smart agriculture might indeed be useful for smart village development, Hargreaves et al. [97] explained that specific policies grounded in the values of rural areas are needed to help them transform into smart villages. This transformation must, therefore, bring effective utilization and management of resources within smart villages. The idea of transformation within the context of smart villages mostly draws attention to digital transformation, which is very important [98]. Another result of technological change is the social changes it brings [99,100].

Given the forgone discussion on how smart village development can be spurred from ideas borrowed from smart- and climate-smart agriculture, we argued that the development of a smart village has to be a gradual process. This is because the development must systematically and strategically prioritize the most important aspects, such as clean energy management and agriculture, bearing in mind the sustainability of the process.

5. Current Lessons & Future Research Direction

Overall, this study revealed a number of lessons from smart-agriculture and climate-smart agriculture ideas, which, if adopted in smart villages, would achieve the following goals.

- Improvement and optimization of existing smart village projects/processes in terms of precision and speed.
- Increased efficiency and productivity, which can lead to increased income/profit on ventures embarked upon by smart-village dwellers.
- Better planning brought about by efficient forecasting and prediction systems, which help to guide against potential dangers, and to take proactive steps in planning and preparation for such eventualities.
- Offer of cheaper and equally effective data gathering avenues for easy detection of challenges and problems.
- Reduced dependence on external funding, and a drive for self-sufficiency encouraged by innovation.

While the above lessons are specific to smart village development, there are specific shortcomings of climate-smart agriculture that must be noted [91,92], and which ought not to be adopted in smart village development.

1. Sain et al. [101] used cost-benefit analysis to analyze variability and uncertainty of some CSA parameters. It was observed that while CSA is generally promising, not all CSA parameters were indeed profitable in the long run [101].
2. With CSA comes IoT, Blockchain, and artificial intelligence in agricultural operations. As such, there is the challenge of helping rural farmers understand the operation of smart farm inputs, and interpretation of data gathered from the farms using CSA tools [102]. The situation might be worse in rural Africa, where farmers rarely have any level of formal education.
3. Interoperability is another serious challenge for adopting CSA. An example is described by Kalatzis et al. [103] in the use of gaiasense^{TM} farming solution.
4. The cost of acquiring smart farming implements cannot be overlooked when listing some known challenges of CSA [104]. Smart sensors for instance are generally expensive [105].

As a result of some of the aforementioned challenges of CSA, future studies might look at the challenges posed by the adoption of "climate-smart" agriculture, prior to the full adoption of its fundamental aspects, as described in this study. This is because research by Taylor [86] pointed out certain foundation faults in the description of CSA by the World Bank group. There is also a fundamental problem in how CSA handles climate justice [87]. Another aspect opened to future research is the development of the climate-smart villages, as used in some studies [69]. While this has been achieved in some parts of the world today [13], it might be the case that smart-village research is yet to reach a maturity level as to warrant even more terms to be coined from it.

6. Conclusions

The uniqueness of smart village projects around the world means that approaches towards smart village development might also differ. This study showed that smart and CSA are key areas that must be considered in developing a smart village project, and offer several lessons to proponents of smart village ideas, given how these concepts have enjoyed steady conceptualization in the research literature. Another important consideration that must be carefully explored is the tendency of developing smart villages in line with the concepts upon which smart cities are built. Having clarified in Section 1 that smart villages are not extensions of smart cities [106], it is important to understand that the challenges of rural areas differ significantly from those of cities. Hence, smart village development must come with uniquely defined plans and strategies for its development [107].

A major driving force for "smarting up" rural areas is the mass exodus of persons to the cities, as well as inferior services offered in these villages [107]. Nevertheless, an introduction of the smart village concept comes with new opportunities brought about by technology, which is currently touted as the major economic driver of the 21st century. The current study, therefore, tries to adopt the technological ideas of CSA in creating a foundational path for smart village development. To do this, the study carefully analyzes the framework of CSA and proposes that the same be adopted for developing smart villages. It is observed that certain fundamental aspects of technological innovation; productivity, new knowledge, new technology, capacity building, vulnerability reduction, increased profits, etc., are fundamental to the building of smart villages. Nevertheless, these fundamental terms cannot be embedded immediately. Rather, it must follow gradual process that gives priority to the important aspects.

Author Contributions: Conceptualization, A.A., O.F., O.K., A.S. and K.K.; methodology, K.K., P.M., O.K., O.F. and A.S.; software, A.A., M.A. and O.F.; validation, A.S., K.K., P.M. and O.K.; formal analysis, O.F., K.K. and O.K.; investigation, A.A., O.F., M.A.; resources, O.K., K.K. and A.S.; data curation, A.A., O.F. and O.K.; writing—original draft preparation, A.A., O.F., M.A., K.K., P.M., A.S. and O.K.; writing—review and editing, A.A., O.F., K.K. and O.K.; visualization, A.A., O.F. and M.A.; supervision, A.S., P.M., K.K. and O.K.; project administration, O.F., O.K., P.M. and K.K.; funding acquisition, O.K. and K.K. All authors have read and agreed to the published version of the manuscript.

Funding: This work was funded in part by the project (2020/2205), Grant Agency of Excellence, University of Hradec Kralove, Faculty of Informatics and Management, Czech Republic; project at Universiti Teknologi Malaysia (UTM) under Research University Grant Vot-20H04, Malaysia Research University Network (MRUN) Vot 4L876 and the Fundamental Research Grant Scheme (FRGS) Vot5F073, supported under Ministry of Education Malaysia for the completion of the research.

Conflicts of Interest: Authors declare no conflict of interest.

References

1. Mackenzie, D. IEEE Smart Village Sustainable Development Is a Global Mission. *IEEE Syst. Man Cybern. Mag.* **2019**, *5*, 39–41. [CrossRef]
2. Jung, J.G. Smart Communities: Digitally-Inclined and Content-Rich. Available online: http://tfi.com/pubs/ntq/articles/view/98Q1_A3.pdf (accessed on 27 May 2020).
3. European Commission EU Support for Smart Villages. European Commission-European Commission. 4 August 2019. Available online: https://ec.europa.eu/digital-single-market/en/news/eu-support-smart-villages (accessed on 3 June 2020).
4. Maheswari, R.; Azath, H.; Sharmila, P.; Gnanamalar, S.S.R. Smart Village: Solar Based Smart Agriculture with IoT Enabled for Climatic Change and Fertilization of Soil. In Proceedings of the 2019 IEEE 5th International Conference on Mechatronics System and Robots (ICMSR), Singapore, 3–5 May 2019; pp. 102–105. [CrossRef]
5. Azevedo, D. Precision Agriculture and the Smart Village Concept. In *Smart Villages in the EU and Beyond*; Visvizi, A., Lytras, M.D., Mudri, G., Eds.; Emerald Publishing Limited: Bingley, UK, 2019; pp. 83–97.
6. Antonacci, A.; Arduini, F.; Moscone, D.; Palleschi, G.; Scognamiglio, V. Nanostructured (Bio)sensors for smart agriculture. *TrAC Trends Anal. Chem.* **2018**, *98*, 95–103. [CrossRef]
7. Khoa, T.A.; Man, M.M.; Nguyen, T.-Y.; Nguyen, V.; Nam, N.H. Smart Agriculture Using IoT Multi-Sensors: A Novel Watering Management System. *J. Sens. Actuator Netw.* **2019**, *8*, 45. [CrossRef]
8. Mogili, U.R.; Deepak, B.B.V.L. Review on Application of Drone Systems in Precision Agriculture. *Procedia Comput. Sci.* **2018**, *133*, 502–509. [CrossRef]
9. Visvizi, A.; Lytras, M.D. It's Not a Fad: Smart Cities and Smart Villages Research in European and Global Contexts. *Sustainability* **2018**, *10*, 2727. [CrossRef]
10. Zavratnik, V.; Kos, A.; Duh, E.S. Smart Villages: Comprehensive Review of Initiatives and Practices. *Sustainability* **2018**, *10*, 2559. [CrossRef]
11. Larsen, R.S.; Estes, D. IEEE Smart Village Launches SunBlazer IV and Smart Portable Battery Kits Empowering Remote Communities. *IEEE Syst. Man Cybern. Mag.* **2019**, *5*, 49–51. [CrossRef]

12. Larsen, R.S.; Welbourn, D.; Wessner, D.; Podmore, R.; Lacourciere, M.; Larsen, A.; Lee, P.; Moulton, R.; Myers, S.; Niboh, M.; et al. 'Learning beyond the Light Bulb' among Least Developed Countries based on a sustainable PV solar utility model. In Proceedings of the IEEE Global Humanitarian Technology Conference (GHTC 2014), San Jose, CA, USA, 10–13 October 2014; pp. 106–114. [CrossRef]
13. CGIAR Climate-Smart Villages. 8 July 2013. Available online: https://ccafs.cgiar.org/climate-smart-villages (accessed on 3 June 2020).
14. CGIAR Scaling-Out Climate-Smart Village Program in the Vulnerable Areas of Indo-Gangetic Plains of India. 25 July 2019. Available online: https://ccafs.cgiar.org/scaling-out-climate-smart-village-program-vulnerable-areas-indo-gangetic-plains-india (accessed on 3 June 2020).
15. CGIAR Climate Services for Farmers. 2016. Available online: https://ccafs.cgiar.org/themes/climate-services-farmers (accessed on 3 June 2020).
16. European Commission Rural Development. European Commission-European Commission. Available online: https://ec.europa.eu/info/food-farming-fisheries/key-policies/common-agricultural-policy/rural-development_en (accessed on 3 June 2020).
17. *European Commission Bled Declaration: For a Smarter Future of the Rural Areas in EU*; European Commission: Bled, Slovenia, 2018. Available online: https://www.sciteceuropa.eu/smart-villages-rural-development/95112/ (accessed on 17 November 2019).
18. Rural Mountainous Remote Areas & Smart Villages' Group Inaugural Meeting of the New Intergroup SMART Villages for Rural Communities. RUMRA & Smart Villages. 22 July 2019. Available online: https://www.smart-rural-intergroup.eu/inaugural-meeting-of-the-new-intergroup-smart-villages-for-rural-communities/ (accessed on 7 June 2020).
19. *European Network for Rural Development Digital and Social Innovation in Rural Services: Projects Brochure*; The European Agricultural Fund for Rural Development: Luxembourg, 2018. Available online: https://data.europa.eu/doi/10.2762/58984 (accessed on 7 June 2020).
20. Arduini, F.; Cinti, S.; Scognamiglio, V.; Moscone, D. Nanomaterials in electrochemical biosensors for pesticide detection: Advances and challenges in food analysis. *Microchim. Acta* **2016**, *183*, 2063–2083. [CrossRef]
21. Arduini, F.; Cinti, S.; Scognamiglio, V.; Moscone, D.; Palleschi, G. How cutting-edge technologies impact the design of electrochemical (bio)sensors for environmental analysis. A review. *Anal. Chim. Acta* **2017**, *959*, 15–42. [CrossRef]
22. De Oliveira, J.L.; Campos, E.V.R.; Bakshi, M.; Abhilash, P.C.; Fraceto, L.F. Application of nanotechnology for the encapsulation of botanical insecticides for sustainable agriculture: Prospects and promises. *Biotechnol. Adv.* **2014**, *32*, 1550–1561. [CrossRef]
23. Scognamiglio, V. Nanotechnology in glucose monitoring: Advances and challenges in the last 10 years. *Biosens. Bioelectron.* **2013**, *47*, 12–25. [CrossRef] [PubMed]
24. Yu, Z.; Zhao, G.; Liu, M.; Lei, Y.; Li, M. Fabrication of a Novel Atrazine Biosensor and Its Subpart-per-Trillion Levels Sensitive Performance. *Environ. Sci. Technol.* **2010**, *44*, 7878–7883. [CrossRef] [PubMed]
25. Dong, J.; Fan, X.; Qiao, F.; Ai, S.; Xin, H. A novel protocol for ultra-trace detection of pesticides: Combined electrochemical reduction of Ellman's reagent with acetylcholinesterase inhibition. *Anal. Chim. Acta* **2013**, *761*, 78–83. [CrossRef]
26. Shi, H.; Zhao, G.; Liu, M.; Fan, L.; Cao, T. Aptamer-based colorimetric sensing of acetamiprid in soil samples: Sensitivity, selectivity and mechanism. *J. Hazard. Mater.* **2013**, *260*, 754–761. [CrossRef]
27. Safarpour, H.; Safarnejad, M.R.; Tabatabaei, M.; Mohsenifar, A.; Rad, F.; Basirat, M.; Shahryari, F.; Hasanzadeh, F. Development of a quantum dots FRET-based biosensor for efficient detection of Polymyxa betae. *Can. J. Plant Pathol.* **2012**, *34*, 507–515. [CrossRef]
28. Bakhori, N.M.; Yusof, N.A.; Abdullah, A.H.; Hussein, M.Z. Development of a Fluorescence Resonance Energy Transfer (FRET)-Based DNA Biosensor for Detection of Synthetic Oligonucleotide of Ganoderma boninense. *Biosensors* **2013**, *3*, 419. [CrossRef] [PubMed]
29. Ali, M.A.; Jiang, H.; Mahal, N.K.; Weber, R.J.; Kumar, R.; Castellano, M.J.; Dong, L. Microfluidic impedimetric sensor for soil nitrate detection using graphene oxide and conductive nanofibers enabled sensing interface. *Sens. Actuators B Chem.* **2017**, *239*, 1289–1299. [CrossRef]
30. Prasad, B.B.; Jauhari, D.; Tiwari, M.P. Doubly imprinted polymer nanofilm-modified electrochemical sensor for ultra-trace simultaneous analysis of glyphosate and glufosinate. *Biosens. Bioelectron.* **2014**, *59*, 81–88. [CrossRef]

31. Siddiquee, S.; Rovina, K.; Yusof, N.A.; Rodrigues, K.F.; Suryani, S. Nanoparticle-enhanced electrochemical biosensor with DNA immobilization and hybridization of Trichoderma harzianum gene. *Sens. Bio Sens. Res.* **2014**, *2*, 16–22. [CrossRef]
32. Pan, P.; Miao, Z.; Yanhua, L.; Linan, Z.; Haiyan, R.; Pan, K.; Linpei, P. Preparation and Evaluation of a Stable Solid State Ion Selective Electrode of Polypyrrole/Electrochemically Reduced Graphene/Glassy Carbon Substrate for Soil Nitrate Sensing. *Int. J. Electrochem. Sci.* **2016**, *11*, 4779–4793. [CrossRef]
33. Deng, H.-H.; Hong, G.-L.; Lin, F.-L.; Liu, A.-L.; Xia, X.-H.; Chen, W. Colorimetric detection of urea, urease, and urease inhibitor based on the peroxidase-like activity of gold nanoparticles. *Anal. Chim. Acta* **2016**, *915*, 74–80. [CrossRef]
34. Zhao, Y.; Liu, L.; Kong, D.; Kuang, H.; Wang, L.; Xu, C. Dual Amplified Electrochemical Immunosensor for Highly Sensitive Detection of Pantoea stewartii sbusp. stewartii. *ACS Appl. Mater. Interfaces* **2014**, *6*, 21178–21183. [CrossRef]
35. Lin, H.-Y.; Huang, C.-H.; Lu, S.-H.; Kuo, I.-T.; Chau, L.-K. Direct detection of orchid viruses using nanorod-based fiber optic particle plasmon resonance immunosensor. *Biosens. Bioelectron.* **2014**, *51*, 371–378. [CrossRef] [PubMed]
36. Daponte, P.; De Vito, L.; Glielmo, L.; Iannelli, L.; Liuzza, D.; Picariello, F.; Silano, G. A review on the use of drones for precision agriculture. *IOP Conf. Ser. Earth Environ. Sci.* **2019**, *275*, 012022. [CrossRef]
37. Primicerio, J.; Di Gennaro, S.F.; Fiorillo, E.; Genesio, L.; Lugato, E.; Matese, A.; Vaccari, F.P. A flexible unmanned aerial vehicle for precision agriculture. *Precis. Agric.* **2012**, *13*, 517–523. [CrossRef]
38. Burgos, S.; Mota, M.; Noll, D.; Cannelle, B. Use of Very High-Resolution Airborne Images to Analyse 3d Canopy Architecture of a Vineyard. *ISPAr* **2015**, *XL3*, 399–403. [CrossRef]
39. Berni, J.A.J.; Zarco-Tejada, P.J.; Suarez, L.; Fereres, E. Thermal and Narrowband Multispectral Remote Sensing for Vegetation Monitoring From an Unmanned Aerial Vehicle. *IEEE Trans. Geosci. Remote Sens.* **2009**, *47*, 722–738. [CrossRef]
40. Nagpure, S.; Ingale, S.; Pahurkar, S.; Bobade, A.M.; Ghosal, M.; Dhope, T. Smart Agriculture Using IOT. *HELIX* **2019**, *9*, 5081–5083. [CrossRef]
41. Mat, I.; Kassim, M.R.M.; Harun, A.N.; Yusoff, I.M. Smart Agriculture Using Internet of Things. In Proceedings of the 2018 IEEE Conference on Open Systems (ICOS), Langkawi Island, Malaysia, 21–22 November 2018; pp. 54–59. [CrossRef]
42. Prathibha, S.R.; Hongal, A.; Jyothi, M.P. IOT Based Monitoring System in Smart Agriculture. In Proceedings of the 2017 International Conference on Recent Advances in Electronics and Communication Technology (ICRAECT), Bangalore, India, 16–17 March 2017; pp. 81–84. [CrossRef]
43. Doshi, J.; Patel, T.; Bharti, S.K. Smart Farming using IoT, a solution for optimally monitoring farming conditions. *Procedia Comput. Sci.* **2019**, *160*, 746–751. [CrossRef]
44. Pathak, A.; AmazUddin, M.; Abedin, M.J.; Andersson, K.; Mustafa, R.; Hossain, M.S. IoT based Smart System to Support Agricultural Parameters: A Case Study. *Procedia Comput. Sci.* **2019**, *155*, 648–653. [CrossRef]
45. Patil, A.S.; Tama, B.A.; Park, Y.; Rhee, K.-H. A Framework for Blockchain Based Secure Smart Green House Farming. In *Advances in Computer Science and Ubiquitous Computing*; Springer: Singapore, 2018; pp. 1162–1167. [CrossRef]
46. Lin, J.; Shen, Z.; Zhang, A.; Chai, Y. Blockchain and IoT based Food Traceability for Smart Agriculture. In Proceedings of the 3rd International Conference on Crowd Science and Engineering, Singapore, 28–31 July 2018; pp. 1–6. [CrossRef]
47. Fox, D.G.; van Amburgh, M.E.; Tylutki, T.P. Predicting Requirements for Growth, Maturity, and Body Reserves in Dairy Cattle. *J. Dairy Sci.* **1999**, *82*, 1968–1977. [CrossRef]
48. Halachmi, I.; Guarino, M.; Bewley, J.; Pastell, M. Smart Animal Agriculture: Application of Real-Time Sensors to Improve Animal Well-Being and Production. *Annu. Rev. Anim. Biosci.* **2019**, *7*, 403–425. [CrossRef] [PubMed]
49. Wallace, M.A.; Stouffer, J.R.; Westervelt, R.G. Relationships of ultrasonic and carcass measurements with retail yield in beef cattle. *Livest. Prod. Sci.* **1977**, *4*, 153–164. [CrossRef]
50. Mizrach, A.; Flitsanov, U.; Maltz, E.; Spahr, S.L.; Novakofski, J.E.; Murphy, M.R. Ultrasonic assessment of body condition changes of the dairy cow during lactation. *Trans. Am. Soc. Agric. Eng.* **1999**, *42*, 805–812. [CrossRef]
51. Bercovich, A.; Edan, Y.; Alchanatis, V.; Moallem, U.; Parmet, Y.; Honig, H.; Maltz, E.; Antler, A.; Halachmi, I. Development of an automatic cow body condition scoring using body shape signature and Fourier descriptors. *J. Dairy Sci.* **2013**, *96*, 8047–8059. [CrossRef]

52. Halachmi, I.; Polak, P.; Roberts, D.J.; Klopcic, M. Cow body shape and automation of condition scoring. *J. Dairy Sci.* **2008**, *91*, 4444–4451. [CrossRef]
53. Bewley, J.M.; Peacock, A.M.; Lewis, O.; Boyce, R.E.; Roberts, D.J.; Coffey, M.P.; Kenyon, S.J.; Schutz, M.M. Potential for Estimation of Body Condition Scores in Dairy Cattle from Digital Images. *J. Dairy Sci.* **2008**, *91*, 3439–3453. [CrossRef]
54. Halachmi, I.; Klopčič, M.; Polak, P.; Roberts, D.J.; Bewley, J.M. Automatic assessment of dairy cattle body condition score using thermal imaging. *Comput. Electron. Agric.* **2013**, *99*, 35–40. [CrossRef]
55. Spoliansky, R.; Edan, Y.; Parmet, Y.; Halachmi, I. Development of automatic body condition scoring using a low-cost 3-dimensional Kinect camera. *J. Dairy Sci.* **2016**, *99*, 7714–7725. [CrossRef]
56. Yanmaz, L.; Okumus, Z.; Dogan, E. Instrumentation of thermography and its applications in horses. *J. Anim. Vet. Adv.* **2007**, *6*, 858–862.
57. Hurnik, J.F.; Boer, S.D.; Webster, A.B. Detection of Health Disorders in Dairy Cattle Utilizing a Thermal Infrared Scanning Technique. *Can. J. Anim. Sci.* **1984**, *64*, 1071–1073. [CrossRef]
58. Steensels, M.; Maltz, E.; Bahr, C.; Berckmans, D.; Antler, A.; Halachmi, I. Towards practical application of sensors for monitoring animal health; design and validation of a model to detect ketosis. *J. Dairy Res.* **2017**, *84*, 139–145. [CrossRef] [PubMed]
59. Steensels, M.; Antler, A.; Bahr, C.; Berckmans, D.; Maltz, E.; Halachmi, I. A decision-tree model to detect post-calving diseases based on rumination, activity, milk yield, BW and voluntary visits to the milking robot. *Animal* **2016**, *10*, 1493–1500. [CrossRef] [PubMed]
60. Cornou, C.; Lundbye-Christensen, S. Classifying sows' activity types from acceleration patterns: An application of the Multi-Process Kalman Filter. *Appl. Anim. Behav. Sci.* **2008**, *111*, 262–273. [CrossRef]
61. Halachmi, I.; Meir, Y.B.; Miron, J.; Maltz, E. Feeding behavior improves prediction of dairy cow voluntary feed intake but cannot serve as the sole indicator. *Animal* **2016**, *10*, 1501–1506. [CrossRef] [PubMed]
62. Broell, F.; Noda, T.; Wright, S.; Domenici, P.; Steffensen, J.F.; Auclair, J.P.; Taggart, C.T. Accelerometer tags: Detecting and identifying activities in fish and the effect of sampling frequency. *J. Exp. Biol.* **2013**, *216*, 1255–1264. [CrossRef]
63. Marchioro, G.F.; Cornou, C.; Kristensen, A.R.; Madsen, J. Sows' activity classification device using acceleration data–A resource constrained approach. *Comput. Electron. Agric.* **2011**, *77*, 110–117. [CrossRef]
64. Glennon, T.; O'Quigley, C.; McCaul, M.; Matzeu, G.; Beirne, S.; Wallace, G.G.; Stroiescu, F.; O'Mahoney, N.; White, P.; Diamond, D. 'SWEATCH': A Wearable Platform for Harvesting and Analysing Sweat Sodium Content. *Electroanalysis* **2016**, *28*, 1283–1289. [CrossRef]
65. Nograles, A.H.H.; Caluyo, F.S. Wireless system for pregnancy detection in cows by monitoring temperature changes in body. In Proceedings of the 2013 IEEE 9th International Colloquium on Signal Processing and its Applications, Kuala Lumpur, Malaysia, 8–10 March 2013; pp. 11–16. [CrossRef]
66. Zia, A.I.; Syaifudin, A.M.; Mukhopadhyay, S.C.; Yu, P.L.; Al-Bahadly, I.H.; Kosel, J.; Gooneratne, C. Sensor and instrumentation for progesterone detection. In Proceedings of the 2012 IEEE International Instrumentation and Measurement Technology Conference Proceedings, Graz, Austria, 13–16 May 2012; pp. 1220–1225. [CrossRef]
67. Ferrari, S.; Silva, M.; Guarino, M.; Aerts, J.M.; Breckmans, D. Cough sound analysis to identify respiratory infection in pigs. *Comput. Electron. Agric.* **2008**, *64*, 318–325. [CrossRef]
68. Campbell, B.M.; Thornton, P.; Zougmoré, R.; van Asten, P.; Lipper, L. Sustainable intensification: What is its role in climate smart agriculture? *Curr. Opin. Environ. Sustain.* **2014**, *8*, 39–43. [CrossRef]
69. Aggarwal, P.K.; Jarvis, A.; Campbell, B.M.; Zougmoré, R.B.; Khatri-Chhetri, A.; Vermeulen, S.; Loboguerrero Rodriguez, A.M.; Sebastian, L.; Kinyangi, J.; Bonilla Findji, O.; et al. The climate-smart village approach: Framework of an integrative strategy for scaling up adaptation options in agriculture. *Ecol. Soc.* **2018**, *23*. [CrossRef]
70. Food and Agriculture Organization. *Climate Smart Agriculture: Policies, Practices and Financing For Food Security, Adaptation and Mitigation*; Food and Agriculture Organization of the United Nations: Rome, Italy, 2010.
71. Food and Agriculture Organization. *Climate-Smart Agriculture Sourcebook*; Food and Agriculture Organization of the United Nations: Rome, Italy, 2013.
72. Engel, S.; Muller, A. Payments for environmental services to promote 'climate-smart agriculture'? Potential and challenges. *Agric. Econ.* **2016**, *47*, 173–184. [CrossRef]

73. Rosenstock, T.S.; Lamanna, C.; Chesterman, S.; Bell, P.; Arslan, A.; Richards, M.; Rioux, J.; Akinleye, A.O.; Champalle, C.; Cheng, Z.; et al. *The Scientific Basis of Climate-Smart Agriculture: A Systematic Review Protocol*; CGIAR Research Program on Climate Change; CCAFS Working Paper; Agriculture and Food Security (CCAFS): Copenhagen, Denmark, 2016; p. 138. Available online: https://cgspace.cgiar.org/bitstream/handle/10568/70967/http://CCAFSWP138.pdf (accessed on 17 November 2019).
74. Steenwerth, K.L.; Hodson, A.K.; Bloom, A.J.; Carter, M.R.; Cattaneo, A.; Chartres, C.J.; Hatfield, J.L.; Henry, K.; Hopmans, J.W.; Horwath, W.R. Climate-smart agriculture global research agenda: Scientific basis for action. *Agric. Food Secur.* **2014**, *3*, 11. [CrossRef]
75. Hellin, J.; Fisher, E. Building pathways out of poverty through climate smart agriculture and effective targeting. *Dev. Pract.* **2018**, *28*, 974–979. [CrossRef]
76. Rao, N.H. Big Data and Climate Smart Agriculture-Status and Implications for Agricultural Research and Innovation in India. *Proc. Indian Natl. Sci. Acad.* **2018**, *84*, 625–640. [CrossRef]
77. Alliance, R. What Is Climate-Smart Agriculture? *Rainforest Alliance: 2020.* Available online: https://www.rainforest-alliance.org/articles/what-is-climate-smart-agriculture (accessed on 5 June 2020).
78. Scherr, S.J.; Shames, S.; Friedman, R. From climate-smart agriculture to climate-smart landscapes. *Agric. Food Secur.* **2012**, *1*, 12. [CrossRef]
79. Xiong, W.; van der Velde, M.; Holman, I.P.; Balkovic, J.; Lin, E.; Skalský, R.; Porter, C.; Jones, J.; Khabarov, N.; Obersteiner, M. Can climate-smart agriculture reverse the recent slowing of rice yield growth in China? *Agric. Ecosyst. Environ.* **2014**, *196*, 125–136. [CrossRef]
80. William, J.R. The EPIC model. In *Computer Models in Watershed Hydrology*; Singh, V.P., Ed.; Water Resources Publication, LLC: Highlands Ranch, CO, USA, 2012; pp. 90–100.
81. Jones, J.W.; Hoogenboom, G.; Porter, C.H.; Boote, K.J.; Batchelor, W.D.; Hunt, L.A.; Wilkens, P.W.; Singh, U.; Gijsman, A.J.; Ritchie, J.T. The DSSAT cropping system model. *Eur. J. Agron.* **2003**, *18*, 235–265. [CrossRef]
82. Adoghe, A.U.; Popoola, S.I.; Chukwuedo, O.M.; Airoboman, A.E.; Atayero, A.A. Smart Weather Station for Rural Agriculture using Meteorological Sensors and Solar Energy. In Proceedings of the World Congress on Engineering, London, UK, 5–7 July 2017; Volume I; pp. 1–4. Available online: http://eprints.covenantuniversity.edu.ng/8584/#.Xt0880VKiM8 (accessed on 7 June 2020).
83. Tenzin, S.; Siyang, S.; Pobkrut, T.; Kerdcharoen, T. Low cost weather station for climate-smart agriculture. In Proceedings of the 2017 9th International Conference on Knowledge and Smart Technology (KST), Chonburi, Thailand, 1–4 February 2017; pp. 172–177. [CrossRef]
84. Doyle, L.; Oliver, L.; Whitworth, C. Design of a Climate Smart Farming System in East Africa. In Proceedings of the 2018 IEEE Global Humanitarian Technology Conference (GHTC), San Jose, CA, USA, 18–21 October 2018; pp. 1–6. [CrossRef]
85. Chandra, A.; McNamara, K.E.; Dargusch, P. Climate-smart agriculture: Perspectives and framings. *Clim. Policy* **2018**, *18*, 526–541. [CrossRef]
86. Taylor, M. Climate-smart agriculture: What is it good for? *J. Peasant Stud.* **2018**, *45*, 89–107. [CrossRef]
87. Budiman, I. Climate-smart agriculture policy and (in)justice for smallholders in developing countries. *Future Food J. Food Agric. Soc.* **2019**, *7*, 31–41. [CrossRef]
88. Chandra, A.; McNamara, K.E. Climate-Smart Agriculture in Southeast Asia: Lessons from Community-Based Adaptation Programs in the Philippines and Timor-Leste. In *Resilience: The Science of Adaptation to Climate Change*; Zommers, Z., Alverson, K., Eds.; Elsevier Science BV: Amsterdam, The Netherlands, 2018; pp. 165–179.
89. Safdar, T.; Heap, B. *Energy and Agriculture for Smart Villages in India*; Technical Report 7; International Crop Research Institute for the Semi-Arid Tropics: Telangana, India, 2016; Available online: https://e4sv.org/wp-content/uploads/2017/01/Energy-and-Agriculture-for-Smart-Villages-in-India.compressed.pdf (accessed on 8 July 2020).
90. Khatri-Chhetri, A.; Aryal, J.P.; Sapkota, T.B.; Khurana, R. Economic benefits of climate-smart agricultural practices to smallholder farmers in the Indo-Gangetic Plains of India. *Curr. Sci.* **2016**, *110*, 1251–1256. [CrossRef]
91. Bullard, W.E. Effects of Land Use on Water Resources. *J. (Water Pollut. Control Fed.)* **1996**, *38*, 645–659.
92. Visvizi, A.; Lytras, M.D. Sustainable Smart Cities and Smart Villages Research: Rethinking Security, Safety, Well-being, and Happiness. *Sustainability* **2020**, *12*, 215. [CrossRef]
93. Katara, S.K. Envisioning Smart Villages through Information and Communication Technologies–A Framework for Implementation in India. In *Digital Transformation and Global Society*; Springer: Cham, Switzerland, 2016; pp. 463–468. [CrossRef]

94. Hüseyin, Ş.; Aysun, Ş. Digital Farming and Productivity Effect: 'The Smart Village' In Turkey. *Ann. Fac. Econ.* **2019**, *1*, 371–379.
95. Naldi, L.; Nilsson, P.; Westlund, H.; Wixe, S. What is smart rural development? *J. Rural Stud.* **2015**, *40*, 90–101. [CrossRef]
96. Anderson, A.; Loomba, P.; Orajaka, I.; Numfor, J.; Saha, S.; Janko, S.; Johnson, N.; Podmore, R.; Larsen, R. Empowering Smart Communities Electrification, education, and sustainable entrepreneurship In IEEE Smart Village initiatives. *IEEE Electrif. Mag.* **2017**, *5*, 6–16. [CrossRef]
97. Hargreaves, T.; Hielscher, S.; Seyfang, G.; Smith, A. Grassroots innovations in community energy: The role of intermediaries in niche development. *Glob. Environ. Chang.* **2013**, *23*, 868–880. [CrossRef]
98. Plevák, O. Smart Village is a Remedy for Outflow of People from Rural Areas. Available online: https://www.euractiv.com/section/agriculture-food/news/smart-village-is-a-remedy-for-outflow-of-people-from-rural-areas/ (accessed on 9 June 2020).
99. Del Sesto, S.L. Technology and social change: William Fielding Ogburn revisited. *Technol. Forecast. Soc. Chang.* **1983**, *24*, 183–196. [CrossRef]
100. Fenni, M.Y.D.; Ezziyyani, M. The Influence of Technology and Social Networks on Social Change. In *Advanced Intelligent Systems for Sustainable Development (AI2SD'2019)*; Springer: Cham, Switzerland, 2020; pp. 431–438. [CrossRef]
101. Sain, G.; Loboguerrero, A.M.; Corner-Dolloff, C.; Lizarazo, M.; Nowak, A.; Martínez-Barón, D.; Andrieu, N. Costs and benefits of climate-smart agriculture: The case of the Dry Corridor in Guatemala. *Agric. Syst.* **2017**, *151*, 163–173. [CrossRef]
102. Villa-Henriksen, A.; Edwards, G.T.C.; Pesonen, L.A.; Green, O.; Sørensen, C.A.G. Internet of Things in arable farming: Implementation, applications, challenges and potential. *Biosyst. Eng.* **2020**, *191*, 60–84. [CrossRef]
103. Kalatzis, N.; Marianos, N.; Chatzipapadopoulos, F. IoT and data interoperability in agriculture: A case study on the gaiasenseTM smart farming solution. In Proceedings of the 2019 Global IoT Summit (GIoTS), Aarhus, Denmark, 17–21 June 2019; pp. 1–6. [CrossRef]
104. McCarthy, N.; Lipper, L.; Zilberman, D. Economics of Climate Smart Agriculture: An Overview. In *Climate Smart Agriculture: Building Resilience to Climate Change*; Lipper, L., McCarthy, N., Zilberman, D., Asfaw, S., Branca, G., Eds.; Springer International Publishing: Cham, Switzerland, 2018; pp. 31–47.
105. García, L.; Parra, L.; Jimenez, J.M.; Lloret, J.; Lorenz, P. IoT-Based Smart Irrigation Systems: An Overview on the Recent Trends on Sensors and IoT Systems for Irrigation in Precision Agriculture. *Sensors* **2020**, *20*, 1042. [CrossRef] [PubMed]
106. Garner, C. Smart Villages: A New Concept for Rural Development; SciTech Europa: 17 May 2019. Available online: https://www.scitecheuropa.eu/smart-villages-rural-development/95112/ (accessed on 17 November 2019).
107. Komorowski, Ł.; Stanny, M. Smart Villages: Where Can They Happen? *Land* **2020**, *9*, 151. [CrossRef]

Publisher's Note: MDPI stays neutral with regard to jurisdictional claims in published maps and institutional affiliations.

© 2020 by the authors. Licensee MDPI, Basel, Switzerland. This article is an open access article distributed under the terms and conditions of the Creative Commons Attribution (CC BY) license (http://creativecommons.org/licenses/by/4.0/).

Article

Urban Green Infrastructure Monitoring Using Remote Sensing from Integrated Visible and Thermal Infrared Cameras Mounted on a Moving Vehicle

Sigfredo Fuentes *, Eden Tongson and Claudia Gonzalez Viejo

Digital Agriculture, Food and Wine Sciences Group, Faculty of Veterinary and Agricultural Sciences, School of Agriculture and Food, Parkville, VIC 3010, Australia; eden.tongson@unimelb.edu.au (E.T.); cgonzalez2@unimelb.edu.au (C.G.V.)
* Correspondence: sfuentes@unimelb.edu.au

Abstract: Climate change forecasts higher temperatures in urban environments worsening the urban heat island effect (UHI). Green infrastructure (GI) in cities could reduce the UHI by regulating and reducing ambient temperatures. Forest cities (i.e., Melbourne, Australia) aimed for large-scale planting of trees to adapt to climate change in the next decade. Therefore, monitoring cities' green infrastructure requires close assessment of growth and water status at the tree-by-tree resolution for its proper maintenance and needs to be automated and efficient. This project proposed a novel monitoring system using an integrated visible and infrared thermal camera mounted on top of moving vehicles. Automated computer vision algorithms were used to analyze data gathered at an Elm trees avenue in the city of Melbourne, Australia (n = 172 trees) to obtain tree growth in the form of effective leaf area index (*LAIe*) and tree water stress index (TWSI), among other parameters. Results showed the tree-by-tree variation of trees monitored (5.04 km) between 2016–2017. The growth and water stress parameters obtained were mapped using customized codes and corresponded with weather trends and urban management. The proposed urban tree monitoring system could be a useful tool for city planning and GI monitoring, which can graphically show the diurnal, spatial, and temporal patterns of change of *LAIe* and TWSI to monitor the effects of climate change on the GI of cities.

Keywords: urban tree management; tree monitoring; computer vision; tree water stress index; leaf area index

Citation: Fuentes, S.; Tongson, E.; Gonzalez Viejo, C. Urban Green Infrastructure Monitoring Using Remote Sensing from Integrated Visible and Thermal Infrared Cameras Mounted on a Moving Vehicle. *Sensors* 2021, *21*, 295. https://doi.org/10.3390/s21010295

Received: 26 October 2020
Accepted: 30 December 2020
Published: 4 January 2021

Publisher's Note: MDPI stays neutral with regard to jurisdictional claims in published maps and institutional affiliations.

Copyright: © 2021 by the authors. Licensee MDPI, Basel, Switzerland. This article is an open access article distributed under the terms and conditions of the Creative Commons Attribution (CC BY) license (https://creativecommons.org/licenses/by/4.0/).

1. Introduction

Green infrastructure (GI) has become a priority in most cities worldwide and has been recognized as an essential element in urban planning and development. The urban green infrastructure, which includes natural vegetation, parks, street trees, green roofs, and small gardens, provides various benefits to the environment, community, and the economy. The GI of a city contributes valuable benefits such as regulation and reduction of temperature during heatwaves [1] through plant transpiration [2], while green roofing decreases albedo [3]. GI improves air quality [2] and reduces flood risk [4,5] and stormwater pollution [6], among other environmental benefits. Beyond improving the ecosystem, GI has been linked to improving people's physical and mental health [1]. Within the major challenges in urban cities are extreme heat and the urban heat island effect (UHI). UHI mainly occurs within cities with a higher proportion of concrete in relation to their green infrastructure (GI) [7]. In these cities, the ambient temperature increase can be multiplied by a 1.4–15 factor depending on circumstances within and surrounding a particular city environment [8,9]. UHI may worsen in the future, corresponding to the predicted climate change.

The maintenance of GI in cities may pose a challenge for city councils due to the number and complexity of tree and plant species, especially in cities classified as forest cities, such as Melbourne, Australia [1,2]. The City of Melbourne established a goal as part of the Urban Forestry Strategy to plant 3000 trees per year as one of the primary strategies for the climate adaptation program. The city council aimed to increase canopy to 140,000 trees in 2040, twice the coverage of the existing number of trees that it manages at present [3]. Trained arborists maintain public trees through routine inspection and assessment, which can also be requested through public reporting. Trees situated in heavily accessed areas such as parks and boulevards are inspected annually, while other locations are inspected at least once in 2 years. The city eliminates about 800 tree stands due to various reasons, including threats to safety. A considerable population of trees managed by the City of Melbourne are a century old and may pose higher risk of decline and, consequently, safety risk.

The manual inspection of urban trees by arborists is quite time demanding and inefficient. With the large-scale greening plans in urban cities such as Melbourne, it is highly impractical and nearly impossible to monitor GI to achieve high temporal and spatial resolution through the current manual practice. Other methods use wireless sensor networks, including the monitoring of soil moisture, temperature, light, humidity, and pressure [10], using internet of things (IoT) systems [4], real-time controls [5], and plant/tree-based sensors, such as sap flow probes [6,11]. These methods produce accurate and high temporal data resolution. However, one major disadvantage is that this method can only be applied to a few representative plants, as installing sensors to numerous trees is costly. Furthermore, the sensor networks require frequent monitoring and high maintenance, requiring specialized personnel with specific technical skills.

More spatially representative approaches for GI monitoring of cities are based on satellite remote sensing on relevant vegetation indices (VIs) related to growth and water status [12]. This has been applied in China for 70 major cities [13], and in Sweden [14] using Sentinel-2 and Landsat-8 satellites [13], as well as in Croatia using World View 1, 2, and 3 with high resolution visible and multispectral bands [15]. The use of satellite imagery can monitor large areas from a single image or stitched up images incorporating several square kilometers, which is a major advantage. Also, information can be readily available, and sources can be either free or low-cost (i.e., Landsat and Sentinel satellites). However, disadvantages can include low resolution of information per pixel, reaching 0.5 m for panchromatic imagery, and between 2 m to 30 m per pixel for multispectral imagery. Higher spatial resolution imagery may be expensive, such as those from the World View satellites. Furthermore, satellite revisit time to the same spot (i.e., in cities) may be between 10–15 days, and data quality depends on how clear the skies are.

To address the problem of low temporal and spatial resolution, airborne, and unmanned aerial vehicles (UAV) have been implemented to monitor GI in cities [16–19]. However, the use of airborne remote sensing comes with a cost, requiring a pilot and skilled personnel to operate the instrumentation, process the information, and deliver interpreted information to relevant city council personnel for GI management and decision making. Some services such as Nearmap (Nearmap, Barangaroo, NSW, Australia) offer high-resolution visible images with a high temporal resolution for major cities and coastlines [20,21]. However, the application of visible images is mainly for monitoring of growth parameters for trees [22]. The recent popularity of UAV has also expanded its application in remote sensing, with the accommodation of various camera and sensor payloads aside from visible such as multispectral camera and LIDAR [23–26]. With UAV, the main challenge is the implementation in countries with strict civil aviation regulations, such as many cities in Europe, the United States, and Australia, among others. In Australia, for example, the Civil Aviation Safety Authority (CASA) has a very strict regulation to fly drones within 30 m in proximity to people, making flights in heavily populated cities, such as Melbourne (Victoria, Australia) virtually impossible [27].

This paper proposes a novel GI monitoring approach based on prototype integrated visible and infrared thermal cameras to automatically obtain different VIs based on growth and tree water status parameters on a tree-by-tree scale. The integrated cameras are mounted on top of moving vehicles, circulated through one of the most important and historical Elm tree avenues in Melbourne (Royal Parade), Australia. The integrated system, which is composed of low-cost instrumentation, could be mounted on top of public transport vehicles, such as buses and trams, city council vehicles, and rubbish trucks. Public transport vehicles allow for incursion to potentially every street at multiple times in a day, offering a high temporal and spatial resolution of trees monitoring. The proposed system enables automated data acquisition, analysis, and mapping and does not require specialized personnel. It can also offer diurnal monitoring of trees to assess in real-time the effects of weather anomalies, such as heatwaves, floods, and heavy winds, among others, and the detection of pest and disease incidence. The novel technology (integrated cameras) and application could potentially be an accurate, cost-effective, and user-friendly tool for city councils, and they could base management strategies with high reliability on the system proposed, such as tree lopping, detection of encroachment of tree branches on power lines, and deterioration of old trees.

Only a few studies are based on the implementation of cameras for upward-looking imagery and analysis without any automation. These have been restricted to the 3D modeling of trees using handheld cameras and point cloud analysis [28] and the analysis of trees' thermal characteristics [29]. Other applications using different technologies, such as hyperspectral cameras [30] and low-cost electronic noses (e-noses) [31], using the methodology proposed can be implemented to obtain more information from trees and their environment, such as diagnosis of vegetation health.

This study has been based on the integration of previously developed technology from our research group for the automated analysis of visible and infrared thermal imagery for different crops such as eucalyptus trees [22], grapevines [32–35], kiwi plants [36], apple trees [37], cherry trees [38–40], and cocoa plants [41], among others.

2. Materials and Methods

Data acquisition was performed in Melbourne, Australia, mainly based on an integrated visible and infrared thermal camera developed and processed using customized computer vision algorithms.

2.1. Urban Site and Tree Material Description

The monitoring site (Figure 1) was located along the iconic Royal Parade avenue in the city of Melbourne, Australia, that starts at Grattan Street (−37°48′02.27″ S; 144°57′26.27″ E; 33 m.a.s.l.) finishing on Park Street (−37°46′41.45″ S; 144°57′36.56″ E; 46 m.a.s.l.), and vice versa. The trees are planted along a nature strip separating the main road and the access road in both directions. The main roads (North and South bound) are divided by a median strip containing the Route 19 tram lane. Each way corresponds to 2.52 km, with a total distance of 5.04 km both ways. There are 172 deciduous trees considered in the monitoring, composed of different Elm species (*Ulmus* spp.) planted in 1900 and 1997 [42]. The trees are irrigated using sub-surface irrigation, and tree lopping management is performed regularly by the Melbourne city council.

2.2. Climate and Weather Information Description

The climate in Melbourne is classified as subtropical oceanic with mild winters and pleasant to hot summers. Windy conditions are common, and weather changes can occur within the same day. The average temperatures between November and January are between 22 and 26 °C, with minimum temperatures between 11 and 14 °C. The yearly average precipitation is 670 mm, with an even distribution throughout the year of around 50 mm per month. Sunshine hours are higher between September and March (between 6–9 h). Specific weather data available for the trial site and the monitoring period were acquired from the Bureau of Meteorology, measured from a meteorological station located in Melbourne Olympic Park (Number: 086338) at 3.4 km from Royal Parade. The weather information extracted from this station was: maximum daily temperature (°C), rain (mm), and solar radiation (MJ m^{-2}). Monitoring was performed from November 2016 (late spring) to January 2017 (summer).

Figure 1. Location monitored using a moving vehicle from (**A**) start of Royal Parade from Grattan Street to (**B**) Park Street (2.52 km), and vice versa. There were 172 trees monitored, consisting of different species of Elm trees (*Ulmus* spp.).

2.3. Integrated Visible and Thermal Infrared Camera System

The integrated camera system (Figure 2) consisted of a visible RGB video camera and a thermal infrared camera FLIR AX8™ (FLIR Systems, Wilsonville, OR, USA) with a resolution of 90 × 60 pixels, connected to a web-based system that can simultaneously capture and store the videos and infrared thermal images (IRTIs) to be further downloaded for analysis or transmitted to cloud storage and processing system. The thermal camera had a spectral range of 7.5–13 µm, an accuracy of ±2 °C, and an emissivity of 0.985. The IRTI capture rate was every second. The RGB video camera is connected to a Raspberry Pi Camera Module V2.1 (Raspberry Pi Foundation, Cambridge, UK; Figure 2A), board, and memory card. This device has an 8-megapixel sensor with a resolution of 640 × 360 pixels, 4:3 aspect ratio, and 30 frames per second (fps). Videos were recorded within the unit in H.264 video compression format and automatically converted into Motion Pictures Expert Group-4 (.mp4) files. The camera was fitted with a 3-axis gimbal to minimize movements when acquiring the data (Figure 2A); an integrated temperature, relative humidity, and solar radiation sensors within a 3D printed Stevenson screen (Figure 2A); and a magnetic GPS tracker (Figure 2B). The integrated camera was mounted on top of a car (Figure 2B) with a height between the camera and the tree canopies of approximately 5 m.

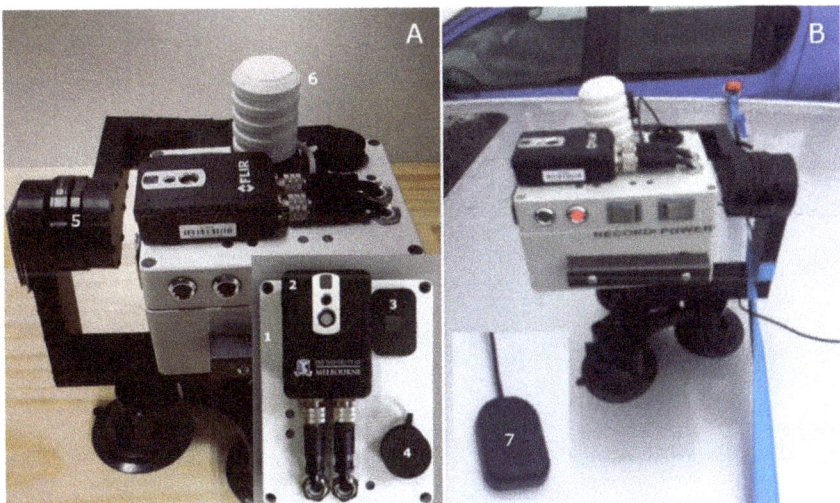

Figure 2. The integrated camera system. (**A**) The system is composed of weather and shock resistant case (1) that holds the Raspberry Pi boards and battery; thermal infrared camera FLIR AX8™ (2); the visible Red, Green, and Blue (RGB) Raspberry Pi Camera Module V2.1 (3); power mount receptacle (4) to charge the internal battery; 3-axis gimbal (5) to provide stability to the camera; integrated temperature, relative humidity, and solar radiation sensors (Stevenson screen, 6). (**B**) Example of the mounting procedure of the integrated camera on top of a vehicle. The camera was also integrated with a magnetic GPS tracker (7).

The radiometric data from the thermal infrared camera were obtained every second while traveling through the Royal Parade. They were recorded as in comma-separated values (.csv) format files and the visible RGB images in Joint Photographic Experts Group (.jpg). Both sets of data were obtained using the Sense Batch software (SENSE Software, Warszawa, Mazowsze, Poland). The data were analyzed using customized codes developed and updated using Matlab® R2020b (Mathworks Inc., Natick, MA, USA).

The camera's integrated sensors consisted of an AM2302 (wired DHT22) temperature-humidity sensor (Guangzhou Aosong Electronics Co., Ltd., Guangzhou, China). This sensor can obtain new data from it once every 2 s (0.5 Hz), which is accurate for 0–100% humidity readings with 2–5% accuracy and −40 to 80 °C temperature readings with ±0.5 °C accuracy. The SP-510-SS upward-Looking Thermopile Pyranometer (Apogee Instruments, Inc., Logan, UT, USA) has a sensitivity of 0.05 mV per W m^{-2}, with a measurement range between 0 to 2000 W m^{-2} (net shortwave irradiance) and repeatability of <1%. The detector response time is 0.5 s with a field of view of 180° and spectral range of 385–2105 nm, directional (Cosine) response less than 30 W m^{-2} at 80° solar zenith, temperature response: <5% from −15 to 45 °C at the operating environment: −50 to 80 °C, and 0 to 100% relative humidity.

2.4. Image Pre-Processing and Computer Vision Algorithms

Every frame corresponding to a canopy from the visible (RGB) video and infrared thermal images were analyzed using the computer vision algorithms described in Sections 2.4.1 and 2.4.2, respectively. Figure 3 shows an example of a visible (RGB) frame and corresponding infrared thermal image from an Elm tree canopy along the Royal Parade.

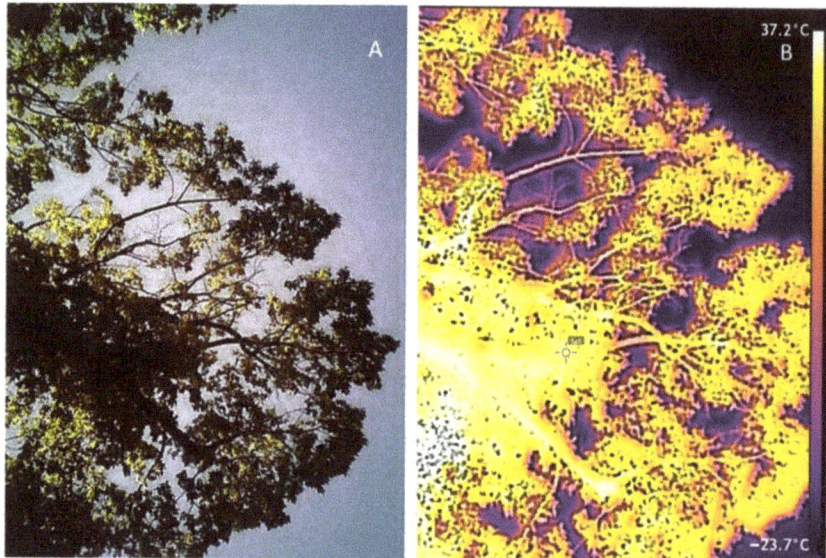

Figure 3. Example of a visible (RGB) image (**A**) and the corresponding infrared thermal image (**B**) from an Elm tree taken using the integrated camera on top of a moving vehicle.

The pre-processing of the RGB images consisted of the binarization (Figure 4A) using the blue channel from the RGB images by selecting the lowest part of the histogram curve (valley) detected automatically between the pixels corresponding to the canopy material (first peak) and the background or sky (second peak) (Figure 4B). After binarization, each image was automatically subdivided into a 5 × 5 sub-images to perform gap analysis. A large gap (lg) per sub-image was considered when there was over 75% of sky. Total pixels (tp) corresponded to a fixed value related to the resolution of the camera used. This pre-analysis has been described in detail in Fuentes et al. [22].

The pre-processing of the thermal images was performed in batch after each measurement campaign using the SENSE Batch software (Sense Software, Warszawa, Mazowsze, Poland), which extracts radiometric data per pixel in a comma-separated file (.csv) in the form of a matrix processed in Matlab (Figure 4C). Leaf material was selected by simple automatic elimination of temperatures below 0 + °C since this separates the sky from the canopy material (Figure 4D). From the segmented image, the canopy temperature was automatically extracted (T_{canopy}) as entry parameter for the TWSI and Ig calculation (Equations (7) and (8)).

2.4.1. Canopy Architecture and Growth Parameters

Videos from the visible camera were processed automatically using a customized code written in Matlab® R2020b to analyze frames following a computational process proposed by Fuentes et al. (2008) [22].

Canopy architecture parameters were obtained using the following algorithms considering the fractions of foliage projective cover (f_f), crown cover (f_c), and crown porosity (Φ), which were calculated using the following computational algorithms proposed by Fuentes et al. (2008) [22]:

$$f_f = 1 - \frac{tg}{tp} \qquad (1)$$

$$f_c = 1 - \frac{lg}{tp} \qquad (2)$$

$$\phi = 1 - \frac{f_f}{f_c}, \tag{3}$$

where lg = large gap pixels, tg = total pixels in all gaps, and tp = total gap pixels.

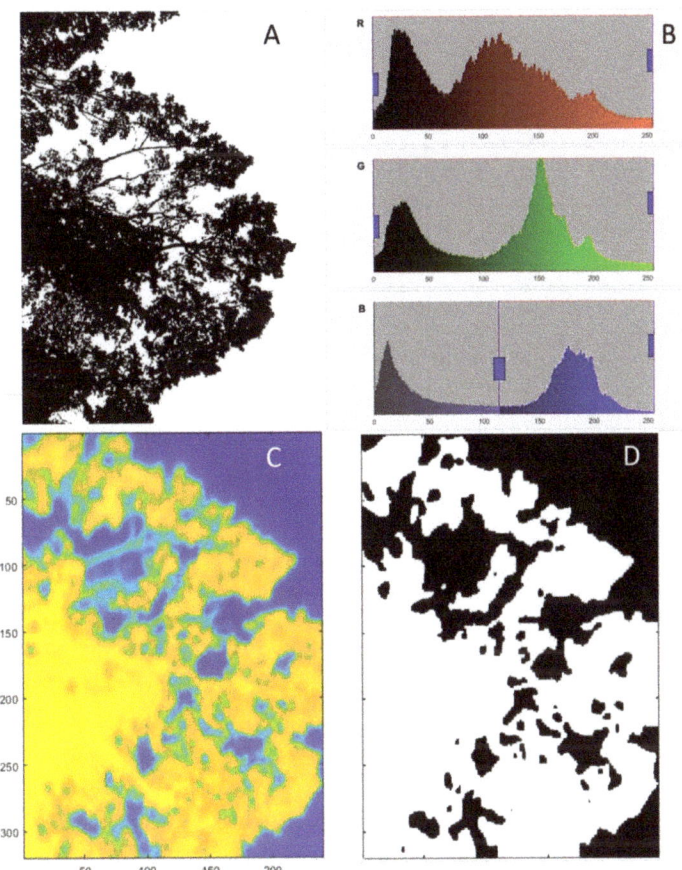

Figure 4. Example of the automated pre-processing of visible (RGB) images transformed to binary images (**A**) using the blue channel as filter (**B**) and the corresponding infrared thermal image (**C**) filtered to create a mask (**D**) to account for leaf material.

LAI (adimensional) is calculated from Beer's Law, defined as the total one-sided area of leaf tissue per unit 3 ground surface area [43]. Hence, the LAI values describe m² of leaf area per m² of soil.

$$LAI = -f_c \frac{\ln \phi}{k} \tag{4}$$

where k = coefficient of light extinction (k = 0.5), which is applicable for tall trees [22], and the clumping index at the zenith, $\Omega(0)$, was calculated as follows:

$$\Omega(0) = \frac{(1-\phi)\ln(1-f_f)}{\ln(\phi)/f_f}. \tag{5}$$

The clumping index is a correction factor in obtaining effective LAI (LAI_e), also adimensional, which is the product of:

$$LAI_e = LAIx\Omega(0). \tag{6}$$

Equation (5) describes the non-random distribution of canopy elements. If $\Omega(0) = 1$ means that the canopy displays random dispersion, then for $\Omega(0)>$ or <1, the canopy is defined as clumped.

2.4.2. Infrared Thermal Image Analysis

A tree water stress index (TWSI) was derived from the common crop water stress index (CWSI) [32] used in agriculture, which is a normalized value (0–1) and, therefore adimensional, and it was calculated using the following equation after determining T_{dry} and T_{wet} [44]:

$$\text{TWSI} = \frac{T_{canopy} - T_{wet}}{T_{dry} - T_{wet}} \tag{7}$$

where T_{canopy} is the actual canopy temperature extracted from the thermal image at determined positions, and T_{dry} and T_{wet} are the reference temperatures (in °C) obtained using the statistical temperature distribution discrimination described in published research [39].

An infrared index (I_g), which is adimensional and proportional to leaf conductance and water vapor transfer (g_s), can be obtained using the relationship as follows [45]:

$$I_g = \frac{T_{canopy} - T_{wet}}{T_{dry} - T_{wet}} = g_s \left(r_{aw} + \left(\frac{s}{\gamma}\right) r_{HR}\right) \tag{8}$$

where r_{aw} = boundary layer resistance to water vapor, γ = psychrometric constant, and s = slope of the curve relating saturation vapor pressure to temperature [45,46].

For automated analysis, the leaf energy balance approached was implemented using integrated sensors within the camera described in Figure 2A as [32]:

$$Tdry - Ta = \frac{r_{HR} R_{ni}}{\rho c_p} \tag{9}$$

where Ta is the air temperature measured at the same positions and time as infrared thermography acquisition, r_{RH} = the parallel resistance to heat and radiative transfer, R_{ni} is the net isothermal radiation (the net radiation that would be received by an equivalent surface at air temperature), ρ is the density of air, and c_p is the specific heat capacity of air. This formula uses the concept of isothermal radiation and assumes a dry surface with the same aerodynamic and radiative properties, in which the sensible heat loss will equal the net radiation absorbed [47].

$$Twet - Ta = \frac{r_{HR} r_{aW} \gamma R_{ni}}{\rho c_p [\gamma(r_{aW}) + s r_{HR}]} - \frac{r_{HR} \delta e}{\gamma(r_{aW}) + s r_{HR}} \tag{10}$$

The thresholds $Twet$ and $Tdry$ are references that can be leaves painted with water ($Twet$) and use petroleum jelly ($Tdry$) to obtain through infrared thermography the maximum and minimum temperatures to be found within a specific canopy at the time of measurements [32]. The leaf energy balance approach allows the implementation of an automated procedure to obtain these thresholds using the sensors incorporated in the integrated camera proposed (Figure 2).

2.5. Survey, Automated Detection of Trees Location, Data Extraction, and Mapping

Acquisition of images was performed on four dates: twice in November 2016 (17 and 19 November), followed by 19 December 2016 and 16 January 2017. The image surveys were all performed at 1–2 pm during maximum atmospheric demand (maximum vapor pressure deficit), a common practice in agriculture to assess plant water status for irrigation assessment requirements.

For the 172 Elm trees monitored in this study, the GPS location was extracted from Google Earth Pro (Googleplex, Mountain View, CA, USA). The tree positions were used as anchors to automatically extract information from procedures previously explained for canopy architecture and infrared thermal-based parameters. The automated extraction consisted of identifying the nearest coordinates registered in the integrated camera to the anchored GPS for specific trees.

Once the data were extracted, they were mapped using a customized code written in Matlab® R2020b to produce: (i) geo-located icons (circles) with relative sizes to denote changes in growth ($LAIe$), and (ii) geo-located circles with a different color to represent different TWSI values. The process can be used to map any parameter extracted using Equations (1)–(10).

3. Results
3.1. Weather Data within the Period of Measurement and Calculated Parameters

Figure 5 shows the weather information acquired from the closest meteorological station from the trial site. The first two dates of measurement (A: 17 November 2016 and B: 29 November 2016) had maximum rain events of 12.6 and 17 mm of rain in the previous week, and maximum temperatures of 31.4 and 20 °C, respectively. These dates had high solar radiation (28.8 MJ m^{-2} and 31 MJ m^{-2}, respectively). In the last two measurement surveys (19 December 2016 and 16 January 2017), there were no or minimal rain events (0 mm and 4 mm, respectively) within two weeks preceding the measurements. Both dates had high maximum temperatures and solar radiation values (30.02 and 32.7 °C and 29.3 and 30.5 MJ m^{-2}, respectively).

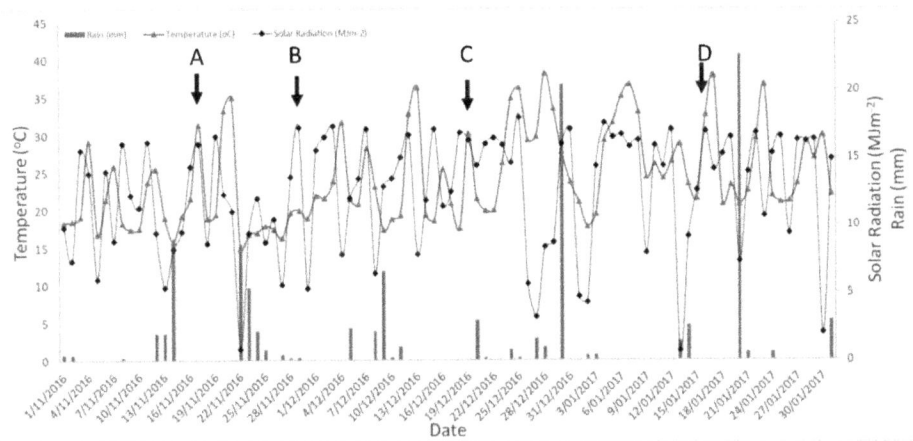

Figure 5. Meteorological data showing daily maximum temperature (°C) solar radiation (MJ m^{-2}) and rain (mm) in the Royal Parade for four different dates studied: (**A**) 17 November 2016, (**B**) 29 November 2016, (**C**) 19 December 2016, and (**D**) 16 January 2017.

Table 1 shows the main canopy and tree water status parameters for all the measurement survey days. There was considerable variation in LAI and $LAIe$ from a minimum of 0.61 and 0.41, found the last date of measurement, to maximum values of 5.98 for LAI in the first date of measurement and 4.97 $LAIe$ for the second date of measurement. Furthermore, the lowest Tc values, TD, and TWSI corresponded to the second measurement date.

Table 1. Growth and water stress parameters obtained using the proposed urban tree monitoring system, measured at Royal Parade in Melbourne, Australia, for four measurement surveys between 2016 and 2017. Parameters are presented with maximum, minimum, means, and standard deviation values (SD) for leaf area index (*LAI*, adimensional), effective *LAI* (*LAIe*, adimensional), canopy temperature of trees (Tc, °C), temperature depression (TD, °C), thermal infrared index (Ig, adimensional), and tree water stress index (TWSI, adimensional).

Parameter/Date	17 November 2016				29 November 2016				19 December 2016				16 January 2017			
	Min	Max	Mean	SD	Min	Max	Mean	SD	Min	Max	Mean	SD	Min	Max	Mean	SD
LAI	0.81	5.98	2.67	±1.17	0.86	5.33	2.58	±0.90	0.63	4.88	2.70	±0.88	0.61	4.11	1.86	±0.57
LAIe	0.48	3.56	1.59	±0.70	0.80	4.97	2.41	±0.84	0.47	3.67	2.03	±0.66	0.41	2.72	1.23	±0.38
Tc	25.9	30.7	28.2	±1.21	16.5	21.5	19.3	±1.10	23.6	30.3	27.9	±1.05	23.7	36.6	31.5	±1.99
TD	0.7	5.5	3.2	±1.21	−1.5	3.4	0.7	±1.10	−0.1	6.6	2.3	±1.05	−3.9	9.1	1.3	±1.99
Ig	0.19	0.93	0.43	±0.12	0.26	1.36	0.66	±0.17	0.20	1.07	0.39	±0.12	0.18	1.19	0.45	±0.15
TWSI	0.52	0.84	0.70	±0.06	0.42	0.79	0.61	±0.06	0.48	0.84	0.73	±0.06	0.46	0.84	0.70	±0.07

3.2. Comparative Analysis of Main Extracted Parameters from Trees

Figure 6A compares growth parameters (LAIe) for the 172 trees monitored with the TWSI for the different measurement dates. The trends followed apparent curvilinear relationships with the last two dates (19 December 2016 and 16 January 2017) with lower LAIe and higher TWSI than the earliest dates (17 November 2016 and 29 November 2016). Figure 6B shows the comparison between the Ig and TD parameters related to stomatal conductance from trees. There was contrasting behavior of these parameters for the first two dates with lower Ig and higher TD for the first and flat distribution of the whole range of Ig values with low TD close to the 0 values. On the contrary, the last two dates had similar behavior with low Ig and TD values ranging from −3 to around 3 °C.

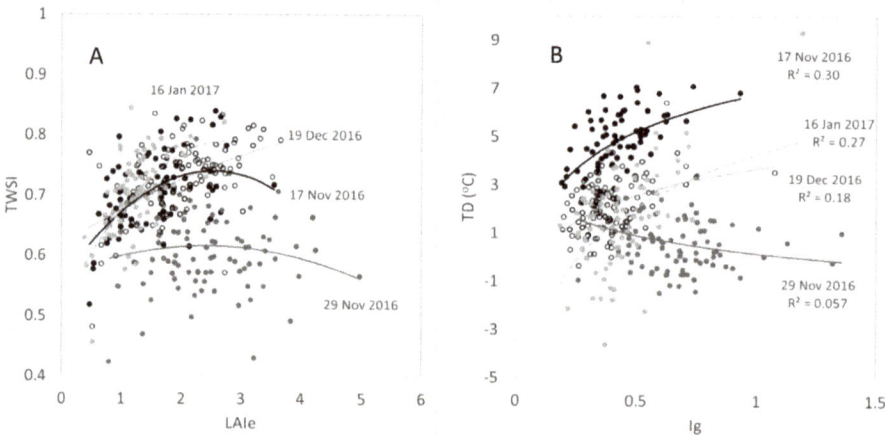

Figure 6. Comparison between effective leaf area index (LAIe, dimensionless) and tree water stress index (TWSI) (**A**), and between the infrared thermal index (Ig) and temperature depression (TD, °C) (**B**) for 172 elm trees monitored along the Royal Parade in Melbourne, Australia for four different dates between 2–16 and 2017, using the proposed urban tree monitoring system.

3.3. Main Growth and Tree Water Stress Parameters Map

Figures 7 and 8 show the proposed urban tree monitoring system's main outputs, displaying the main parameters extracted per tree along Royal Parade in four measurement dates. Figure 7 shows the LAIe for different trees with the relative size of circles corresponding to trees changing according to growth differences between dates. Figure 8 shows changes in color of circles representing the trees relative to the TWSI for different dates.

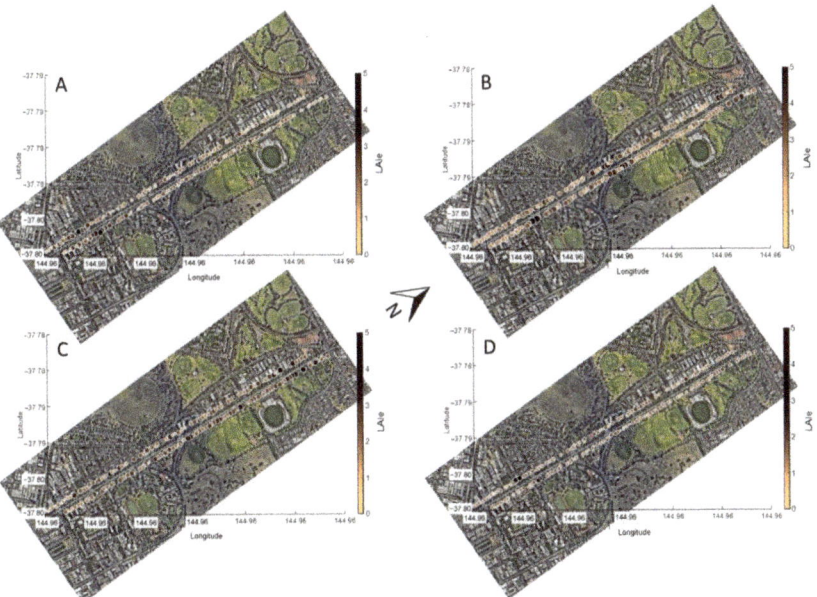

Figure 7. Mapping of effective leaf area index (*LAIe*) along the Royal Parade (5.04 Km) of 172 trees using the proposed urban tree monitoring system for four different dates: (**A**) 17 November 2016, (**B**) 29 November 2016, (**C**) 19 December 2016, and (**D**) 16 January 2017. Different colors and relative circle sizes correspond to the *LAIe* scale.

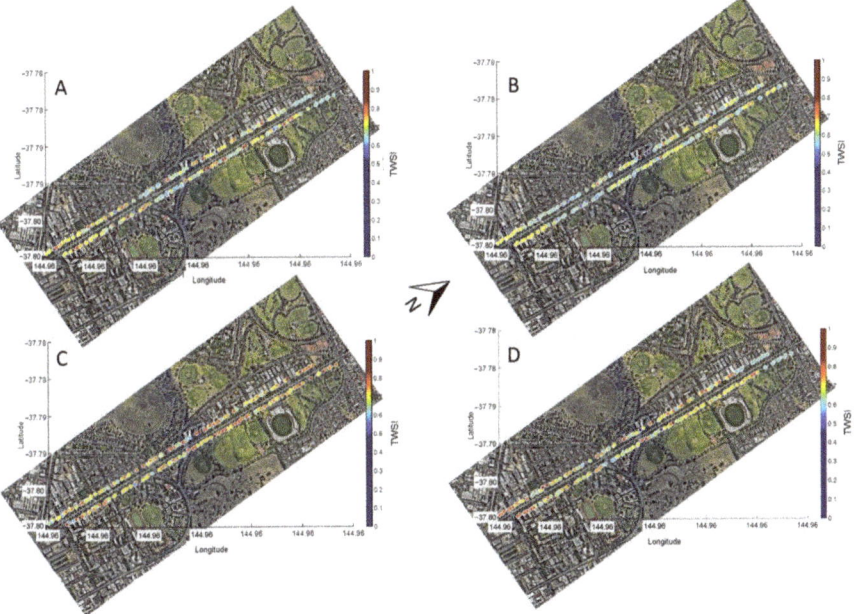

Figure 8. Mapping of tree water stress index (TWSI) along the Royal Parade (5.04 Km) of 172 trees using the proposed urban tree monitoring system for four different dates: (**A**) 17 November 2016, (**B**) 29 November 2016, (**C**) 19 December 2016, and (**D**) 16 January 2017. Different colors correspond to the TWSI scale.

4. Discussion

The proposed urban tree monitoring system that uses an integrated camera on moving vehicles can automatically provide information on trees' growth and water status changes, which can serve as a powerful decision-making tool for city councils for tree management (i.e., supply water requirement at appropriate times and tree lopping for power lines encroachment and public safety management). The reliability of the system is based on the growth and canopy architecture parameters and algorithms used, which have been successfully implemented for other trees such as eucalyptus [22], and tree crops such as cherry trees [38,40], apple trees [37], and grapevines [33,48–50]. Tree water stress algorithms have been used to describe the water status of many trees and crops [32,51–53]. Furthermore, most of the tree canopies were visible in the field of view of canopies for both visible and infrared thermal images (Figures 3 and 4), making the analysis representative of the whole tree. Furthermore, since images are upward-looking, the monitored parts of the trees were the under canopy and were shaded, which has been regarded as the most consistent and representative part to monitor using infrared thermal imagery [46,54].

The sensitivity of the growth and physiological parameters obtained and their variations are specifically shown in Figure 6 and compared between the trees measured (individually) and temporally (within dates). The variation is sensible in response to weather conditions and changes related to atmospheric demand (temperature) and water availability (rain). These trends and their sensitivity are further supported by the mapping of the processed data in the form of *LAI* (Figure 7) and TWSI (Figure 8). The parameters obtained are in accordance with weather information acquired within the measurement dates. The first two dates (17 November 2016 and 29 November 2016) corresponded to milder weather, with cooler weather for the second date with a maximum temperature of 20 °C, followed by rain events. For the second date, lower atmospheric demands produced a flat response for TWSI and TD. In the case of TD, lower TD values, close to 0 °C, are related to low stomata opening and transpiration (Figure 6B). However, they were not associated with higher TWSI (Figure 6A). The rest of the dates (19 December 2016 and 16 January 2017) have more significant increases in TWSI with higher atmospheric demands (evapotranspiration), as shown by higher maximum temperatures and solar radiation. The highest and more significant determination coefficient (Figure 6B) between Ig and TD was found for the dates with higher atmospheric demand (first, third, and fourth dates), which was expected since these parameters are related to stomata aperture [32,54].

Another advantage of the proposed system is that it allows the automatic mapping of data obtained from surveys on a tree-per-tree scale (Figures 7 and 8). For growth parameters, such as *LAIe* (Figure 7), some of the trees with higher growth showed decreased *LAIe* from the first to the second date of measurement, which may be related to continuous tree lopping management from the council (Figure 7A,B). However, the lowest and most consistent *LAIe* values were found in the last date of measurement (January 2017), which corresponded to one of the hottest months in summer and the starting of the senescence stage for the Elm trees (Figure 7D), in comparison to the previous dates. For TWSI, the parameter trends followed water availability from rain events during the last weeks and maximum temperatures with the highest values corresponding to the warmest dates in December 2016 and January 2017 (Figure 8C,D).

The integrated cameras could be mounted on public transport of cities, such as buses and trams. The installation of the system on trams is ideal, being on rails are on a fixed route, which can offer more precise data acquisition and more reliable comparative analysis. Furthermore, at least along the Royal Parade route (Route 19), a particular tram can pass through the same spot every 80–90 min, which can acquire at least 13 data points in one day from 5 am to midnight. To access more places within the city, such as suburb streets, cameras could be installed on rubbish trucks and buses, which have more extensive access to residential areas. This layout of the trams' path is similar to many European cities since they have similar designs.

The diurnal data collected may be more relevant for infrared thermal parameters to assess tree water status changes throughout the day compared to changes in growth, which are expected to be minimal, while data related to changes in leaf or branch angle due to water stress after sunset could be relevant to assess night-time water loss by trees, as this phenomenon is relevant to other tree species and crops [55–59]. Continuous daily data of water stress may also offer insights of tree behavior within heat waves [60], pest and disease interactions [61,62], windy days, and mortality estimates [63]. The volume of data that can be gathered through the proposed system allows the implementation of machine learning modeling and artificial intelligence to promptly detect problems for management and mitigation, avoiding damage to infrastructure and the public due to unpredicted fallen trees or big branches.

The system has been proposed, and data analysis can be deployed as a user-friendly digital platform producing maps with tree water status and growth maps depicted in Figures 7 and 8. Users can click on any individual tree and obtain numerical and other management information as it is already set up for planting date and basic information of trees by the Melbourne city council [3]. Furthermore, since the conception of the integrated camera idea, on which this paper was based, FLIR has released an integrated visible 4K video camera and a high resolution infrared thermal imaging: FLIR Duo Pro (FLIR Systems, Wilsonville, OR, USA). This camera is intended to be mounted as a payload for UAV vehicles. It can also be used to acquire data to obtain the analysis proposed in this paper mounted on vehicles as per Figure 2B. The downside will be the costs of using these cameras if many vehicles are required for this purpose. It is thought that the higher resolution from the FLIR camera will not impact with statistical significance results obtained with the low-cost camera system presented in this paper. The latter is supported by previous research that has compared different resolutions of visible and thermal infrared cameras for growth and water status assessment on trees with no significant differences for the parameters studied [39,64], which can be explained by the short height between the camera and the canopies included for these type of studies, which is between 3–5 m.

This study was based on an extensive avenue in which there were a predominant tree species. Hence, further studies should be conducted for different tree species to account for the variety that exists in a normal urban green infrastructure environment. Even though the algorithms used in this study have been proven to be robust for other horticultural tree species, specific calibrations should be made to consider different canopy architectures and sensitivity/tolerance to different water stress levels.

5. Conclusions

The urban green infrastructure could be automatically monitored using a low-cost integrated camera system mounted on top of moving vehicles. Specifically, the main advantages of the system described in this paper compared to similar studies to monitor the green infrastructure in urban environments are: (i) low-cost instrumentation required to integrate visible and infrared thermal cameras; (ii) the system can be mounted on public transport such as buses, trams, and city council vehicles with the extra advantage when considering garbage trucks since they can access every street of a city if extensive monitoring is required; (iii) it could provide high spatial and temporal data resolution, which is related to the frequency of public transport through the same trees; (iv) algorithms implemented are robust and have been successfully tested on a wide variety of horticultural trees; (v) the system does not require special permits or trained pilots, such as the case of UAVs, and they also do not have restrictions due to privacy issues since they monitor urban infrastructure in an upward-looking fashion above the pedestrian level. These operational, cost-effectiveness, accuracy, and privacy-related advantages of the system proposed can be compared to those of manual measurements of green infrastructure, using sensors and IoT on sentinel trees, remote sensing using satellites, UAVs, or the airborne instrumentation (Nearmap) discussed in this paper. Furthermore, the high volume of data collected (spatial and temporal) using the system proposed in this paper could allow the implementation of machine learning algorithms and artificial intelligence (AI) to obtain further vegetation indices of trees to manage the cities' green infrastructure efficiently, to maximize resources, and to minimize detrimental effects of climate change and risk to infrastructure and people.

Author Contributions: Conceptualization, S.F. and E.T.; methodology, S.F.; software, S.F.; validation, S.F., E.T. and C.G.V.; formal analysis, S.F., E.T. and C.G.V.; investigation, S.F.; resources, S.F. and E.T.; data curation, S.F., E.T. and C.G.V.; writing—original draft preparation, S.F.; writing—review and editing, E.T. and C.G.V.; visualization, S.F.; funding acquisition, S.F. All authors have read and agreed to the published version of the manuscript.

Funding: This research was partially funded by the Melbourne Network Society, belonging to The University of Melbourne.

Institutional Review Board Statement: Not Applicable.

Informed Consent Statement: Not Applicable.

Data Availability Statement: Data and intellectual property belong to The University of Melbourne; any sharing needs to be evaluated and approved by the University.

Conflicts of Interest: The authors declare no conflict of interest.

References

1. Phillips, C.; Atchison, J. Seeing the trees for the (urban) forest: More-than-human geographies and urban greening. *Aust. Geogr.* **2020**, *51*, 155–168. [CrossRef]
2. Han, J.; Tang, B.; Hou, S. Spatial Pattern Characteristics and Influencing Factors of National Forest Cities in China. *J. Landsc. Res.* **2019**, *11*, 35–40.
3. City of Melbourne. Green Our City Strategic Action Plan. Available online: https://data.melbourne.vic.gov.au/Environment/Trees-with-species-and-dimensions-Urban-Forest-/fp38-wiyy/data (accessed on 26 October 2020).
4. Lv, Z.; Hu, B.; Lv, H. Infrastructure monitoring and operation for smart cities based on IoT system. *IEEE Trans. Ind. Inform.* **2019**, *16*, 1957–1962. [CrossRef]
5. Lewellyn, C.; Wadzuk, B.; Traver, R. Performance optimization of a green infrastructure treatment train using real-time controls. Proceedings of International Low Impact Development Conference 2016: Mainstreaming Green Infrastructure, Portland, ME, USA, 29–31 August 2016; pp. 123–130.
6. Matasov, V.; Marchesini, L.B.; Yaroslavtsev, A.; Sala, G.; Fareeva, O.; Seregin, I.; Castaldi, S.; Vasenev, V.; Valentini, R. IoT Monitoring of Urban Tree Ecosystem Services: Possibilities and Challenges. *Forests* **2020**, *11*, 775. [CrossRef]
7. Mohajerani, A.; Bakaric, J.; Jeffrey-Bailey, T. The urban heat island effect, its causes, and mitigation, with reference to the thermal properties of asphalt concrete. *J. Environ. Manag.* **2017**, *197*, 522–538. [CrossRef]
8. Wouters, H.; De Ridder, K.; Poelmans, L.; Willems, P.; Brouwers, J.; Hosseinzadehtalaei, P.; Tabari, H.; Broucke, S.V.; van Lipzig, N.P.; Demuzere, M. Heat stress increase under climate change twice as large in cities as in rural areas: A study for a densely populated midlatitude maritime region. *Geophys. Res. Lett.* **2017**, *44*, 8997–9004. [CrossRef]
9. Chapman, S.; Watson, J.E.; Salazar, A.; Thatcher, M.; McAlpine, C.A. The impact of urbanization and climate change on urban temperatures: A systematic review. *Landsc. Ecol.* **2017**, *32*, 1921–1935. [CrossRef]
10. Le, T.; Wang, L.; Haghani, S. Design and implementation of a DASH7-based wireless sensor network for green infrastructure. In Proceedings of the World Environmental and Water Resources Congress 2019: Emerging and Innovative Technologies and International Perspectives, Pittsburgh, PA, USA, 19–23 May 2019; pp. 118–129.
11. Jones, T.S. Advances in Environmental Measurement Systems: Remote Sensing of Urban Methane Emissions and Tree Sap Flow Quantification. Ph.D. Thesis, Harvard University, Cambridge, MA, USA, 2019.
12. Xue, J.; Su, B. Significant remote sensing vegetation indices: A review of developments and applications. *J. Sens.* **2017**, *2017*. [CrossRef]
13. Kuang, W.; Dou, Y. Investigating the Patterns and Dynamics of Urban Green Space in China's 70 Major Cities Using Satellite Remote Sensing. *Remote Sens.* **2020**, *12*, 1929. [CrossRef]
14. Furberg, D.; Ban, Y.; Mörtberg, U. Monitoring Urban Green Infrastructure Changes and Impact on Habitat Connectivity Using High-Resolution Satellite Data. *Remote Sens.* **2020**, *12*, 3072. [CrossRef]
15. Gašparović, M.; Medak, D.; Miler, M. Geospatial monitoring of green infrastructure–case study Zagreb, Croatia. In Proceedings of the International Conference SGEM Vienna GREEN 2017, Vienna, Austria, 27–29 November 2017.
16. Perc, M.N.; Cirella, G.T. Evaluating green infrastructure via unmanned aerial systems and optical imagery indices. In *Sustainable Human–Nature Relations*; Springer: Berlin/Heidelberg, Germany, 2020; pp. 171–184.
17. Dimitrov, S.; Georgiev, G.; Georgieva, M.; Gluschkova, M.; Chepisheva, V.; Mirchev, P.; Zhiyanski, M. Integrated assessment of urban green infrastructure condition in Karlovo urban area by in-situ observations and remote sensing. *One Ecosyst.* **2018**, *3*, e21610. [CrossRef]
18. Bartesaghi-Koc, C.; Osmond, P.; Peters, A. Mapping and classifying green infrastructure typologies for climate-related studies based on remote sensing data. *Urban For. Urban Green.* **2019**, *37*, 154–167. [CrossRef]

19. Koc, C.B.; Osmond, P.; Peters, A.; Irger, M. Understanding land surface temperature differences of local climate zones based on airborne remote sensing data. *IEEE J. Sel. Top. Appl. Earth Obs. Remote Sens.* **2018**, *11*, 2724–2730.
20. Lumiatti, G.; Carley, J.T.; Drummond, C.D.; Vos, K. Use of emerging remote sensing technologies for measuring long-term shoreline change and coastal management. In Proceedings of the Australasian Coasts and Ports 2019 Conference: Future directions from 40 [degrees] S and beyond, Hobart, Australia, 10–13 September 2019; p. 797.
21. Evans, S.M.; Griffin, K.J.; Blick, R.A.; Poore, A.G.; Vergés, A. Seagrass on the brink: Decline of threatened seagrass Posidonia australis continues following protection. *PLoS ONE* **2018**, *13*, e0190370. [CrossRef] [PubMed]
22. Fuentes, S.; Palmer, A.R.; Taylor, D.; Zeppel, M.; Whitley, R.; Eamus, D. An automated procedure for estimating the leaf area index (*LAI*) of woodland ecosystems using digital imagery, MATLAB programming and its application to an examination of the relationship between remotely sensed and field measurements of *LAI*. *Funct. Plant Biol.* **2008**, *35*, 1070–1079. [CrossRef] [PubMed]
23. Ritter, B. Use of Unmanned Aerial Vehicles (UAV) for Urban Tree Inventories. Master's Thesis, Clemson University, Clemson, SC, USA, 2014.
24. Näsi, R.; Honkavaara, E.; Blomqvist, M.; Lyytikäinen-Saarenmaa, P.; Hakala, T.; Viljanen, N.; Kantola, T.; Holopainen, M. Remote sensing of bark beetle damage in urban forests at individual tree level using a novel hyperspectral camera from UAV and aircraft. *Urban For. Urban Green.* **2018**, *30*, 72–83. [CrossRef]
25. Wei, L.; Huang, C.; Wang, Z.; Wang, Z.; Zhou, X.; Cao, L. Monitoring of Urban Black-Odor Water Based on Nemerow Index and Gradient Boosting Decision Tree Regression Using UAV-Borne Hyperspectral Imagery. *Remote Sens.* **2019**, *11*, 2402. [CrossRef]
26. Miyoshi, G.T.; Arruda, M.d.S.; Osco, L.P.; Marcato Junior, J.; Gonçalves, D.N.; Imai, N.N.; Tommaselli, A.M.G.; Honkavaara, E.; Gonçalves, W.N. A Novel Deep Learning Method to Identify Single Tree Species in UAV-Based Hyperspectral Images. *Remote Sens.* **2020**, *12*, 1294. [CrossRef]
27. Molnar, A.; Parsons, C. Unmanned Aerial Vehicles (UAVs) and law enforcement in Australia and Canada: Governance through 'privacy' in an era of counter-law? In *National Security, Surveillance and Terror*; Springer: Berlin/Heidelberg, Germany, 2016; pp. 225–247.
28. Miller, J.; Morgenroth, J.; Gomez, C. 3D modelling of individual trees using a handheld camera: Accuracy of height, diameter and volume estimates. *Urban For. Urban Green.* **2015**, *14*, 932–940. [CrossRef]
29. Lee, S.; Moon, H.; Choi, Y.; Yoon, D.K. Analyzing thermal characteristics of urban streets using a thermal imaging camera: A case study on commercial streets in Seoul, Korea. *Sustainability* **2018**, *10*, 519. [CrossRef]
30. Hernández-Clemente, R.; Hornero, A.; Mottus, M.; Penuelas, J.; González-Dugo, V.; Jiménez, J.; Suárez, L.; Alonso, L.; Zarco-Tejada, P.J. Early diagnosis of vegetation health from high-resolution hyperspectral and thermal imagery: Lessons learned from empirical relationships and radiative transfer modelling. *Curr. For. Rep.* **2019**, *5*, 169–183. [CrossRef]
31. Viejo, C.G.; Fuentes, S.; Godbole, A.; Widdicombe, B.; Unnithan, R.R. Development of a low-cost e-nose to assess aroma profiles: An artificial intelligence application to assess beer quality. *Sens. Actuators B Chem.* **2020**, *308*, 127688. [CrossRef]
32. Fuentes, S.; De Bei, R.; Pech, J.; Tyerman, S. Computational water stress indices obtained from thermal image analysis of grapevine canopies. *Irrig. Sci.* **2012**, *30*, 523–536. [CrossRef]
33. Fuentes, S.; Poblete-Echeverría, C.; Ortega-Farias, S.; Tyerman, S.; De Bei, R. Automated estimation of leaf area index from grapevine canopies using cover photography, video and computational analysis methods. *Aust. J. Grape Wine Res.* **2014**, *20*, 465–473. [CrossRef]
34. Fuentes, S.; Tongson, E.J.; De Bei, R.; Viejo, C.G.; Ristic, R.; Tyerman, S.; Wilkinson, K. Non-Invasive Tools to Detect Smoke Contamination in Grapevine Canopies, Berries and Wine: A Remote Sensing and Machine Learning Modeling Approach. *Sensors* **2019**, *19*, 3335. [CrossRef]
35. Baofeng, S.; Jinru, X.; Chunyu, X.; Yuyang, X.; Fuentes, S. Digital surface model applied to unmanned aerial vehicle based photogrammetry to assess potential biotic or abiotic effects on grapevine canopies. *Int. J. Agric. Biol. Eng.* **2016**, *9*, 119–130.
36. Xue, J.; Fan, Y.; Su, B.; Fuentes, S. Assessment of canopy vigor information from kiwifruit plants based on a digital surface model from unmanned aerial vehicle imagery. *Int. J. Agric. Biol. Eng.* **2019**, *12*, 165–171. [CrossRef]
37. Poblete-Echeverría, C.; Fuentes, S.; Ortega-Farias, S.; Gonzalez-Talice, J.; Yuri, A.J. Digital Cover Photography for Estimating Leaf Area Index (*LAI*) in Apple Trees Using a Variable Light Extinction Coefficient. *Sensors* **2015**, *15*, 2860–2872. [CrossRef]
38. Mora, M.; Avila, F.; Carrasco-Benavides, M.; Maldonado, G.; Olguín-Cáceres, J.; Fuentes, S. Automated computation of leaf area index from fruit trees using improved image processing algorithms applied to canopy cover digital photograpies. *Comput. Electron. Agric.* **2016**, *123*, 195–202. [CrossRef]
39. Carrasco-Benavides, M.; Antunez-Quilobrán, J.; Baffico-Hernández, A.; Ávila-Sánchez, C.; Ortega-Farías, S.; Espinoza, S.; Gajardo, J.; Mora, M.; Fuentes, S. Performance Assessment of Thermal Infrared Cameras of Different Resolutions to Estimate Tree Water Status from Two Cherry Cultivars: An Alternative to Midday Stem Water Potential and Stomatal Conductance. *Sensors* **2020**, *20*, 3596. [CrossRef]
40. Carrasco-Benavides, M.; Mora, M.; Maldonado, G.; Olguín-Cáceres, J.; von Bennewitz, E.; Ortega-Farías, S.; Gajardo, J.; Fuentes, S. Assessment of an automated digital method to estimate leaf area index (*LAI*) in cherry trees. *N. Z. J. Crop Hortic. Sci.* **2016**, *44*, 247–261. [CrossRef]
41. Fuentes, S.; Gonzalez Viejo, C.; Wang, X.; Torrico, D.D. Aroma and quality assessment for vertical vintages using machine learning modelling based on weather and management information. In Proceedings of the 21st GiESCO International Meeting, Thessaloniki, Greece, 23–28 June 2019; pp. 23–28.
42. City of Melbourne. Trees, with Species and Dimensions (Urban Forest). Available online: https://data.melbourne.vic.gov.au/Environment/Trees-with-species-and-dimensions-Urban-Forest-/fp38-wiyy/data (accessed on 26 October 2020).

43. Watson, D.J. Comparative physiological studies on the growth of field crops: I. Variation in net assimilation rate and leaf area between species and varieties, and within and between years. *Ann. Bot.* **1947**, *11*, 41–76. [CrossRef]
44. Nyakatya, M.; McGeoch, M. Temperature variation across Marion Island associated with a keystone plant species (Azorella selago Hook.(Apiaceae)). *Polar Biol.* **2008**, *31*, 139–151. [CrossRef]
45. Jones, H.G. Use of infrared thermometry for estimation of stomatal conductance as a possible aid to irrigation scheduling. *Agric. For. Meteorol.* **1999**, *95*, 139–149. [CrossRef]
46. Jones, H.G.; Stoll, M.; Santos, T.; Sousa, C.; Chaves, M.M.; Grant, O.M. Use of infrared thermography for monitoring stomatal closure in the field: Application to grapevine. *J. Exp. Bot.* **2002**, *53*, 2249–2260. [CrossRef] [PubMed]
47. Jones, H.G. *Plants and Microclimate: A Quantitative Approach to Environmental Plant Physiology*; Cambridge University Press: Cambridge, UK, 1992.
48. De Bei, R.; Fuentes, S.; Collins, C. Vineyard variability: Can we assess it using smart technologies? *IVES Tech. Rev. Vine Wine* **2019**. [CrossRef]
49. De Bei, R.; Fuentes, S.; Gilliham, M.; Tyerman, S.; Edwards, E.; Bianchini, N.; Smith, J.; Collins, C. VitiCanopy: A free computer App to estimate canopy vigor and porosity for grapevine. *Sensors* **2016**, *16*, 585. [CrossRef]
50. De Bei, R.; Kidman, C.; Wotton, C.; Shepherd, J.; Fuentes, S.; Gilliham, M.; Tyerman, S.; Collins, C. Canopy architecture is linked to grape and wine quality in Australian Shiraz 2018. In Proceedings of the Web of Conferences—XII International Terroir Congress Zaragoza 2018, Zaragoza, Spain, 18–22 June 2018.
51. Zovko, M.; Boras, I.; Švaić, S. Assessing plant water status from infrared thermography for irrigation management. In Proceedings of the 14th Quantitative Infrared Thermography Conference, Berlin, Germany, 25–29 June 2018.
52. Vidal, D.; Pitarma, R. Infrared Thermography Applied to Tree Health Assessment: A Review. *Agriculture* **2019**, *9*, 156. [CrossRef]
53. Pitarma, R.; Crisóstomo, J.; Ferreira, M.E. Contribution to Trees Health Assessment Using Infrared Thermography. *Agriculture* **2019**, *9*, 171. [CrossRef]
54. Jones, H.G. Thermal imaging and infrared sensing in plant ecophysiology. In *Advances in Plant Ecophysiology Techniques*; Springer: Berlin/Heidelberg, Germany, 2018; pp. 135–151.
55. Fricke, W. Night-time transpiration–favouring growth? *Trends Plant Sci.* **2019**, *24*, 311–317. [CrossRef]
56. Zhao, C.; Si, J.; Feng, Q.; Yu, T.; Li, P.; Forster, M.A. Nighttime transpiration of Populus euphratica during different phenophases. *J. For. Res.* **2019**, *30*, 435–444. [CrossRef]
57. Fuentes, S.; Mahadevan, M.; Bonada, M.; Skewes, M.A.; Cox, J. Night-time sap flow is parabolically linked to midday water potential for field-grown almond trees. *Irrig. Sci.* **2013**, *31*, 1265–1276. [CrossRef]
58. Fuentes, S.; De Bei, R.; Collins, M.; Escalona, J.; Medrano, H.; Tyerman, S. Night-time responses to water supply in grapevines (*Vitis vinifera* L.) under deficit irrigation and partial root-zone drying. *Agric. Water Manag.* **2014**, *138*, 1–9. [CrossRef]
59. Zeppel, M.J.; Lewis, J.D.; Medlyn, B.; Barton, C.V.; Duursma, R.A.; Eamus, D.; Adams, M.A.; Phillips, N.; Ellsworth, D.S.; Forster, M.A. Interactive effects of elevated CO2 and drought on nocturnal water fluxes in Eucalyptus saligna. *Tree Physiol.* **2011**, *31*, 932–944. [CrossRef] [PubMed]
60. Teskey, R.; Wertin, T.; Bauweraerts, I.; Ameye, M.; McGuire, M.A.; Steppe, K. Responses of tree species to heat waves and extreme heat events. *Plant Cell Environ.* **2015**, *38*, 1699–1712. [CrossRef]
61. Jactel, H.; Petit, J.; Desprez-Loustau, M.L.; Delzon, S.; Piou, D.; Battisti, A.; Koricheva, J. Drought effects on damage by forest insects and pathogens: A meta-analysis. *Glob. Chang. Biol.* **2012**, *18*, 267–276. [CrossRef]
62. Meineke, E.K.; Dunn, R.R.; Sexton, J.O.; Frank, S.D. Urban warming drives insect pest abundance on street trees. *PLoS ONE* **2013**, *8*, e59687. [CrossRef]
63. Barigah, T.S.; Charrier, O.; Douris, M.; Bonhomme, M.; Herbette, S.; Améglio, T.; Fichot, R.; Brignolas, F.; Cochard, H. Water stress-induced xylem hydraulic failure is a causal factor of tree mortality in beech and poplar. *Ann. Bot.* **2013**, *112*, 1431–1437. [CrossRef]
64. Pekin, B.; Macfarlane, C. Measurement of crown cover and leaf area index using digital cover photography and its application to remote sensing. *Remote Sens.* **2009**, *1*, 1298–1320. [CrossRef]

Article

Machine-Learning Classification of a Number of Contaminant Sources in an Urban Water Network

Ivana Lučin [1,2,*], Luka Grbčić [1,2], Zoran Čarija [1,2] and Lado Kranjčević [1,2]

[1] Faculty of Engineering, University of Rijeka, Vukovarska 58, 51000 Rijeka, Croatia; lgrbcic@riteh.hr (L.G.); zcarija@riteh.hr (Z.Č.); lado.kranjcevic@riteh.hr (L.K.)
[2] Center for Advanced Computing and Modelling, University of Rijeka, Radmile Matejčić 2, 51000 Rijeka, Croatia
* Correspondence: ilucin@riteh.hr; Tel.: +385-51-651-418

Abstract: In the case of a contamination event in water distribution networks, several studies have considered different methods to determine contamination scenario information. It would be greatly beneficial to know the exact number of contaminant injection locations since some methods can only be applied in the case of a single injection location and others have greater efficiency. In this work, the Neural Network and Random Forest classifying algorithms are used to predict the number of contaminant injection locations. The prediction model is trained with data obtained from simulated contamination event scenarios with random injection starting time, duration, concentration value, and the number of injection locations which varies from 1 to 4. Classification is made to determine if single or multiple injection locations occurred, and to predict the exact number of injection locations. Data was obtained for two different benchmark networks, medium-sized network Net3 and large-sized Richmond network. Additionally, an investigation of sensor layouts, demand uncertainty, and fuzzy sensors on model accuracy is conducted. The proposed approach shows excellent accuracy in predicting if single or multiple contaminant injections in a water supply network occurred and good accuracy for the exact number of injection locations.

Keywords: water distribution networks; water network contamination; machine learning; random forest; neural network

1. Introduction

Contamination in water distribution networks can occur due to deliberate or unintentional intrusions and it is of extreme importance to determine the contamination event parameters so it can be detected which parts of water distribution networks have been exposed to the contaminant and needed measures can be conducted. This is considered to be an inverse problem since injection location, injection starting time, injection duration, and contaminant chemical concentration value needs to be predicted based on sensor measurements. Numerical simulations are used to determine these parameters, but model limitations need to be taken into consideration. EPANET [1] is the most commonly used software for water distribution network simulations and uses an advective approach which cannot efficiently analyze contaminant dispersion in the networks. Piazza et al. [2] conducted experiments where it was shown that dispersive and diffusive processes must be incorporated in the transport model for less turbulent fluid flows to achieve more accurate results than the pure advection model. Also, EPANET assumes complete mixing in all network junctions, which can be valid only in the case of a single outlet or if there is considerable distance between two junctions. Therefore, EPANET extension EPANET-BAM [3] was proposed which uses experimentally calibrated mixing model parameter to more accurately model mixing in network junctions. A number of studies investigated mixing behavior for different conditions, both experimentally and numerically, to further enhance these simpler 1D numerical models [4–9].

Huang and McBean [10] investigated a data mining approach for identifying possible sources of intrusion where single and multiple injection scenarios were considered. In the case of multiple injection scenario, the method provided a limited number of nodes with the probability of them being the true contamination source. However, in their work, it is not predicted what is the true number of injection locations. In Wang and Harrison [11] a Bayesian approach was coupled with Support Vector Regression to provide a probability distribution of water network nodes being contaminant sources. However, a single injection is assumed, and it is noted that multiple contaminant sources should be considered in future work where the likelihood evaluation needs to be adjusted. Seth et al. [12] investigated the efficiency of three different methods for source detection; Bayesian probability-based method, backtracking method (using contaminant status algorithm), and optimization-based method where accuracy in case of multiple injection locations was investigated for two and three contamination injection locations. It was noted that the Bayesian method is designed only for a single contamination location while the contaminant status algorithm used in De Sanctis et al. [13] provides a list of possible solutions that narrow down search space for the optimization method; however, it also does not identify the possible number of injection locations. In Lučin et al. [14] a new search space reduction method was proposed, which can eliminate a considerable number of source nodes for both single and multiple injection locations, but with considerably greater reduction for single injection scenario. A number of different optimization approaches were considered to determine the contamination source, an overview of proposed methods can be found in Adedoja et al. [15]. Optimization approach can be easily extended to consider multiple contamination sources, as mentioned in [16–18].

If considering the optimization approach with multiple injection locations, with each additional source of contamination, the complexity of search space increases with an increase of optimization variables. Since the number of injection locations is not known, as a precaution, multiple injection locations should be allowed, since optimization can set variables to zero (which eliminates that source node and eliminates the number of injection locations), but it cannot add additional variables (injection locations) during the optimization process. In this way, in the case of a single injection location, optimization can eliminate other source nodes (all contamination parameters would be set to 0). However, this considerably increases the complexity of the considered problem since unnecessary fitness function evaluations would be conducted due to greater search space. Thus, it would be greatly beneficial to determine the number of injection locations before the optimization algorithm is employed. Also, if it is known that a single injection event occurred, a number of methods can be used more efficiently to reduce the complexity of the problem. For example, the machine learning approach provides probabilities for each network node being the true contamination scenario, which greatly reduces the number of suspect nodes and helps in quicker detection of true contamination location. However, in the case of multiple injections, different likelihood evaluation is needed which increases the complexity of the machine learning approach. Prediction of the number of contamination sources has previously been conducted for air pollution in Wade and Senocak [19], but to authors knowledge was not conducted for water distribution network contamination scenarios.

Machine learning tools have been increasingly used in contamination detection, where Random Forest has been used for groundwater source of contamination detection [20] and source detection in a river [21]. In Grbčić et al. [22] Random Forest algorithm was used to predict contamination event parameters in water distribution networks and in Grbčić et al. [23] new machine learning-based algorithm was proposed. A great advantage of prediction models is that they can be constructed before an accident occurs, so when a contamination event is detected prediction can be made even for large networks in a computationally efficient way. Thus, the proposed model which predicts number of injection locations can be used prior to conducting approaches that search for contamination parameters, without influencing the reaction time needed to contain the contamination event. However, in accident situations hydraulic conditions can greatly differ from those on which model was trained, thus, a wrong prediction could be made. This can be handled with the preparation of multiple prediction models with different hydraulic conditions or by using a prediction model that achieves great accuracy with the small number of inputs so time for prediction also becomes negligible considering the benefit of search space reduction when redundant optimization parameters are not used.

In this paper, the Random Forest and Artificial Neural Network classifier are used to predict the number of contamination sources based on contamination sensor measurements in the water distribution network. Sensor measurements of contamination needed for model teaching are obtained from contamination scenarios simulated using EPANET2 with Monte Carlo generated contamination parameters. An investigation was conducted for two different sized benchmark water distribution networks with different sensor layouts, to examine the efficiency of the proposed machine learning approach. Investigation of demand uncertainty and fuzzy sensors is also estimated.

2. Materials and Methods

2.1. Benchmark Water Supply Networks

Prediction of the number of injection sources is conducted for two benchmark different sized networks. Investigated networks are Net3 EPANET2 example consisting of 92 nodes and Richmond network consisting of 865 nodes, obtained from The Centre for Water Systems (CWS) at the University of Exeter [24]. For the Net3 network, two different sensor layouts are investigated. In first layout four sensors were placed in network nodes 117, 143, 181, and 213 as in [25] and in second layout four sensor were placed in network nodes 115, 119, 187, and 209 as in [26]. Additionally, an investigation of the number of sensors was conducted. For the first layout, two sensors were placed in network nodes 117 and 181, and for the second layout sensors were placed in network nodes 119 and 209. For Richmond network five sensors were placed in network nodes 93, 352, 428, 600, and 672 where sensor layout was taken from [27]. Layout with three sensors placed in network nodes 93, 428, and 672 was also considered. Considered networks with sensor layouts can be seen in Figures 1 and 2.

Contamination scenarios are simulated using EPANET2 version 2.0.12. where for both networks, simulation time is 24 h with a hydraulic time step of 10 min, quality time step 5 min, pattern time step 10 min and report time step 1 h. For all conducted simulations, the EPANET2 flow paced method is used for the contaminant injection. Contamination scenario parameters are chosen randomly. The number of injection locations is chosen from 1 to 4 nodes. The starting time and duration of contamination injection are chosen from 0 to 24 h. Concentration was randomly chosen from 10 to 2000 mg/L. For contamination scenarios with multiple injection locations starting time, duration, and concentration was kept the same for every injection location.

Prior to simulating multiple injection scenario, independent simulations for each randomly chosen node as a source of contamination are conducted. If contamination is not registered for the investigated node with chosen contamination parameters, that node is eliminated as source location and only nodes for which contamination was detected in at least one sensor are kept as a source of contaminant. For example, if four source nodes are randomly chosen to be the source of contamination, but only two source nodes influence sensor detection of contaminant, the same time series of sensor measurements would be obtained for two, three, and four injection locations since the latter two do not influence contamination measurements. If four sources are given to the prediction model as input, where contamination can be measured only from two sources, that would significantly reduce the accuracy of the prediction model. Thus, only nodes which contribute to the contamination measurements in sensors are considered for multiple injection scenario.

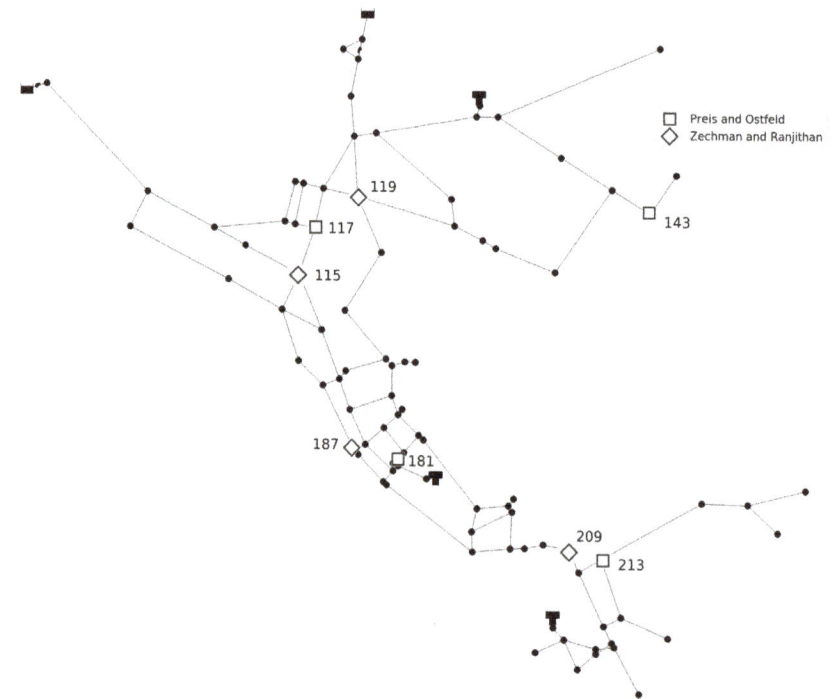

Figure 1. Net3 network with sensor layouts.

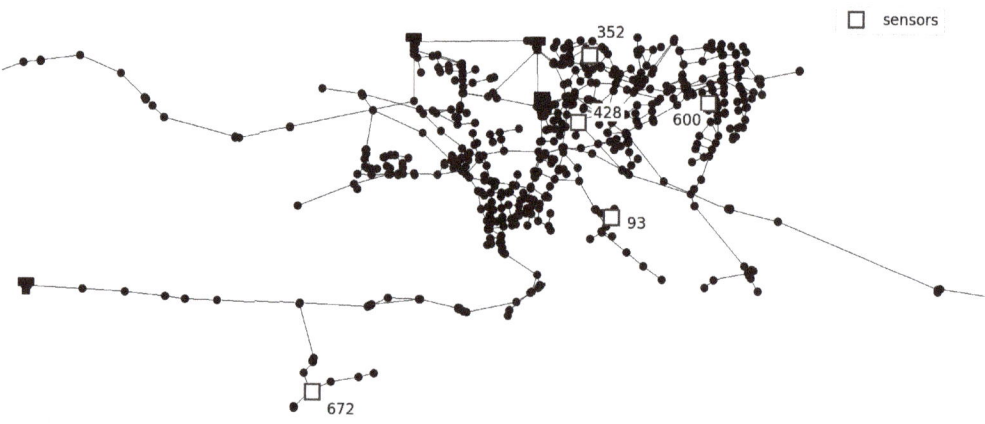

Figure 2. Richmond network detail with sensor layout.

An example of the proposed methodology can be seen for arbitrarily chosen Net3 contamination scenario in Figure 3. Randomly chosen contamination scenario parameters are 3 source nodes (159, 151 and 123), with contamination value of 200 mg/L, starting time 13 h and 20 min and injection duration 2 h. Sensor measurements for chosen contamination scenario can be seen in Figure 4. It can be observed that for source node 151 contamination scenario remains undetected in all sensors placed in the water distribution network, thus for multiple sources scenario only source nodes 123 and 159 are further considered.

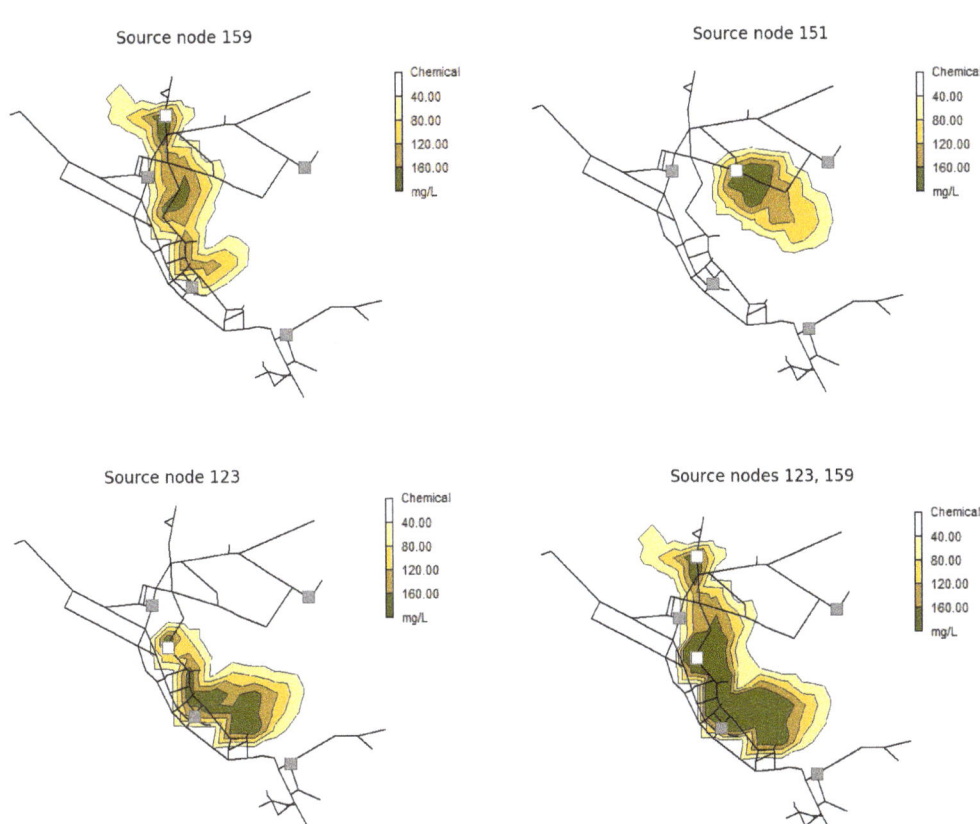

Figure 3. Contours of chemical for randomly chosen Net3 contamination scenario 90 min after injection starting time. Contamination from source node 151 remains undetected, so the source node is not included for multiple injections scenario.

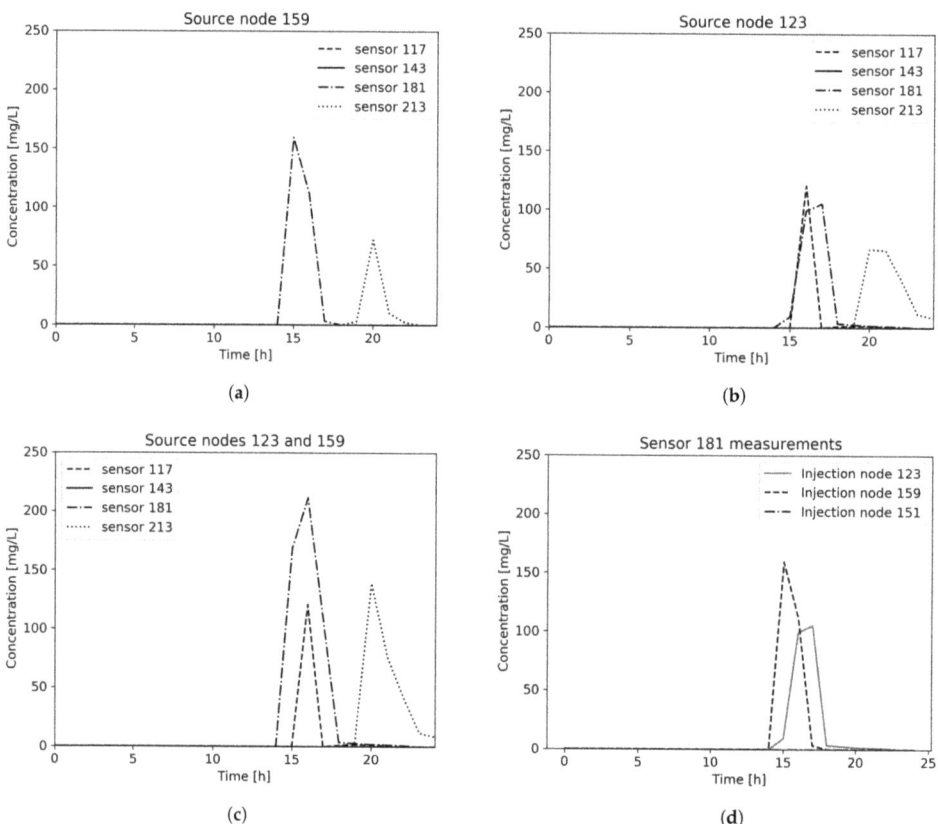

Figure 4. Sensor measurements for Net3 contamination scenario with (**a**) injection node 159, (**b**) injection node 123, (**c**) injection nodes 123 and 159 and (**d**) contamination measurements in the sensor in node 181.

2.2. Demand Uncertainty and Sensor Type

To investigate demand uncertainty, for both Net3 and Richmond networks, for every network node first it was randomly chosen if demand will be altered or not. If node base demand was to be altered, the percentage from 0–5% is randomly chosen for each network node, to reduce or increase base demand by the chosen percentage, resulting in a random demand span of 10%. To further investigate influence of demand uncertainty, the percentage from 0–10% is randomly chosen to reduce or increase base demand, resulting in a random demand span of 20%. All network demand patterns were kept the same, only base demand was changed. This method was conducted for every contamination scenario, thus resulting in different hydraulic conditions for each contamination scenario.

For sensor type influence, fuzzy sensor measurements were made where sensor detection was considered either low, medium, or high. Chemical concentration value C in range $0 < C < 300$ mg/L was considered low, in range $300 < C < 1000$ mg/L was considered medium and high if $C > 1000$ mg/L. Prediction model input features were defined as 0 if no contaminant was detected, 1 for low measurements, 2 and 3 for medium and high measurements, respectively.

2.3. Machine Learning Classifiers

Two different machine learning classifiers, Random Forest and Artificial Neural Network were used to compare the efficiency of the proposed method. Random Forest algorithm [28], based on multiple decision trees is used, with 250 estimators (trees) with a maximum depth of 30 and the minimum number of samples required to split an internal node 8. An artificial neural network with three hidden layers with 100 nodes in each layer, with hyperbolic tangent activation function and Adam solver for weight optimization is used. Proposed parameters were chosen with the grid search hyperparameter optimization method, while other parameters, which are not mentioned, are kept constant. Implementation in the Python library Scikit-learn [29] version 0.20.3 is used for both classifiers. Obtained data was split 70% for teaching and 30% for model testing. Flowchart of the prediction model can be seen in Figure 5. Data generation and prediction model training was done using the supercomputing resources at the Center for Advanced Computing and Modelling, University of Rijeka.

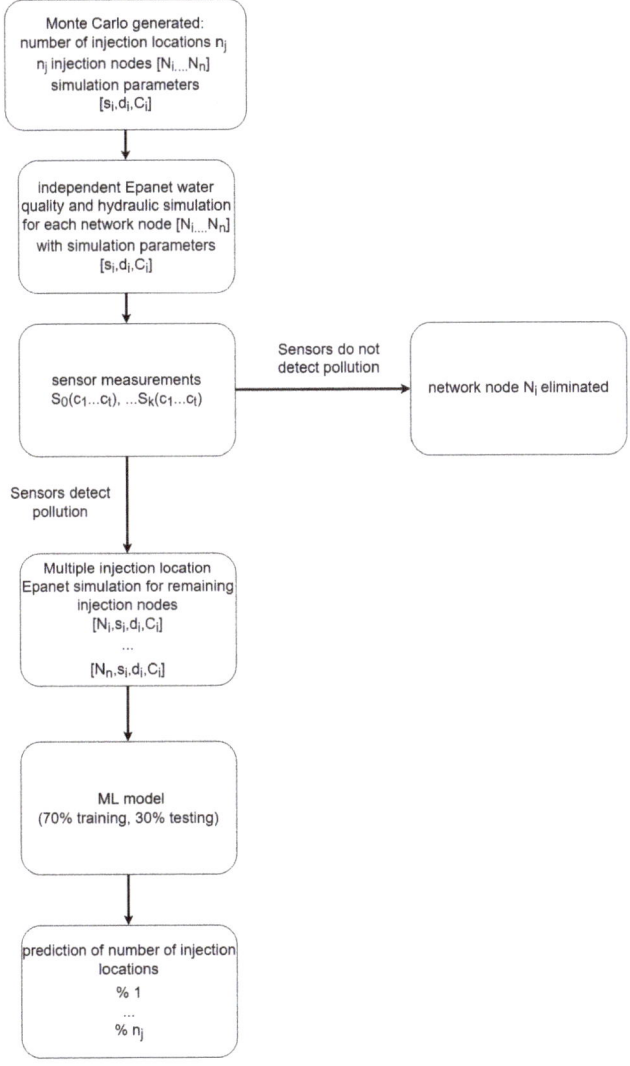

Figure 5. Flowchart of Machine Learning algorithm for prediction of number of contamination sources.

Input data for the prediction model is the time series of sensor measurements. For both Net3 and Richmond network, 25 features per sensor are obtained, which resulted in 100 features for Net3 and 125 features for Richmond network. The output of the machine learning model is the number of injection locations where two different prediction models are used. The first prediction model was used to predict the exact number of injection locations, i.e., 4 different classes are predicted. In the second model it is predicted only if single or multiple injections occurred, i.e., 2, 3 and 4 injection locations are treated as same, multiple injections class, thus only 2 different classes are predicted (single and multiple injections). To further increase the accuracy of the latter prediction model, the threshold value is introduced. Only if the model predicts a single source scenario with a probability greater than the chosen threshold value, single source prediction is made. In other cases, the scenario is treated as multiple sources. Threshold values of 50%, 60%, 70%, 80%, 90%, and 95% are investigated.

3. Results

3.1. Model Accuracy

The influence of input data on prediction model accuracy is investigated for both benchmark networks where data ranged from 50,000 to 500,000 inputs (Figure 6). An investigation is conducted for prediction model with 2 categories (model predicts only if single or multiple injection locations are present) and with 4 categories (model predicts an exact number of injection locations). For each model and each number of inputs, 20 runs were conducted to take into consideration the influence of random seed. For the Net3 network second sensor layout with sensors placed in nodes 115, 119, 187, and 209 was considered. For Net 3 results are presented for both RF and NN prediction models. Standard deviation ranged from 0.63% for 50,000 to 0.33% for 500,000 inputs for NN model, and from 0.33% for 50,000 to 0.1% for 500,000 inputs. It can be observed that the RF model has slightly better accuracy for all investigated models. Also, due to the faster execution time of the RF model, for all further analyses, only RF results will be presented. For Richmond network, standard deviation ranged from 0.28% for 50,000 inputs to 0.12% for 500,000 inputs which indicates the stability of the model. Presented results are an average of all 20 runs.

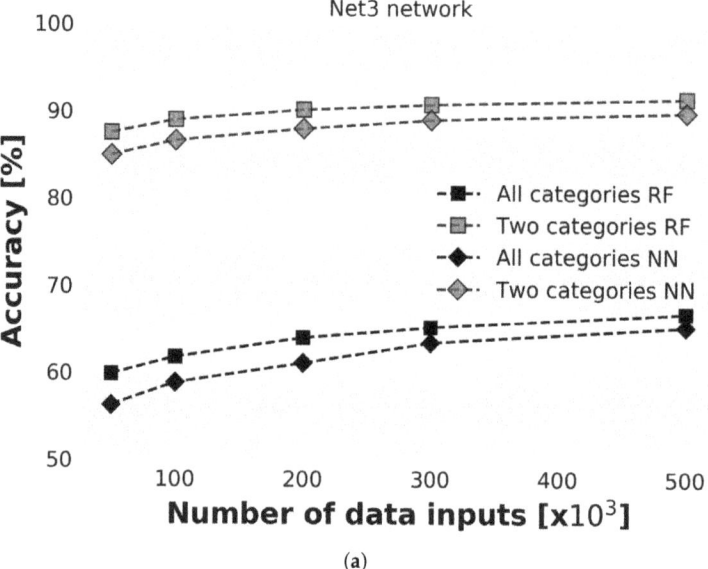

Figure 6. Cont.

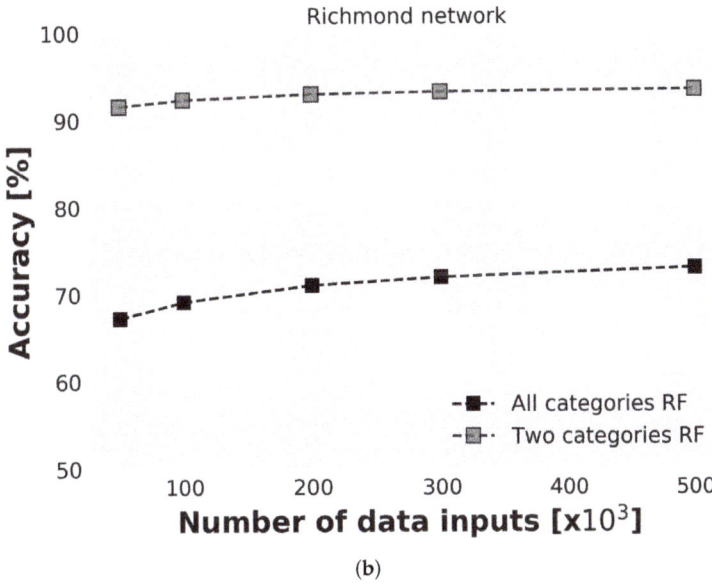

Figure 6. Accuracy of prediction models for different number of inputs for (**a**) Net3 network and (**b**) Richmond network.

It can be observed that even for a small number of input data considerable accuracy can be achieved. For model with 2 categories even with 50,000 inputs accuracy of the model is above 85% for both considered networks. After 200,000 inputs accuracy of the models for both networks tend to only slightly increase with the further increase of the number of input data. For 500,000 inputs accuracy of the Net3 network is 66.83% and for Richmond network 72.96%. When simplification is made, and the model only needs to predict single or multiple injection locations, accuracy significantly increases and for 500,000 inputs for the Net3 network is 91.46% and for the Richmond network 93.4%.

3.2. Threshold Influence

To further increase the accuracy of the prediction model, the threshold value is introduced for the model which predicts 2 categories. Detailed results are presented for models with 500,000 inputs for Net3 (Tables 1 and 2) and Richmond network (Tables 3 and 4). Presented results are the average of values obtained from 20 runs. As expected, with the increase in threshold value accuracy of the prediction model increases. However, with a greater threshold value, a greater number of single injection scenarios, as a precaution, are classified as multiple sources, thus a smaller number of true single injection scenarios are detected. For both networks, when the threshold value is 95%, a very low percentage of correct prediction of single source scenarios can be observed when prediction model parameters chosen with grid search optimization method (250 estimators, maximum depth 30, minimum samples for split 8) were used (Tables 1 and 3). Thus, different prediction model parameters (180 estimators, maximum depth 80, minimum samples for split 10) were also investigated to test its influence on model accuracy when threshold values are considered. In Tables 2 and 4 it can be observed that for the greatest threshold value (95%) correct prediction of single sources scenarios greatly increases, and is around 30% of the total number of single source scenarios. As threshold value decreases, similar percentages are observed for both models, which indicates that model accuracy is similar for different RF parameters. However, when greater prediction certainty is expected, model parameters must be carefully considered.

For both networks, accuracy with threshold value 95% is above 99.5%. It can be observed from Table 2 that for Net3 only 36% of total number of single source scenarios are correctly predicted where for Richmond network (Table 4) that value is 37%. For threshold value 50% for Net3 94.5% of single injection scenarios are correctly predicted; however, the number of wrong predictions increases. The same can be observed for the Richmond network where for threshold value 50%, 97.8% of single injection scenarios are correctly predicted but the percentage of wrong single injection scenarios increases from 0.8% to 12.7%.

The problem remains with scenarios that are wrongly predicted even for a threshold value of 95%. With further increase of threshold value, the number of wrongly predicted scenarios would decrease, but only because ultimately all scenarios would be classified as multiple sources (this can also be observed in Tables 1 and 3 for first chosen RF parameters). Thus, optimum threshold value should be chosen to both provide a reasonable number of single injection scenario predictions but with a high model accuracy. In-depth analysis of scenarios where the model wrongly predicts a single injection scenario with a high threshold value should be conducted. Also, it should be investigated how much accuracy of the model can be further increased with a larger number of inputs and with the usage of different classifiers.

Table 1. Influence of threshold value on model accuracy for Net3 network (250 estimators, maximum depth 30, minimum samples for split 8). Percentage indicates number of predicted simulations based on total number of single source scenarios.

Threshold Value	Accuracy	Single Source Scenarios	Correct Prediction	Wrong Prediction
95%	99.98%	48,682	3307 (6.8%)	36 (0.07%)
90%	99.73%	48,682	15,388 (31.6%)	405 (0.8%)
80%	98.6%	48,682	34,717 (71.3%)	2085 (4.3%)
70%	97.5%	48,682	41,204 (84.6%)	3683 (7.6%)
60%	96.7%	48,682	44,334 (91.1%)	4914 (10.1%)
50%	95.7%	48,682	46,388 (95.3%)	6390 (13.1%)

Table 2. Influence of threshold value on model accuracy for Net3 network (180 estimators, maximum depth 80, minimum samples for split 10). Percentage indicates number of predicted simulations based on total number of single source scenarios.

Threshold Value	Accuracy	Single Source Scenarios	Correct Prediction	Wrong Prediction
95%	99.7%	48,783	17,458 (35.8%)	508 (1%)
90%	99.4%	48,783	25,426 (52.1%)	863 (1.8%)
80%	98.9%	48,783	35,197 (72.2%)	1667 (3.4%)
70%	98.2%	48,783	40,640 (83.3%)	2636 (5.4%)
60%	96.7%	48,783	43,977 (90.2%)	3737 (7.7%)
50%	95.7%	48,783	46,091 (94.5%)	5072 (10.4%)

Table 3. Influence of threshold value on model accuracy for Richmond network (250 estimators, maximum depth 30, minimum samples for split 8). Percentage indicates number of predicted simulations based on total number of single source scenarios.

Threshold Value	Accuracy	Single Source Scenarios	Correct Prediction	Wrong Prediction
95%	99.9%	52,911	375 (0.7%)	5 (0.001%)
90%	99.8%	52,911	10,889 (20.6%)	303 (0.6%)
80%	97.9%	52,911	37,463 (70.8%)	3076 (5.8%)
70%	95.8%	52,911	49,149 (92.9%)	6269 (11.9%)
60%	94.8%	52,911	51,427 (97.2%)	7819 (14.8%)
50%	93.9%	52,911	52,198 (98.65%)	9178 (17.3%)

Table 4. Influence of threshold value on model accuracy for Richmond network (180 estimators, maximum depth 80, minimum samples for split 10). Percentage indicates number of predicted simulations based on total number of single source scenarios.

Threshold Value	Accuracy	Single Source Scenarios	Correct Prediction	Wrong Prediction
95%	99.7%	52,941	19,499 (36.8%)	435 (0.8%)
90%	99.3%	52,941	30,305 (57.2%)	1085 (2.1%)
80%	98.3%	52,941	42,000 (79.3%)	2567 (4.9%)
70%	97.3%	52,941	47,654 (90%)	4061 (7.7%)
60%	96.4%	52,941	50,433 (95.3%)	5433 (10.3%)
50%	95.5%	52,941	51,775 (97.8%)	6703 (12.7%)

3.3. Sensor Layout

The influence of sensor layout was tested for both Net3 and Richmond networks. 20 runs were conducted for the model with 500,000 inputs and average accuracy for all runs can be seen in Table 5. It can be observed that for the same number of sensors, their layout influences the accuracy of prediction models. This is expected, since the same behavior can be seen when the detection rate of contamination event is investigated for different sensor layouts. In the paper by Ostfeld et al. [30] for the same network and the same number of sensors detection likelihood of contamination event greatly differs for different sensor layouts. Results show that the prediction model for 2 categories (predicts single or multiple injections) is less influenced by sensor layout and all sensor layouts have accuracy around 90% or higher.

Interestingly, greater model accuracy can be observed when a smaller number of sensors is placed for Net3 layout with sensors in nodes 117, 143, 181, and 213 and for Richmond network. However, it can be explained with the fact that a greater number of contamination events remain undetected. i.e., with the greater number of sensors, contamination events from the greater number of network nodes are detected, resulting in more combinations when considering multiple injection locations. When sensor placement is sparser, a smaller number of network nodes can be detected when the contamination event occurs, resulting in a smaller number of combinations for multiple injection locations and consequently providing better model accuracy with 500,000 inputs.

Table 5. Influence of sensor layout for Net3 and Richmond networks on prediction model accuracy.

	Sensors Locations	Accuracy	
		4 Categories	2 Categories
Net3	117, 143, 181, 213	71%	94%
	115, 119, 187, 209	67%	91%
	117, 181	75%	89%
	119, 209	63%	89%
Richmond	93, 352, 428, 600, 672	73%	93%
	93, 428, 672	83%	92%

3.4. Demand Uncertainty and Fuzzy Sensors

Influence of demand uncertainty and fuzzy sensors was investigated for Net3 network with 4 sensors in nodes 117, 143, 181 and 213 and for Richmond network with 5 sensors in nodes 93, 352, 428, 600 and 672. 20 runs were conducted for RF models with 500,000 inputs and average accuracy can be observed in Table 6. When demand uncertainty is considered the accuracy of RF models slightly decreases for both networks. The influence of fuzzy sensors is more prominent, where the greater reduction in prediction accuracy can be observed for the Net3 network. When considering both demand uncertainty and fuzzy sensors in the same model, accuracy further slightly decreases. However, it can be observed that for both networks model which predicts 2 categories has accuracy above 90% for all cases. This shows that the proposed model could be applied in a real case scenario.

Table 6. Influence of demand uncertainty and fuzzy sensors for Net3 and Richmond network on prediction model accuracy.

	Net3	
	4 Categories	2 Categories
perfect sensors	71%	94%
demand uncertainty (±5%)	69%	93%
demand uncertainty (±10%)	69%	93%
fuzzy sensors	65%	91%
demand uncertainty (±5%) and fuzzy sensors	64%	90%
demand uncertainty (±10%) and fuzzy sensors	63%	90%
	Richmond	
	4 Categories	2 Categories
perfect sensors	73%	93%
demand uncertainty (±5%)	72%	93%
demand uncertainty (±10%)	72%	93%
fuzzy sensors	72%	93%
demand uncertainty (±5%) and fuzzy sensors	71%	93%
demand uncertainty (±10%) and fuzzy sensors	71%	92%

4. Discussion

Accuracy of prediction models for both networks has similar results with small differences, which shows that the proposed methodology could be successfully applied to other networks. Further investigation should be conducted for large size water distribution networks and different sensor placements, to fully investigate the robustness of the proposed method. Also, it must be noted that simplification was used in this study, where all source nodes had the same parameters (injection starting time, duration, and concentration value), thus, it should be investigated how the model predicts if those parameters are different for each injection node.

Although slightly, with the increase of input data model accuracy still increases, so in further study a greater number of data inputs should be investigated. Also, in the proposed scenarios report time step was chosen to be 1 h, resulting in 25 features per sensor. It should be investigated if a greater number of features, i.e., smaller report time step would increase model accuracy and if similar model accuracy could be achieved with a smaller number of contamination readings. The optimal number of features and inputs should be investigated to achieve great accuracy but with reasonable execution time. However, to obtain a greater number of inputs a greater amount of time is needed, so the model should be trained before the actual contamination event occurs. In that case, the model would be trained with simulation results with average demand patterns. This surely would mean that true contamination event will have different demands which would influence the accuracy of the prediction model. Investigation of demand uncertainty with arbitrarily chosen demand variation spans showed that small differences of base demands slightly influence prediction model accuracy. However, it must be taken into consideration that when base demand variation is defined with percentage, small demand variation is achieved when base demand is small and greater demand variation only when base demand is greater. Greater difference in demands should be further investigated since the usual variability of consumption can be greater than considered in this paper. Different machine learning models, with different expected demand patterns, can be prepared for contamination event so prediction can be obtained instantaneously. However, in case of contamination event, greater oscillations in the hydraulics of water distribution network could occur, such as pipe burst or some other unplanned event, which would greatly influence change in demand patterns. Thus, it would be beneficial to investigate other algorithms that could increase accuracy with a smaller number of input data. In that case, input data can be obtained after the contamination event occurred, in a reasonable amount of time. That would be greatly beneficial since the simulation model can then be calibrated with sensor measurements from the field and input data would be more precise. The proposed method can be easily coupled with other machine learning approaches since inputs obtained for this model can also be used for teaching model that predicts injection location.

Investigation of different sensor layouts, demand uncertainty, and fuzzy sensors showed that sensor layout and type of sensors have the greatest impact on prediction model accuracy. Demand uncertainty slightly decreases model accuracy. However, model accuracy can be greatly reduced when a real case event is considered since both demand uncertainty and measurement errors can be greater than considered in this work. Thus, a threshold value is introduced which can help increase model accuracy. Greater threshold value increases model accuracy; however, it also leads to a greater number of single injection scenarios classified as multiple injections. It is also observed that prediction models are not very sensitive to model parameters; however, when threshold value is used, i.e., model prediction certainty is evaluated, model parameters are very important for method efficiency. Thus, the investigation of different machine learning approaches should be further investigated to increase model accuracy.

When observing presented results it must be taken into consideration that numerical model simplifications are made, where EPANET was used which assumes complete mixing in all network junctions and uses pure advection transport model. Also, in the presented study benchmark networks are used, and numerical simulations are conducted for only 24 h, where more than 24 h are needed to obtain stable contamination scenario results. However, the functionality of the presented machine learning approach is not dependent on the numerical model setup, and it is assumed that the same numerical approach that is chosen for the optimization process is to be also chosen for the prediction model preparation. In this way, all discrepancies due to numerical model simplifications would be also present in the optimization and as such are not the result of using the proposed machine learning approach. Furthermore, network uncertainties were not considered regarding internal pipe diameter and pipe roughness which should be considered in the further research.

5. Conclusions

In this paper, the machine learning approach is presented which helps identify the number of injection locations based on sensor measurements. Random Forest classifier and Neural Network classifier are used on medium-sized benchmark network, where Random Forest classifier provided better accuracy and faster execution time, thus is used for all other investigations. Two different sized benchmark networks are considered, where it is shown that the machine learning approach can be successfully used to predict the number of injection locations. This can help define the number of optimization parameters, where redundant parameters can be avoided which needlessly increase the complexity of the problem. The prediction model shows great accuracy when it predicts only if single or multiple injection locations occurred. The threshold value is proposed which further increases model accuracy since the single injection scenario is assumed only if the model predicts with certainty greater than the threshold value. Lower accuracy is obtained when the exact number of injection locations is predicted. The accuracy of the prediction model is investigated for different sensor layouts and in case of demand uncertainties and fuzzy sensors. Conducted research showed promising results, where exploration of other algorithms and increased number of input data should be investigated to further increase the accuracy of both models.

Author Contributions: Conceptualization, I.L. and L.G.; Data curation, I.L.; Formal analysis, I.L; Investigation, I.L. and L.G.; Methodology, I.L. and L.G.; Resources, Z.Č. and L.K.; Software, I.L.; Supervision Z.Č. and L.K.; Validation, I.L.; Visualization; I.L.; Writing—original draft, I.L.; Writing—

review and editing, L.G., Z.Č. and L.K. All authors have read and agreed to the published version of the manuscript.

Funding: This research received no external funding.

Data Availability Statement: The data presented in this study are available on request from the corresponding author.

Conflicts of Interest: The authors declare no conflict of interest.

References

1. Rossman, L.A. EPANET 2: Users Manual. 2000. Available online: https://epanet.es/wp-content/uploads/2012/10/EPANET_User_Guide.pdf (accessed on 6 September 2020).
2. Piazza, S.; Blokker, E.M.; Freni, G.; Puleo, V.; Sambito, M. Impact of diffusion and dispersion of contaminants in water distribution networks modelling and monitoring. *Water Supply* **2020**, *20*, 46–58. [CrossRef]
3. Ho, C.K.; O'Rear, L., Jr. Evaluation of solute mixing in water distribution pipe junctions. *J. Am. Water Work. Assoc.* **2009**, *101*, 116–127. [CrossRef]
4. Yu, T.; Tao, L.; Shao, Y.; Zhang, T. Experimental study of solute mixing at double-Tee junctions in water distribution systems. *Water Sci. Technol. Water Supply* **2015**, *15*, 474–482. [CrossRef]
5. Yu, T.; Qiu, H.; Yang, J.; Shao, Y.; Tao, L. Mixing at double-Tee junctions with unequal pipe sizes in water distribution systems. *Water Sci. Technol. Water Supply* **2016**, *16*, 1595–1602. [CrossRef]
6. Song, I.; Romero-Gomez, P.; Andrade, M.A.; Mondaca, M.; Choi, C.Y. Mixing at junctions in water distribution systems: An experimental study. *Urban Water J.* **2018**, *15*, 32–38. [CrossRef]
7. Grbčić, L.; Kranjčević, L.; Lučin, I.; Čarija, Z. Experimental and Numerical Investigation of Mixing Phenomena in Double-Tee Junctions. *Water* **2019**, *11*, 1198. [CrossRef]
8. Grbčić, L.; Kranjčević, L.; Družeta, S.; Lučin, I. Efficient Double-Tee Junction Mixing Assessment by Machine Learning. *Water* **2020**, *12*, 238. [CrossRef]
9. Grbčić, L.; Kranjčević, L.; Lučin, I.; Sikirica, A. Large Eddy Simulation of turbulent fluid mixing in double-tee junctions. *Ain Shams Eng. J.* **2020**. [CrossRef]
10. Huang, J.J.; McBean, E.A. Data mining to identify contaminant event locations in water distribution systems. *J. Water Resour. Plan. Manag.* **2009**, *135*, 466–474. [CrossRef]
11. Wang, H.; Harrison, K.W. Improving efficiency of the Bayesian approach to water distribution contaminant source characterization with support vector regression. *J. Water Resour. Plan. Manag.* **2014**, *140*, 3–11. [CrossRef]
12. Seth, A.; Klise, K.A.; Siirola, J.D.; Haxton, T.; Laird, C.D. Testing contamination source identification methods for water distribution networks. *J. Water Resour. Plan. Manag.* **2016**, *142*, 04016001. [CrossRef]
13. De Sanctis, A.E.; Shang, F.; Uber, J.G. Real-time identification of possible contamination sources using network backtracking methods. *J. Water Resour. Plan. Manag.* **2009**, *136*, 444–453. [CrossRef]
14. Lučin, I.; Grbčić, L.; Družeta, S.; Čarija, Z. Source Contamination Detection Using Novel Search Space Reduction Coupled with Optimization Technique. *J. Water Resour. Plan. Manag.* **2020**, *147*, 04020100. [CrossRef]
15. Adedoja, O.; Hamam, Y.; Khalaf, B.; Sadiku, R. Towards development of an optimization model to identify contamination source in a water distribution network. *Water* **2018**, *10*, 579. [CrossRef]
16. Vankayala, P.; Sankarasubramanian, A.; Ranjithan, S.R.; Mahinthakumar, G. Contaminant source identification in water distribution networks under conditions of demand uncertainty. *Environ. Forensics* **2009**, *10*, 253–263. [CrossRef]
17. Liu, L.; Ranjithan, S.R.; Mahinthakumar, G. Contamination source identification in water distribution systems using an adaptive dynamic optimization procedure. *J. Water Resour. Plan. Manag.* **2011**, *137*, 183–192. [CrossRef]
18. Liu, L.; Zechman, E.M.; Mahinthakumar, G.; Ranji Ranjithan, S. Identifying contaminant sources for water distribution systems using a hybrid method. *Civ. Eng. Environ. Syst.* **2012**, *29*, 123–136. [CrossRef]
19. Wade, D.; Senocak, I. Stochastic reconstruction of multiple source atmospheric contaminant dispersion events. *Atmos. Environ.* **2013**, *74*, 45–51. [CrossRef]
20. Rodriguez-Galiano, V.; Mendes, M.P.; Garcia-Soldado, M.J.; Chica-Olmo, M.; Ribeiro, L. Predictive modeling of groundwater nitrate pollution using Random Forest and multisource variables related to intrinsic and specific vulnerability: A case study in an agricultural setting (Southern Spain). *Sci. Total Environ.* **2014**, *476*, 189–206. [CrossRef]
21. Lee, Y.J.; Park, C.; Lee, M.L. Identification of a contaminant source location in a river system using random forest models. *Water* **2018**, *10*, 391. [CrossRef]
22. Grbčić, L.; Lučin, I.; Kranjčević, L.; Družeta, S. Water supply network pollution source identification by random forest algorithm. *J. Hydroinf.* **2020**, *22*, 1521–1535. [CrossRef]
23. Grbčić, L.; Lučin, I.; Kranjčević, L.; Družeta, S. A Machine Learning-based Algorithm for Water Network Contamination Source Localization. *Sensors* **2020**, *20*, 2613. [CrossRef] [PubMed]
24. Centre for Water Systems, U.o.E. Benchmarks. Available online: http://emps.exeter.ac.uk/engineering/research/cws/downloads/benchmarks/ (accessed on 6 November 2019).

25. Preis, A.; Ostfeld, A. A contamination source identification model for water distribution system security. *Eng. Optim.* **2007**, *39*, 941–947. [CrossRef]
26. Zechman, E.M.; Ranjithan, S.R. Evolutionary computation-based methods for characterizing contaminant sources in a water distribution system. *J. Water Resour. Plan. Manag.* **2009**, *135*, 334–343. [CrossRef]
27. Preis, A.; Ostfeld, A. Genetic algorithm for contaminant source characterization using imperfect sensors. *Civ. Eng. Environ. Syst.* **2008**, *25*, 29–39. [CrossRef]
28. Breiman, L. Random forests. *Mach. Learn.* **2001**, *45*, 5–32. [CrossRef]
29. Pedregosa, F.; Varoquaux, G.; Gramfort, A.; Michel, V.; Thirion, B.; Grisel, O.; Blondel, M.; Prettenhofer, P.; Weiss, R.; Dubourg, V.; et al. Scikit-learn: Machine learning in Python. *J. Mach. Learn. Res.* **2011**, *12*, 2825–2830.
30. Ostfeld, A.; Uber, J.G.; Salomons, E.; Berry, J.W.; Hart, W.E.; Phillips, C.A.; Watson, J.P.; Dorini, G.; Jonkergouw, P.; Kapelan, Z.; et al. The battle of the water sensor networks (BWSN): A design challenge for engineers and algorithms. *J. Water Resour. Plan. Manag.* **2008**, *134*, 556–568. [CrossRef]

Article

A Machine Learning-Based Algorithm for Water Network Contamination Source Localization

Luka Grbčić [1,2], Ivana Lučin [1,2], Lado Kranjčević [1,2,*] and Siniša Družeta [1,2]

[1] Department of Fluid Mechanics and Computational Engineering, Faculty of Engineering, University of Rijeka, 51000 Rijeka, Croatia; lgrbcic@riteh.hr (L.G.); ilucin@riteh.hr (I.L.); sinisa.druzeta@riteh.hr (S.D.)
[2] Center for Advanced Computing and Modelling, University of Rijeka, 51000 Rijeka, Croatia
* Correspondence: lado.kranjcevic@riteh.hr

Received: 3 April 2020; Accepted: 30 April 2020; Published: 3 May 2020

Abstract: In this paper, a novel machine learning based algorithm for water supply pollution source identification is presented built specifically for high performance parallel systems. The algorithm utilizes the combination of Artificial Neural Networks for classification of the pollution source with Random Forests for regression analysis to determine significant variables of a contamination event such as start time, end time and contaminant chemical concentration. The algorithm is based on performing Monte Carlo water quality and hydraulic simulations in parallel, recording data with sensors placed within a water supply network and selecting a most probable pollution source based on a tournament style selection between suspect nodes in a network with mentioned machine learning methods. The novel algorithmic framework is tested on a small (92 nodes) and medium sized (865 nodes) water supply sensor network benchmarks with a set contamination event start time, end time and chemical concentration. Out of the 30 runs, the true source node was the finalist of the algorithm's tournament style selection for 30/30 runs for the small network, and 29/30 runs for the medium sized network. For all the 30 runs on the small sensor network, the true contamination event scenario start time, end time and chemical concentration was set as 14:20, 20:20 and 813.7 mg/L, respectively. The root mean square errors for all 30 algorithm runs for the three variables were 48 min, 4.38 min and 18.06 mg/L. For the 29 successful medium sized network runs the start time was 06:50, end time 07:40 and chemical concentration of 837 mg/L and the root mean square errors were 6.06 min, 12.36 min and 299.84 mg/L. The algorithmic framework successfully narrows down the potential sources of contamination leading to a pollution source identification, start and ending time of the event and the contaminant chemical concentration.

Keywords: machine learning; artificial neural networks; random forests; water network pollution; sensor networks; parallel computing

---

1. Introduction

Identifying the source of contamination in a water supply network is an important task since a contamination event is potentially hazardous to the human population in an urban environment. Additionally, a fast identification of a pollution source enables the governing authorities to rapidly react in order to stop the further spread of the contaminant through the water supply network.

Researches have tackled the issue of water supply networks contamination in several ways which include an optimal positioning of water quality sensors in a network [1–3] to facilitate the source identification process and optimally cover all possible intrusion points, rapid contamination event response procedures [4–6] and simulation-optimization methods for contamination source detection and duration based on simulation of the water network contamination event [7–9]. Many researches have incorporated additional uncertainties into the hydraulic simulation process which include uncertain sensor measurements and water demand variability [10–12]. A recent and thorough review of various approaches in tackling the water supply pollution source identification problem can be found in Adedoja et al. [13].

The optimization-simulation approach for finding the source of pollution in a network entails that an optimization algorithm is being coupled with a water supply network hydraulic simulator and the difference between the measured sensor data and the optimization algorithm generated and simulated values (sensor water quality readings through a certain time interval) is being minimized. Through this procedure, the source of contamination, starting time and the end time of the contamination event and the contaminant chemical concentration are obtained.

Probabilistic approaches are also possible for determining the source of contamination in a water supply network and the methods for this approach that were used in previous studies include the Bayesian Belief Networks (BBN) [14–16] and backward probabilistic models [17] which show the ability to predict the source of the contamination with a high probability. Data-driven methods are also a possible tool for the water supply network contamination problem. In [18,19], a database was compiled by massive data mining of hydraulic and water quality simulation contamination events for a fast identification of a source in case of a real event. The sensor readings in case of a contamination event are matched with the simulation events from a database and in [19] a statistical maximum likelihood approach was used for matching. In [20], Monte Carlo (MC) water quality and hydraulic simulations were run in parallel and then used to detect the source of pollution with a certain criteria.

In [21] the logistic regression approach was used to determine top candidates for being the true contamination source with them additionally being explored with local search methods to determine other relevant variables of a contamination event (start time, end time and chemical concentration). The input data for the logistic regression were the sensor readings through time that were constantly updated with new data.

Data-driven models would include using machine learning algorithms to localize the contamination sources in a water supply network. Kim et al. [22] used an Artificial Neural Network (ANN) to find the source of pollution in a small network and Rutkowski and Prokopiuk [23] used a learning vector quantization Neural Network (LVQNN) to locate a zone with a supply network where a potential source of contamination would be located. Wang et al. [24] used Least Squares Support Vector Machines (LS-SVM) to enhance the reliability and accuracy of water sensor contamination detection.

Previous studies have extensively explored Monte-Carlo based methods in air pollution and groundwater pollution source detection problems. In Guo et al. [25] a Markov Chain Monte Carlo (MCMC) sampling method coupled with a Bayesian probabilistic approach was applied to find a source of unsteady atmospheric dispersion which was numerically modeled. Wade and Senocak [26] used Bayesian inference MCMC for the purpose of reconstructing multiple air pollution sources. The study was tested on real-field data and the method includes a ranking of the most probable number of pollution sources which is based on error analysis. In the work by Bashi-Azghadi et al. [27] Probabilistic Support Vector Machines (PSVM) and Probabilistic Neural Networks (PNN) were used to determine an unknown source of pollution in a groundwater system. Vesselinov et al. [28] studied the application of semi-supervised machine learning methods with synthetic and real measured data to identify contamination sources of chemical mixtures in groundwater flows.

In this study we present an algorithm which utilizes the ANN for classification of contamination source in a water supply network and Random Forests (RF) for prediction of contamination start time, end time and chemical concentration. The algorithm is built in a high performance computing (HPC) environment and uses a parallel tournament style selection of most probable contamination source node between a group of nodes. All network nodes are divided in chosen number of tournament groups, where each group is assigned to a single processing core. Monte Carlo (MC) hydraulic and water quality simulations using EPANET2 (Rossman [29]) are run in parallel and the obtained simulation results are used to create models for every tournament group. Network nodes are randomly distributed into tournament groups and the parameters for every simulation (contamination start time, end time and chemical concentration) are randomized. Each network node in a tournament group obtains the same number of results (or MC simulations). Once a node is selected as a winner in the tournament process from existing groups, new tournament groups are created with a reduced number of suspect nodes and this process is repeated until a stopping criterion is satisfied which in this case is the number of set tournament loops which is also a parameter of the algorithm. The whole algorithmic framework was built with a combination of the Python 3.7 programming language and Simple Linux Utility for Resource Management (SLURM) for HPC systems. The ANN classification and the RF regression analysis were done using the Python machine learning library scikit-learn 0.22. The algorithm was tested on two benchmark networks taken from previous studies and it shows good results but also a possibility for improvement with future studies. The algorithm shows a great ability to include the true source node as a top candidate among the remaining source nodes and is good at predicting the other relevant contamination event parameters which can be further used for coupling with optimization algorithms.

2. Materials and Methods

2.1. Water Supply Network Benchmarks

The algorithmic framework was tested on two benchmark networks—the Net3 example from EPANET2 and the Richmond water supply network (Van Zyl [30]).

The Net3 EPANET2 benchmark water supply network consists of 92 nodes and was specifically made for water quality hydraulic simulations. The simulation parameters are set as: the total simulation time is 24 h with a 10 min hydraulic time step and 5 min water quality time step and a 10 min pattern time step. The already optimized water quality sensor layout was set as the one from the work by [7] (network nodes 117, 143, 181 and 213 are set as sensors). The Net3 network layout with the sensor placement can be seen in Figure 1. In each MC simulation of Net3, both the start and the end times (S_m, E_m) of the contamination event were randomly set from 0 to 24 h with an obvious restriction that $E_m > S_m$. The value of C_m was randomly chosen from an interval from 10 to 1000 mg/L and was kept constant throughout the whole contamination scenario. The sensors used in Net3 recorded data during the whole 0–24 h interval for every hour (a total of 25 water quality measurements per sensor) which means that there were 100 input features for the RF regression analysis.

The Richmond network consists of 865 nodes and it was obtained from The Centre for Water Systems (CWS) at the University of Exeter [31]. Simulation time was set as 72 h with a 1 h hydraulic time step, a 5 water quality time step and a 1 pattern time step. The sensor layout was set according to the work by Preis and Ostfeld [7] (nodes 123, 219, 305, 393 and 589 are sensors nodes). The Richmond water supply network layout detail can be seen in Figure 2. The selection of random parameters for each MC Richmond network simulation were defined the same way as for the Net3 network. The Richmond network sensors recorded data during the 0–72 h interval for every hour (a total of 73 measurements per sensor) making a total of 365 input features for the RF regression analysis.

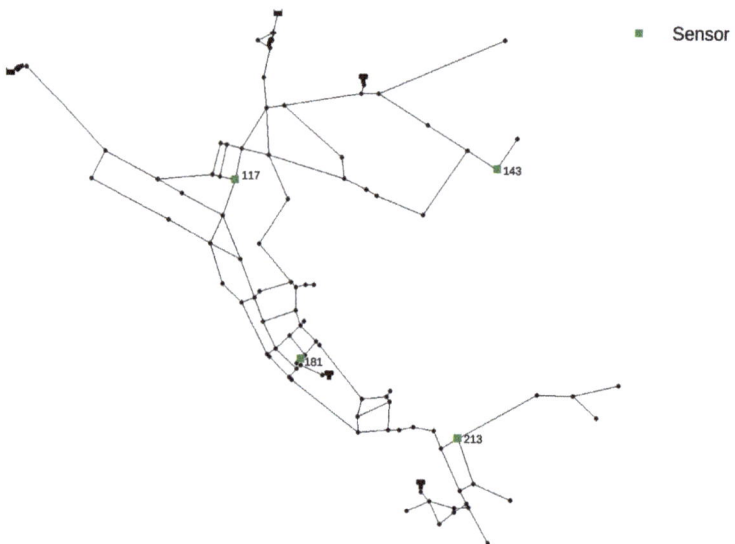

Figure 1. Net3 water supply network layout with sensor positioning by Preis and Ostfeld [7].

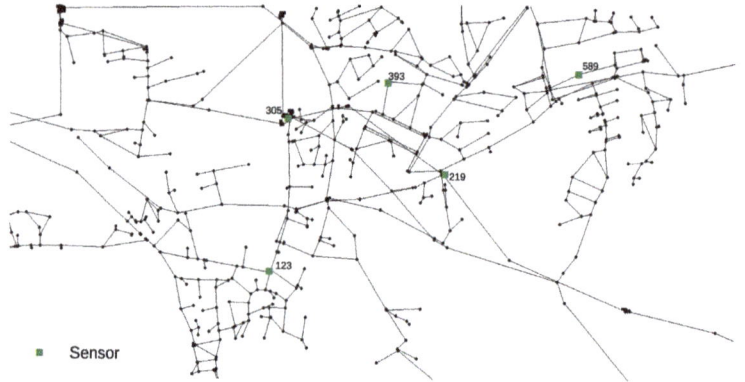

Figure 2. Richmond water supply network layout with sensor positioning by Preis and Ostfeld [7].

2.2. Algorithmic Framework

The algorithm presented in this study was designed to work in a HPC environment in order to detect the water supply network contamination source in a rapid and efficient way. It is based on distributing all potential contamination source nodes of a water supply network into subgroups for which MC simulations would be done in conjunction with machine learning methods to determine which node would best fit to be the true source node of contamination. In this way, each node in a subgroup would be a part of a tournament in which the node with the highest probability of being the contamination source would continue to the next tournament round. After each tournament, the winning nodes would be redistributed in a new tournament group until a predetermined number of tournaments was reached. Each CPU in a HPC system is assigned a tournament node group and after each tournament the total number of used CPUs would decrease since the losing nodes would be discarded and the remaining nodes which are fewer would form new tournament groups. The flowchart of the whole algorithmic framework is shown in Figure 3.

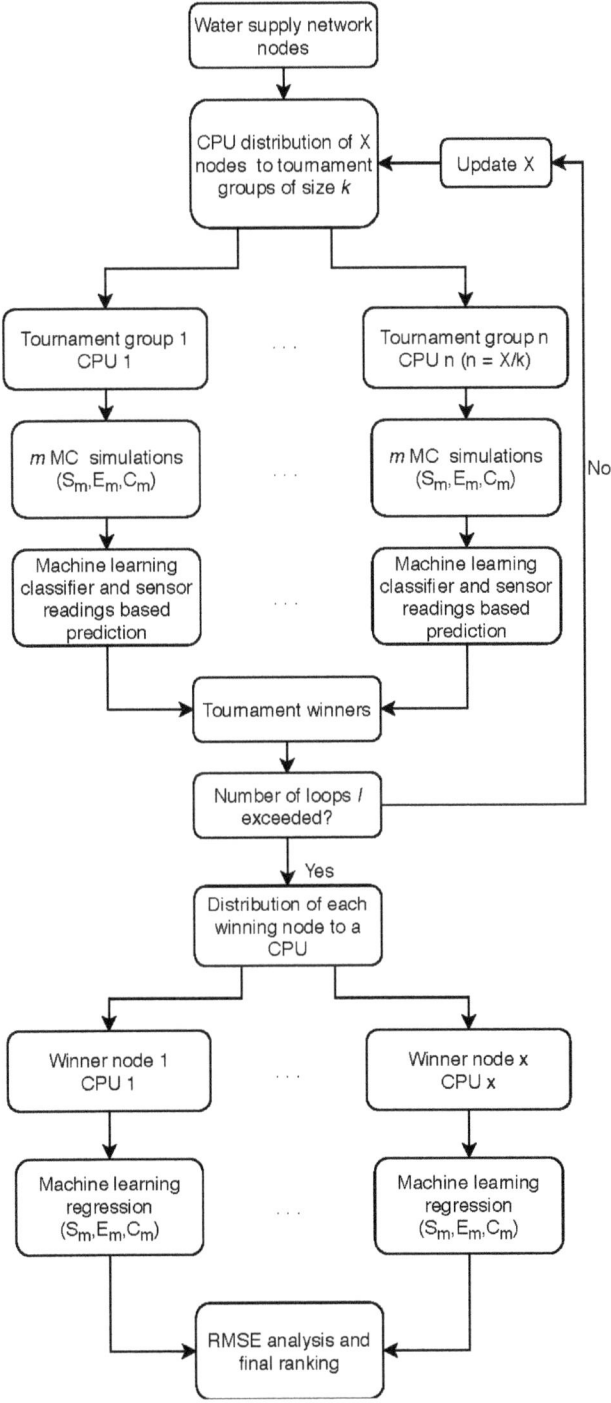

Figure 3. Algorithm flowchart.

As seen in the flowchart, the algorithm is initialized with reading all potential source nodes X in a water supply network and then distributing them in tournament groups of constant size k which is a parameter of the algorithm and can be freely selected. As each tournament group is assigned to a CPU, the number of used CPUs n is determined with n = X/k. After distributing the tournament groups to each CPU, MC simulations are performed m times (m/k times for each suspect node in the tournament group with ideally the modulus of m/k being zero—if not, a node is randomly selected for the additional run) with randomly selected starting S_m and ending E_m times of the contamination event, and the contaminant chemical concentration C_m. Input and output data of each MC simulation are being saved for each suspect node x in every tournament group n.

After the MC simulations are done, the input (sensor readings through time) and output data (source node with used S_m, E_m and C_m) are used for training each tournament group's machine learning (ML) model. The ML model used can be any ML classifying algorithm which supports the prediction of multiple classes. ML output variable set consists of all network nodes that are within that tournament group. After the model was trained with the MC generated data, the sensor readings of the contamination event are being used for the ML prediction of the most probable source node in each tournament group. The nodes with the highest probability in all tournament groups are considered to be the tournament winners. After every used CPU generated a tournament winner, a list of all winners is compiled and if the number of set tournament loops l is not exceeded, the tournament process and distribution is repeated and the number of nodes X is updated (it is equal to the number of winners). In this case, the number of X should be smaller than in the previous tournament loop and consequently the number of used CPUs is reduced since it is dependant on the number of nodes for distribution. It is important to note that each winning node's input and output data is saved from every tournament loop.

If the freely selected algorithm parameter l is exceeded, each winning node is then assigned again to a CPU and with its previously obtained MC input and output data, a ML model is trained and a prediction is performed for the remaining variables of the contamination event (S_m, E_m and C_m). The predicted values of S_m, E_m and C_m of each winning node's ML regression model are then used for simulating the contamination event scenario and the obtained sensor readings are then compared with the real contamination event sensor readings with a RMSE analysis, which in turn creates a ranking where the node with the smallest RMSE is placed at the top.

This whole algorithmic framework was built within the open source workload manager for cluster systems SLURM and the Python 3.7 programming language.

2.3. ANN Classifier

In the previous sub section, the tournament ML classifier was generally defined in the whole algorithmic framework and basically any ML algorithm which can predict multiple classes can be used. In this study, the ANN algorithm is used for classifying the most probable source nodes in a tournament group.

The Multi-layer Perceptron (MLP) type of ANN was used from the Python 3.7 machine learning library scikit-learn 0.22 [32]. The MLP ANN was constructed with both input and output layers and three hidden layers in between. Both first and last hidden layers consisted of 100 neurons, while the middle hidden layer was formed with 500 neurons. The stochastic gradient-based optimizer ADAM for MLP weights optimization was selected through the process of hyper parameter tuning, just as the number of neurons in every ANN layer.

With a preliminary analysis of the possible input variables of the ANN MLP model it was determined that great accuracy of the model can be achieved if only the maximum values of the chemical concentration recorded per sensors through a time interval in the water supply network are used. The preliminary analysis was done through 10 runs and each run when the true source node was in the top 6 of the final nodes was considered successful. The analysis was done on the Net3 benchmark network with the contaminant source node 119 as described in Section 3.1. The true source node was a part of the top 6 ranking suspect nodes for all of the 10 preliminary runs.

This means that the number of neurons at the input layer is equal to the number of sensors used in the water supply network. Furthermore, using the whole time interval of all sensor water quality readings as ANN MLP inputs was tested in the preliminary analysis and it was found that the performance was not better (8 out of 10 runs were successful) than using the maximum values of recorded water quality through time, so naturally, the maximum values per sensor were used as inputs since the number of ML model features is much smaller that way.

The output of the MLP was a list of all tournament group nodes and an assigned probability of each node being the true contamination source after the real contamination event sensor reading was evaluated with the trained model.

In Figure 4 the whole MLP can be seen with $max(C_sn(t))$ being the maximum concentration recorded by the n-th sensor in the network and $N_n\%$ being the probability that the tournament node n is the true source node. All of the MC generated input and output data (of each tournament group) is used to train the ANN model for each tournament group as the goal of the classifier is to predict the most probable contamination source node of each tournament group. The success was assessed by observing the prediction of the whole algorithmic process and not the accuracy of each tournament group ANN model.

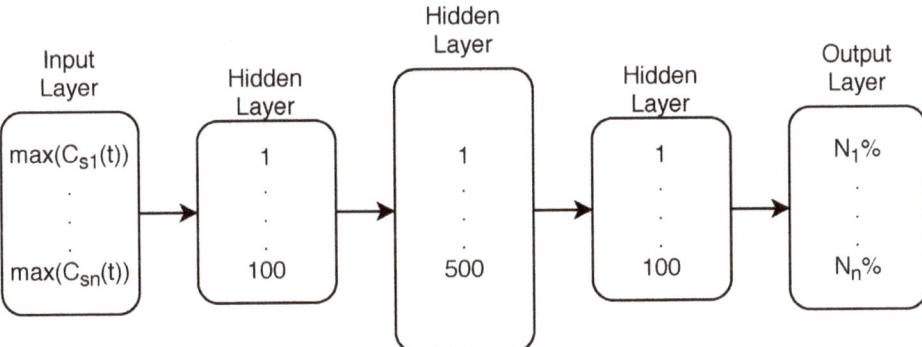

Figure 4. MLP with all layers.

2.4. RF Regression

The ML regression model for each tournament winning node after the parameter l was exceeded in the algorithmic framework can also be done with any ML algorithm which supports multi output regression. The RF algorithm (Breiman [33]) from scikit-learn 0.22 was selected for this purpose. All parameter values of the algorithm were set as default except for the number of estimators (trees in the random forest) which was set to be 200 with the process of hyperparameter optimization.

The input values for the RF regression analysis were sensor water quality readings throughout the whole time interval of the simulation (unlike the inputs used for the MLP ANN) and the output variables were the predicted values of S_m, E_m and C_m for every winner node. A flowchart of the RF regression is seen in Figure 5 with $C_{sn}(t_0...t_{max})$ representing the chemical concentration recorded by the sensor n during simulation time.

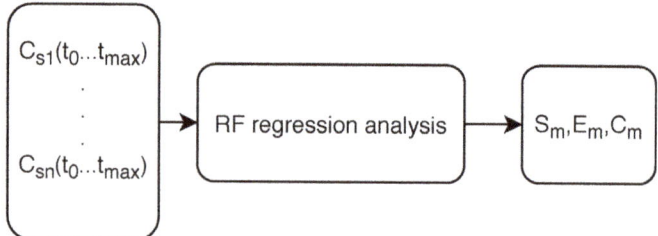

Figure 5. Flowchart of the RF regression analysis.

3. Results and Discussion

3.1. Net3 Network Contamination Scenario

The contamination event scenario for the Net3 benchmark network was chosen to be from the same node (119) as in the one from the work by Preis and Ostfeld [7] and the location can be seen in Figure 6. The contamination event characteristics at source node 119 were freely chosen with the event starting at 14:20 h and lasting until 20:20 h with a constant chemical mass inflow of 813.7 mg/L.

The selected number of algorithm loops l was 3, the number of m (MC simulations for every tournament group) was 200 and the size of a tournament group k was 2, which means that with 92 initial water supply network nodes, the number of used CPUs for every tournament group was 46 and after every loop that number was halved. After three loops, the number of tournament winners was 11, which means that 11 CPUs were used for the RF regression analysis and prediction of other relevant variables.

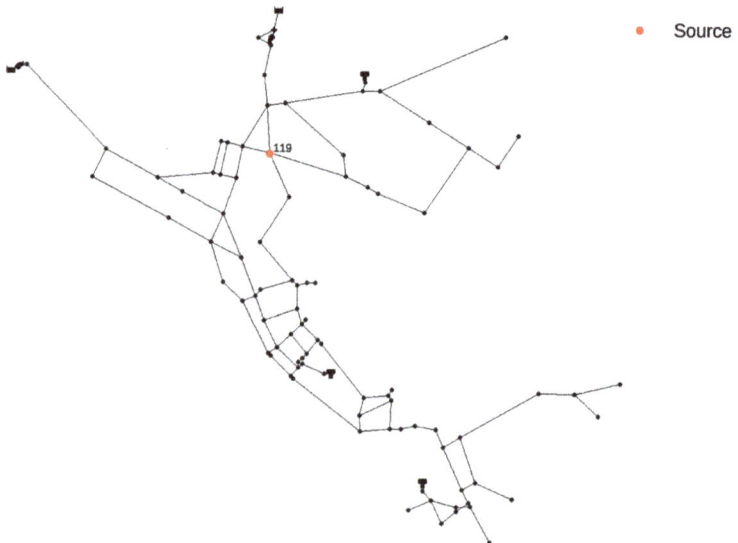

Figure 6. Net3 water supply network contamination source location.

The contamination source search was repeated 30 times since there the algorithm consists of a stochastic component (MC simulations). In 22 out of 30 runs the true source node was the suspect node with the highest probability and in the remaining eight runs the true source node was a part of the final winners list, which means that the ANN classification can successfully narrow down the search space from 92 to 11 nodes in this case. The average run which includes MC simulations, ANN classification and RF regression lasted for 8 min (even though the RF regression lasted only for 8 s).

The algorithm was run (at its initial loop) on 46 Intel Xeon E5 CPUs (two cluster nodes). Out of the other eight runs when the true source node was not ranked first, it was always in the top six of the tournament winners.

In Figure 7 a comparison can be seen between the true contamination event (14:20 h to 20:20 h with 813.7 mg/L) and all of the 30 predicted contamination events for the true source node. It can be seen that the end time prediction of the event is very accurate while the starting time only lacking in accuracy on three runs. The overall RMSE for the starting time for all 30 runs is 48 min, the end time is 4.38 min and the chemical concentration is 18.06 mg/L. The average RMSE For the three of the worst runs with respect to the starting time was 2.47 h. In Table 1, a summary of all runs can be seen through the RMSE analysis and the successful runs represent how many times of the total of 30 runs the true source node was part of the final tournament. The minimum and maximum errors for S_m, E_m and C_m for all 30 runs are presented in Table 2.

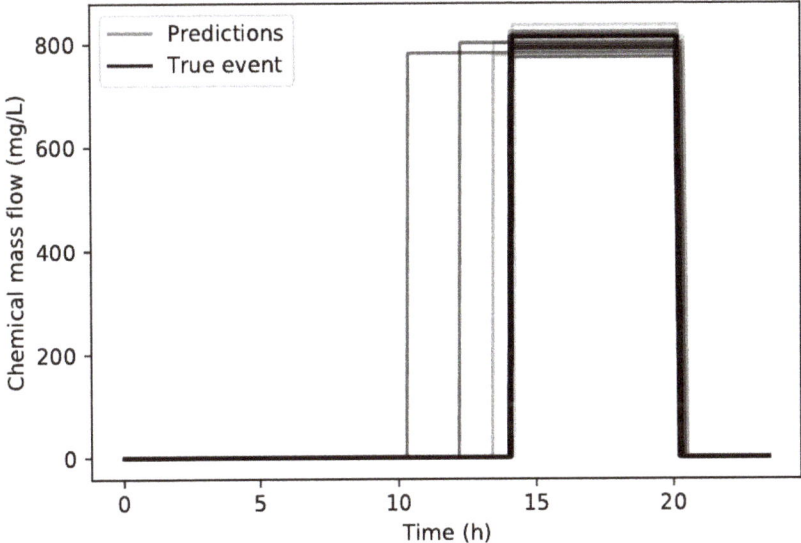

Figure 7. Net3 network true contamination event (black) and predicted contamination events (grey).

In Table 3 the best and worst runs are compared with the true contamination event parameters for S_m, E_m and C_m. The overall best and worst runs are calculated (individual RMSE) by taking into account all of the three variables.

Table 1. Contamination event RMSE analysis for Net3 network of all 30 runs.

Successful Runs	S_m	S_m RMSE	E_m	E_m RMSE	C_m	C_m RMSE
30	14:20 h	48 min	20:20 h	4.38 min	813.7 mg/L	18.06 mg/L

Table 2. Net3 network minimum and maximum errors (S_m, E_m and C_m) for 30 runs.

min S_m	max S_m	min E_m	max E_m	min C_m	max C_m
−3.8 h	+0.2 h	0.0 h	+0.2 h	−0.31 mg/L	−40.11 mg/L

Table 3. Net3 network contamination event results comparison between true, best and worst of the total 30 runs.

Run	S_m	E_m	C_m
True	14:20 h	20:20 h	813.7 mg/L
Best	14:20 h	20:20 h	813.4 mg/L
Worst	14:10 h	20:20 h	773.6 mg/L

In Figure 8 the nodes which were ranked first in the 30 runs can be seen along with a corresponding number of times they were ranked first. It can be observed that the nodes are topologically clustered together. This is expected since due to the multimodal nature of the problem.

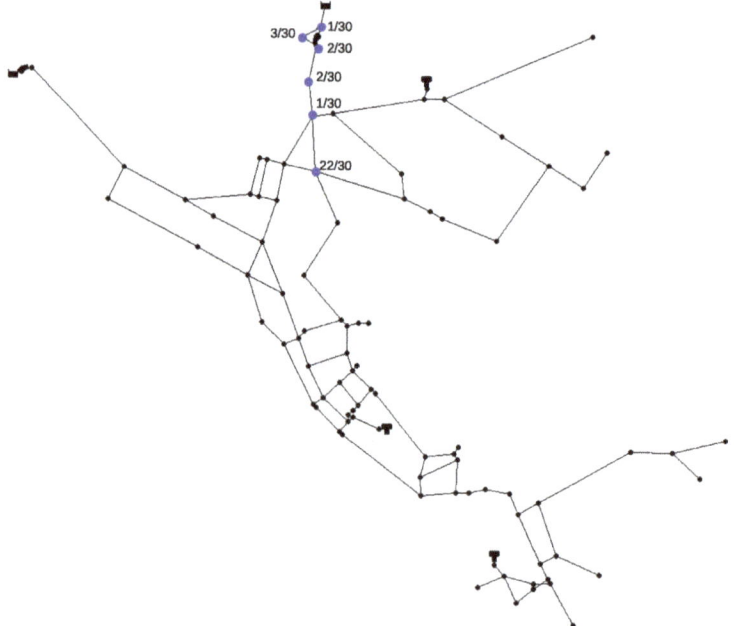

Figure 8. Number of times out of 30 runs for which a node (marked blue) was ranked as first for the Net3 contamination event.

3.2. Richmond Network Contamination Scenario

The Richmond network contamination event scenario was chosen to start at the same node (153) as in the work by Preis and Ostfeld [7] and the location can be seen in Figure 9. The contamination event characteristics at source node 153 were chosen with the event starting at 06:50 h and lasting until 07:40 h with a constant chemical mass inflow of 837 mg/L.

The selected number of algorithm loops l was 1, the number of m (MC simulations for every tournament group) was 2500 and the size of a tournament group k was 4, which means that with 865 initial water supply network nodes, the number of used CPUs for every tournament group was 432. After 1 loop the number of tournament winners was 217, which means that 217 CPUs were used for the RF regression analysis and prediction of other relevant variables.

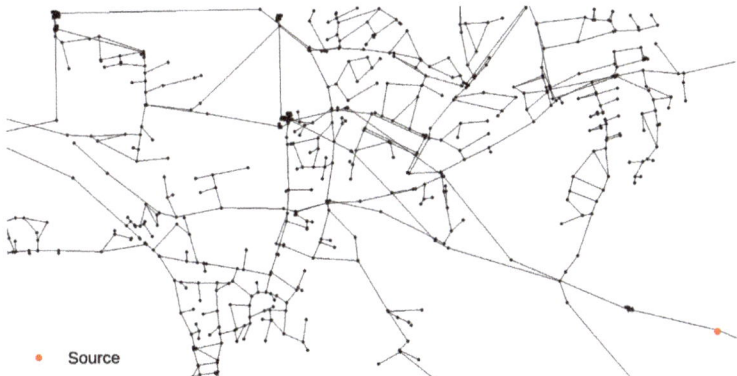

Figure 9. Richmond water supply network contamination source location.

The contamination source search was repeated 30 times just like for the Net3 contamination search. The true source node was ranked first in seven out of 30 runs and it was in the tournament winners list 29 out of 30 times which means that the true source node was not subjected to the RF regression analysis for only one run. The total average run time of the MC simulations and the ANN classification was 41 min and 75 s for the RF regression analysis. The algorithm was run (at its initial loop) on 432 Intel Xeon E5 CPUs. Even though it was ranked first in only seven runs, that was the most number of times a node was ranked first out of the 30 runs. In the 29/30 runs it was in the winners list (which consisted of 217 nodes) it always finished in the top 10 after RF regression was completed.

In Figure 10 the comparison between the true contamination event and the predicted contamination events can be seen. The starting and ending times RMSE for the 29 of the 30 total runs are 6.06 min and 12.36 min respectively, while the chemical concentration RMSE is 299.84 mg/L. The starting and ending times of the predictions are in good agreement with the true values but the chemical concentration value is underestimated by the RF regression.

A RMSE analysis summary of all runs with the number of successful runs can be seen in Table 4. The minimum and maximum errors for all 29 runs for the Richmond network are shown in Table 5 and in Table 6, the best and worst runs comparison is shown in terms of all the predicted variables (S_m, E_m and C_m).

Table 4. Contamination event RMSE analysis for Richmond network of all 30 runs.

Successful Runs	S_m	S_m RMSE	E_m	E_m RMSE	C_m	C_m RMSE
29	06:50 h	6.06 min	07:40 h	12.36 min	837 mg/L	299.84 mg/L

Table 5. Richmond network minimum and maximum errors (S_m, E_m and C_m) for 30 runs.

min S_m	max S_m	min E_m	max E_m	min C_m	max C_m
0.0 h	+0.4 h	−0.4 h	+0.6 h	−122.81 mg/L	−446.81 mg/L

Table 6. Richmond network contamination event results comparison between true, best and worst of the total 30 runs.

Run	S_m	E_m	C_m
True	06:50 h	07:40 h	837 mg/L
Best	06:50 h	07:40 h	714.2 mg/L
Worst	06:50 h	08:00 h	390.2 mg/L

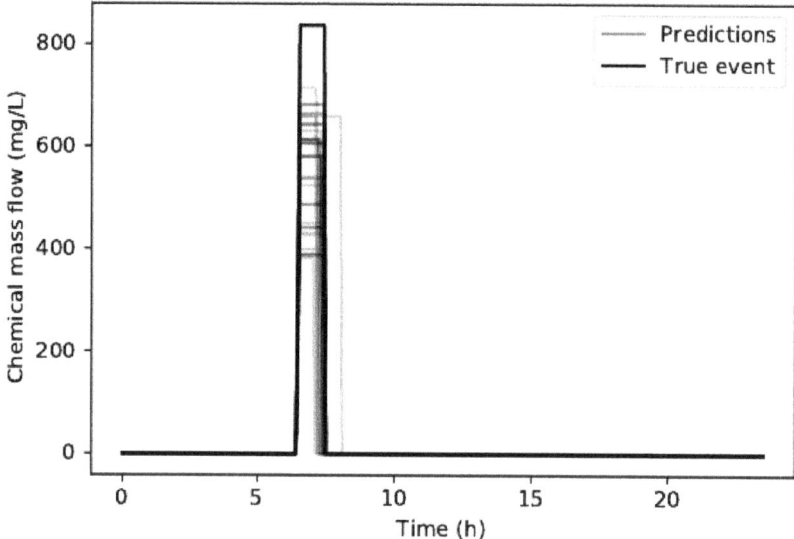

Figure 10. Richmond network true contamination event (black) and predicted contamination events (grey).

3.3. Algorithm Parameters Investigation

In this subsection, an analysis of the influence of algorithm parameters is given. The required number of MC simulations m for each node, the tournament group size k and the number of algorithm loops l are separately explored. The examination of m, k and l was done on the Net3 water supply network with the previously defined contamination event scenario (source at node 119, S_m = 14:20 h, E_m = 20:20 h and C_m = 813.7 mg/L). Each different setup (of m, k and l) was independently run 10 times and a run was considered successful if the node was present in the final ranking after the RF regression and RMSE analysis.

Firstly, the influence of parameter m on the prediction accuracy was explored and a complete summary can be seen in Table 7. The selected tournament group size k for all runs was 2 and the number of loops l was set as 1 during the exploration of parameter m.

Table 7. Influence of the number of MC simulations on the accuracy and efficiency of the algorithm.

m	Successful Runs	Best Rank	Worst Rank	Times Won	S_m RMSE	E_m RMSE	C_m RMSE	Average Run Time
20	9	1	7	2	1.44 h	1.21 h	291.89 mg/L	50 s
40	9	3	8	0	1.76 h	1.11 h	371.93 mg/L	70 s
80	10	1	6	1	0.75 h	0.36 h	283.45 mg/L	120 s
100	10	1	6	4	0.18 h	0.21 h	187.39 mg/L	160 s
200	10	1	6	4	0.58 h	0.18 h	109.40 mg/L	270 s
400	10	1	6	5	0.83 h	0.10 h	39.23 mg/L	420 s
800	10	1	5	6	0.14 h	0.11 h	9.43 mg/L	560 s
1000	10	1	6	9	0.32 h	0.03 h	15.69 mg/L	820 s
1200	10	1	1	10	0.06 h	0.00 h	11.61 mg/L	980 s
2000	10	1	1	10	0.03 h	0.00 h	3.57 mg/L	1400 s
3000	10	1	1	10	0.03 h	0.00 h	4.46 mg/L	2100 s
4000	10	1	1	10	0.00 h	0.00 h	4.09 mg/L	3200 s
5000	10	1	1	10	0.03 h	0.00 h	1.41 mg/L	4000 s
6000	10	1	1	10	0.03 h	0.00 h	3.59 mg/L	4800 s
10,000	10	1	1	10	0.00 h	0.00 h	2.00 mg/L	6300 s

The first column of Table 7 represents the total number of MC simulations per tournament group, which means that since the tournament group size was 2 for each run, each node in the tournament group was the source node in m/2 MC simulations. When observing the analysis of the parameter m in Table 7 it can be seen that the higher the value of the parameter is, both contamination event prediction accuracy (as seen through the decrease S_m, E_m and C_m RMSE) and the average computation time per run (last column) increases. This is expected since more randomly generated data covers more possible scenarios and more input data for the ML model enables a wider and more accurate solution space exploration. Additionally, the best and worst ranks of the true source node are shown and the number of times the true source node won (Times won meaning the rank was 1).

When m was set as 80 the number of successful runs was 10 and the worst possible rank was 6. This means that after 2 min of computation time on average, the search space was reduced from 92 total nodes to 6, which is a reduction of 93.5%. The set of runs when m was 400 could be considered as the first set of runs when the results are acceptable in terms of finding the true source node since it was the winner 50% of the time.

When the value of m is 2000 and above it can be seen that there is not a significant change in prediction accuracy as the RMSE values of S_m and C_m exhibit stability and minor oscillations, while E_m has showed a steady convergence to the same value as the true contamination scenario for all 10 runs.

For further exploration of the tournament group size k, the chosen m was 800 as it exhibited a reasonable computation run time, accuracy in terms of the average RMSE and the number of times the true source node was ranked first. The same scenario was chosen as the one for the investigation of m with the number of tournament loops l set as 1. Five different tournament group sizes k were explored and are summarized in Table 8.

Table 8. Influence of the number of the tournament group size k on the accuracy and efficiency of the algorithm.

k	Successful Runs	Best Rank	Worst Rank	Times Won	S_m RMSE	E_m RMSE	C_m RMSE	CPUs Used
2	10	1	5	6	0.14 h	0.11 h	9.43 mg/L	46
4	10	1	5	7	0.27 h	0.08 h	27.46 mg/L	23
10	10	1	5	5	0.74 h	0.17 h	83.16 mg/L	9
40	5	1	3	1	0.35 h	2.19 h	215.59 mg/L	2
80	4	1	2	2	0.27 h	0.53 h	143.04 mg/L	2

From the results presented in Table 8 it can be observed that when the tournament group is larger, both accuracy and prediction reliability decrease. Furthermore, besides tournament group sizes of 2 and 4, a reasonable result in terms of reliability is achieved with k = 10 with a total search space reduction of 94.6% and even a 50% winning rate in the 10 successful runs. The number of used CPUs for each tournament group size is added and with the given as the last column of the Table 8. Even though a tournament group size of 2 is not that impressive when compared to those of 4 and 10 in the categories of best and worst rank and times won, the achieved overall S_m, E_m and C_m RMSE shows that it is undoubtedly more accurate.

Lastly, the influence of the number of tournament loops l is investigated and a summary of the results is shown in Table 9. The same scenario was used as for the exploration of previous two parameters with k = 2 and with a total number of MC simulations m = 800. An additional column m/L was added to the Table 9 which defines the number of MC simulations m (of a tournament group) per every loop l.

It can be observed that increasing the number of loops l up to a certain value increases the accuracy and reliability of the algorithm. Even though the total number of MC simulations is the same for every run and the computational strain in that sense is similar, adding more loops decreases the number of used CPUs after every tournament loop since losing nodes are omitted and that can be considered as a great advantage.

Table 9. Influence of the number of the tournament loops l on the accuracy and efficiency of the algorithm.

l	Successful Runs	Best Rank	Worst Rank	Times Won	S_m RMSE	E_m RMSE	C_m RMSE	m/L
1	10	1	5	6	0.14 h	0.11 h	9.43 mg/L	800
2	10	1	6	8	0.18 h	0.08 h	23.20 mg/L	400
3	10	1	4	8	0.11 h	0.04 h	15.42 mg/L	267
4	10	1	3	9	0.08 h	0.03 h	14.71 mg/L	200
5	10	1	3	7	0.07 h	0.08 h	14.57 mg/L	160
6	10	1	2	8	0.07 h	0.05 h	23.29 mg/L	134
7	10	1	1	10	0.03 h	0.05 h	14.77 mg/L	115
8	9	1	3	7	0.07 h	0.07 h	25.49 mg/L	100

The value of S_m, E_m and C_m RMSE does not differ much for all tested loops l since the total number of MC simulations is preserved. When the number of loops is set to 8, the successful number of runs dropped as the number of MC simulations per loop (m/L) was not high enough and the true source node was not in the final RF analysis ranking for one run. This was also observed for smaller values of m in Table 7. It can be argued that a higher number of loops positively affects the success of the algorithm in predicting the relevant variables (source node, S_m, E_m and C_m) since there is a higher chance that a main tournament top ranking rival to the true source node is omitted in the process of removing losing nodes after every tournament loop. However, setting l too high could result in unsuccessful runs as well (due to a small m/L) as it can be seen in Table 9.

4. Conclusions

In this paper a novel algorithmic framework for water supply network contamination source node identification is presented and tested on two different benchmark networks. The algorithm is specifically created for massively parallel HPC systems and it utilizes a combination of MC simulations and ML methods to identify the contamination event source node and all the relevant variables such as starting and ending times of the event and the contaminant chemical concentration.

The algorithm is based on running an equal number of MC simulations on a group of nodes in parallel and then selecting the node (via MLP ANN) with the highest probability of being the true contamination source node as the winner of the group (tournament). The number of MC simulations per tournament group is set as a parameter of the algorithm just like the number of tournament loops and the size of a tournament group. After a set of tournament winners is created, the algorithm utilizes a ML regression analysis using RF in parallel and creates a ranking of potential source nodes based on a simple error analysis with the true contamination event sensor data recorded.

The novel algorithmic framework, tested on two realistic and complex benchmark networks cases, displays the capability to narrow down the search space for the source node efficiently, leading to a pollution source identification. The algorithm can be also used in predicting the starting and ending times of the contamination event and the contaminant chemical concentration.

An investigation of algorithm's parameters which are the number of MC simulations, size of a tournament group and number of loops was also conducted. It is demonstrated that increasing the number of MC simulations is beneficial to the algorithm's ability to predict the true source node and relevant variables since more randomly generated data entails a broader solution space coverage, however this comes with an increase in computational time. It was observed that in order to increase the reliability of an accurate prediction the size of a tournament group should be as low as possible, depending on computational resources. Increasing the number of tournament loops shows to be advantageous in prediction accuracy; however, the number of MC simulations for each tournament loop should be high enough in order to preserve the reliability of prediction.

Further research is needed in determining the connection between the network size (initial number of potential source nodes) and the number of MC simulations needed to cover the search space of the contamination event efficiently and thoroughly. Furthermore, additional research should be done regarding the used ML methods investigating the use of different classifiers for the tournament group winner selection and various ML algorithms capable of multi-output regression analysis for the final node ranking. It would be also possible to couple the algorithm with simulation-optimization methods for an even faster convergence towards a true pollution source node detection in a way that only the tournament winners are subjected to the optimization process.

Author Contributions: Conceptualization, L.G., I.L. and S.D.; Data curation, L.G.; Formal analysis, L.G.; Funding acquisition, L.K.; Investigation, L.G. and I.L.; Methodology, L.G. and L.K.; Project administration, L.K.; Resources, L.K.; Software, L.G.; Supervision, S.D.; Validation, L.G.; Visualization, L.G.; Writing—original draft, L.G.; Writing—review & editing, I.L., L.K. and S.D. All authors have read and agreed to the published version of the manuscript.

Funding: This work was supported by the Center for Advanced Computing and Modelling, University of Rijeka.

Conflicts of Interest: The authors declare no conflict of interest.

Abbreviations

The following abbreviations are used in this manuscript:

BBN	Bayesian Belief Networks
MC	Monte Carlo
ANN	Artificial Neural Network
LVQNN	Learning Vector Quantization Neural Network
LS-SVM	Least Squares Support Vector Machines
MCMC	Markov Chain Monte Carlo
PSVM	Probabilistic Support Vector Machines
PNN	Probabilistic Neural Networks
RF	Random Forests
HPC	High Performance Computing
CWS	Center for Water Systems
CPU	Central Processing Unit
ML	Machine Learning
SLURM	Simple Linux Utility for Resource Management
RMSE	Root Mean Square Error
MLP	Multi-layer Perceptron

References

1. Ostfeld, A.; Uber, J.G.; Salomons, E.; Berry, J.W.; Hart, W.E.; Phillips, C.A.; Watson, J.P.; Dorini, G.; Jonkergouw, P.; Kapelan, Z.; others. The battle of the water sensor networks (BWSN): A design challenge for engineers and algorithms. *J. Water Resour. Plan. Manag.* **2008**, *134*, 556–568. [CrossRef]
2. Zhao, Y.; Schwartz, R.; Salomons, E.; Ostfeld, A.; Poor, H.V. New formulation and optimization methods for water sensor placement. *Environ. Model. Softw.* **2016**, *76*, 128–136. [CrossRef]
3. Ung, H.; Piller, O.; Gilbert, D.; Mortazavi, I. Accurate and Optimal Sensor Placement for Source Identification of Water Distribution Networks. *J. Water Resour. Plan. Manag.* **2017**, *143*, 04017032. [CrossRef]
4. Guidorzi, M.; Franchini, M.; Alvisi, S. A multi-objective approach for detecting and responding to accidental and intentional contamination events in water distribution systems. *Urban Water J.* **2009**, *6*, 115–135. [CrossRef]
5. Alfonso, L.; Jonoski, A.; Solomatine, D. Multiobjective optimization of operational responses for contaminant flushing in water distribution networks. *J. Water Resour. Plan. Manag.* **2010**, *136*, 48–58. [CrossRef]
6. Hu, C.; Yan, X.; Gong, W.; Liu, X.; Wang, L.; Gao, L. Multi-objective based scheduling algorithm for sudden drinking water contamination incident. *Swarm Evol. Comput.* **2020**, *55*, p. 100674.
7. Preis, A.; Ostfeld, A. A contamination source identification model for water distribution system security. *Eng. Optim.* **2007**, *39*, 941–947. [CrossRef]
8. Zechman, E.M.; Ranjithan, S.R. Evolutionary computation-based methods for characterizing contaminant sources in a water distribution system. *J. Water Resour. Plan. Manag.* **2009**, *135*, 334–343. [CrossRef]

9. Kranjčević, L.; Čavrak, M.; Šestan, M. Contamination source detection in water distribution networks. *Eng. Rev.* **2010**, *30*, 11–25.
10. Vankayala, P.; Sankarasubramanian, A.; Ranjithan, S.R.; Mahinthakumar, G. Contaminant source identification in water distribution networks under conditions of demand uncertainty. *Environ. Forensics* **2009**, *10*, 253–263. [CrossRef]
11. Xuesong, Y.; Jie, S.; Chengyu, H. Research on contaminant sources identification of uncertainty water demand using genetic algorithm. *Cluster Comput.* **2017**, *20*, 1007–1016. [CrossRef]
12. Yan, X.; Zhu, Z.; Li, T. Pollution source localization in an urban water supply network based on dynamic water demand. *Environ. Sci. Pollut. Res.* **2019**, *26*, 17901–17910. [CrossRef]
13. Adedoja, O.; Hamam, Y.; Khalaf, B.; Sadiku, R. Towards development of an optimization model to identify contamination source in a water distribution network. *Water* **2018**, *10*, 579. [CrossRef]
14. Dawsey, W.J.; Minsker, B.S.; VanBlaricum, V.L. Bayesian belief networks to integrate monitoring evidence of water distribution system contamination. *J. Water Resour. Plan. Manag.* **2006**, *132*, 234–241. [CrossRef]
15. De Sanctis, A.; Boccelli, D.; Shang, F.; Uber, J. Probabilistic approach to characterize contamination sources with imperfect sensors. In *World Environmental and Water Resources Congress 2008: Ahupua'A*; ASCE: Reston, VA, USA 2008; pp. 1–10.
16. Perelman, L.; Ostfeld, A. Bayesian networks for source intrusion detection. *J. Water Resour. Plan. Manag.* **2013**, *139*, 426–432. [CrossRef]
17. Neupauer, R.M.; Ashwood, W.H. Backward probabilistic modeling to identify contaminant sources in a water distribution system. In *World Environmental and Water Resources Congress 2008: Ahupua'A*; ASCE: Reston, VA, USA; 2008; pp. 1–10.
18. Shen, H.; McBean, E. False negative/positive issues in contaminant source identification for water-distribution systems. *J. Water Resour. Plan. Manag.* **2011**, *138*, 230–236. [CrossRef]
19. Huang, J.J.; McBean, E.A. Data mining to identify contaminant event locations in water distribution systems. *J. Water Resour. Plan. Manag.* **2009**, *135*, 466–474. [CrossRef]
20. Eliades, D.; Lambrou, T.; Panayiotou, C.G.; Polycarpou, M.M. Contamination event detection in water distribution systems using a model-based approach. *Procedia Eng.* **2014**, *89*, 1089–1096. [CrossRef]
21. Liu, L.; Zechman, E.M.; Mahinthakumar, G.; Ranjithan, S.R. Coupling of logistic regression analysis and local search methods for characterization of water distribution system contaminant source. *Eng. Appl. Artif. Intell.* **2012**, *25*, 309–316. [CrossRef]
22. Kim, M.; Choi, C.Y.; Gerba, C.P. Source tracking of microbial intrusion in water systems using artificial neural networks. *Water Res.* **2008**, *42*, 1308–1314. [CrossRef]
23. Rutkowski, T.; Prokopiuk, F. Identification of the Contamination Source Location in the Drinking Water Distribution System Based on the Neural Network Classifier. *IFAC-PapersOnLine* **2018**, *51*, 15–22. [CrossRef]
24. Wang, K.; Wen, X.; Hou, D.; Tu, D.; Zhu, N.; Huang, P.; Zhang, G.; Zhang, H. Application of Least-Squares Support Vector Machines for Quantitative Evaluation of Known Contaminant in Water Distribution System Using Online Water Quality Parameters. *Sensors* **2018**, *18*, 938. [CrossRef]
25. Guo, S.; Yang, R.; Zhang, H.; Weng, W.; Fan, W. Source identification for unsteady atmospheric dispersion of hazardous materials using Markov Chain Monte Carlo method. *Int. J. Heat Mass Transf.* **2009**, *52*, 3955–3962. [CrossRef]
26. Wade, D.; Senocak, I. Stochastic reconstruction of multiple source atmospheric contaminant dispersion events. *Atmos. Environ.* **2013**, *74*, 45–51. [CrossRef]
27. Bashi-Azghadi, S.N.; Kerachian, R.; Bazargan-Lari, M.R.; Solouki, K. Characterizing an unknown pollution source in groundwater resources systems using PSVM and PNN. *Expert Syst. Appl.* **2010**, *37*, 7154–7161. [CrossRef]
28. Vesselinov, V.V.; Alexandrov, B.S.; O'Malley, D. Contaminant source identification using semi-supervised machine learning. *J. Contam. Hydrol.* **2018**, *212*, 134–142. [CrossRef]
29. Rossman, L.A. *EPANET 2: Users Manual*; U.S. Environmental Protection Agency: Washington, DC, USA, 2000.
30. Van Zyl, J.E. A Methodology for Improved Operational Optimization of Water Distribution Systems. Ph.D. Thesis, University of Exeter, Exeter, UK, 2001.
31. CWS; UoE. CWS Benchmarks. Available online: http://emps.exeter.ac.uk/engineering/research/cws/downloads/benchmarks/ (accessed on 6 November 2019).

32. Pedregosa, F.; Varoquaux, G.; Gramfort, A.; Michel, V.; Thirion, B.; Grisel, O.; Blondel, M.; Prettenhofer, P.; Weiss, R.; Dubourg, V.; others. Scikit-learn: Machine learning in Python. *J. Mach. Learn. Res.* **2011**, *12*, 2825–2830.
33. Breiman, L. Random forests. *Mach. Learn.* **2001**, *45*, 5–32. [CrossRef]

© 2020 by the authors. Licensee MDPI, Basel, Switzerland. This article is an open access article distributed under the terms and conditions of the Creative Commons Attribution (CC BY) license (http://creativecommons.org/licenses/by/4.0/).

Article

Extraction of Land Information, Future Landscape Changes and Seismic Hazard Assessment: A Case Study of Tabriz, Iran

Ayub Mohammadi [1], Sadra Karimzadeh [1,2,3,*], Khalil Valizadeh Kamran [1,*] and Masashi Matsuoka [3]

1. Department of Remote Sensing and GIS, University of Tabriz, Tabriz 5166616471, Iran; mohammadi.ayub@tabrizu.ac.ir
2. Institute of Environment, University of Tabriz, Tabriz 5166616471, Iran
3. Department of Architecture and Building Engineering, Tokyo Institute of Technology, Yokohama 226-8502, Japan; matsuoka.m.ab@m.titech.ac.jp
* Correspondence: karimzadeh.s.aa@m.titech.ac.jp (S.K.); valizadeh@tabrizu.ac.ir (K.V.K.)

Received: 7 October 2020; Accepted: 7 December 2020; Published: 8 December 2020

Abstract: Exact land cover inventory data should be extracted for future landscape prediction and seismic hazard assessment. This paper presents a comprehensive study towards the sustainable development of Tabriz City (NW Iran) including land cover change detection, future potential landscape, seismic hazard assessment and municipal performance evaluation. Landsat data using maximum likelihood (ML) and Markov chain algorithms were used to evaluate changes in land cover in the study area. The urbanization pattern taking place in the city was also studied via synthetic aperture radar (SAR) data of Sentinel-1 ground range detected (GRD) and single look complex (SLC). The age of buildings was extracted by using built-up areas of all classified maps. The logistic regression (LR) model was used for creating a seismic hazard assessment map. From the results, it can be concluded that the land cover (especially built-up areas) has seen considerable changes from 1989 to 2020. The overall accuracy (OA) values of the produced maps for the years 1989, 2005, 2011 and 2020 are 96%, 96%, 93% and 94%, respectively. The future potential landscape of the city showed that the land cover prediction by using the Markov chain model provided a promising finding. Four images of 1989, 2005, 2011 and 2020, were employed for built-up areas' land information trends, from which it was indicated that most of the built-up areas had been constructed before 2011. The seismic hazard assessment map indicated that municipal zones of 1 and 9 were the least susceptible areas to an earthquake; conversely, municipal zones of 4, 6, 7 and 8 were located in the most susceptible regions to an earthquake in the future. More findings showed that municipal zones 1 and 4 demonstrated the best and worst performance among all zones, respectively.

Keywords: remote sensing; GIS; Markov chain; land use; urban information; Tabriz City

1. Introduction

Rapid urbanization, deforestation and increasing population have led to global environmental changes [1]. Because of this, large areas of agricultural land are being converted into urban land and industrial estates, which are prone to land degradation [1,2]. Typically, urbanization influences climate and water quality [3], which can result in changes in local climate. One of the main means by which to understand the relationship between humans and their environment is recording the changes occurring where they live [3,4]. Useful information can be obtained from the pattern and direction of land cover changes, from which better planning for sustainable development is possible. Landsat satellite imageries are widely used medium-scale data for land surface change analysis [2,5].

The low temporal baseline of these data is considered as a weak point that sometimes results in the omission of some dynamic changes [3]. Urban sprawl significantly changes landscapes across urban areas [6], which is usually associated with changes among vegetation, built-up areas and bare lands in the near or remote future. For seismic hazard assessment caused by an earthquake, urban information is urgently needed [7]. Normally, field checks and digital interpretation using the technology of remote sensing (RS) are among the common ways to extract urban information [1,7,8]. Geographic information systems (GIS) together with remote sensing technology can provide useful information for decision-makers [9]. RS data have successfully been applied for mapping and measuring the area and the extent of land cover. Satellite image performance has now been improved to a ground resolution of less than 1 m to be acquired. Physical assessment of urban areas from remotely sensed data enables comparative analysis of a city's extent within a region [3]. In order to record all changes, selecting the proper temporal baseline of RS data is very important [10]. The Landsat satellite imagery (medium spatial resolution RS data) has widely been utilized to quantitate urbanization across the world [2,11].

Predicted land use changes can help land use planners in mitigating the negative impacts on the environment. In recent years, the city and the surrounding areas have been several times shocked by large and small earthquakes. In developed countries, building inventory information is provided by local and central institutes [7]. Nevertheless, in developing countries like Iran, this is quite the opposite, so researchers should work and provide information to different organizations. Normally, this kind of study requires great effort and considerable financial support. There are many faults in Iran, a few of which have recently been activated and claimed many lives and also caused a great deal of damage to properties [12]; because of the possibility of the recurrence of such events in the near or remote future, fear still exists among inhabitants in these areas. Typically, because of the high buildings, this fear is most common among people in large cities like Tabriz.

Efforts have been made to carry out land cover information extraction using RS data and techniques. Urbanization influences ecosystems, but in order to determine and understand these impacts, precise and accurate information about the land cover's temporal and spatial changes is essential [2,3]. Built-up areas are very important for many studies, including those considering buildings' age, seismic hazard assessment, and prediction, so as to enable the optimal updating of built-up areas. Maximum likelihood, as one of the best methods for classification [11,13], was employed for extracting land cover using four cloud-free Landsat data. Accordingly, building inventory and urban sprawl information are important factors for damage estimation [7,14,15]. In order to determine if older buildings have been constructed under older seismic hazard standards, a map of the ages of buildings is needed [7,12]. For a fast and effective response during an earthquake, an urgent evaluation is needed [7], but, at present, sufficient information is provided to be used in the future to mitigate possible damages in the study area. Previous research simply studied the land cover detection of the city, but a comprehensive study on land cover change detection, future potential land cover, municipal performance evaluation and seismic hazard assessment is missing for Tabriz, which is one of the largest and most important cities in Iran. Therefore, this study presents a more comprehensive examination of the study area compared to the previous research, making it highly significant.

Many studies conducted land cover change detection and prediction using different models of fuzzy logic modeling [16–20]; geo-statistical methods [21–24]; Markov-CA [5,10,25–32]; cellular automata models [33–36]; propagating aleatory and epistemic uncertainty [37,38]; artificial neural network [39–43]; Hopfield neural network [44–48]; supervised back-propagation neural network [49,50]; self-adaptive cellular based deep learning [51–54]; analytical hierarchy process [27,55,56]; geographic information system (GIS)-based hybrid site condition [15,57,58]; recurrent neural network [59–62]; change vector analysis [63–67]; and different satellite imageries of both SAR and optical [1–3,6,7,9–11,13,14,68–76].

Previous studies on Tabriz City were not comprehensive, failing to include seismic hazard assessment and municipality performance. In recent years, the rapidly increasing population of the city has necessitated the construction of high-rise buildings and conversion of agricultural land into built-up areas. A comprehensive study that exploits the advantages of both remote sensing and GIS can turn satellite data into an actionable level that can be used for proper environmental planning. For the study area, a comprehensive evaluation from the point of view of satellite datasets and GIS is lacking. To fill this gap, GIS layers of various spatial information, SAR images from Sentinel-1 and optical images from Landsat missions were gathered to perform a new study of land cover change detection, future potential landscape, distribution of buildings by age, municipal performance evaluation and building damage assessment in Tabriz.

2. Description of the Study Area

The study area (with a population of over 1.7 million and area of 321.03 km^2) is the metropolitan area of Tabriz, East-Azerbaijan, Iran, defined by its 10 municipal regions (Figure 1). Tabriz City is located in Azarbaijan geological zone, which is surrounded in the northern region by Alborz, in the south by Semnan and in the west by the Tabriz–Urumiyeh Faults [77]; therefore, it is considered as an area susceptible to earthquakes. Tabriz City continues to the Pontic highlands in Turkey [77,78]. The central and western regions of Iran are comparable with the Azarbaijan geological zone [79], where there are a few important faults [15,78–80]. The Tabriz fault is the most important one near to the city of Tabriz, extending in the northwest–southeast direction from the Zanjan zone and continuing to the northern mountains of Tabriz City [78]. It has been selected because: (1) it is the 5th large city of Iran which is located near the Tabriz fault, and in recent years, the land cover has rarely been updated; (2) it is a hub for cities in the northwest and west of the country (mainly due to better facilities including more job opportunities, higher quality education and more health centers, more people tend to migrate to the city); and (3) for such cities with this level of importance, future land cover prediction and seismic hazard assessment is vital. Seasons in Tabriz are regular, and it has a continental and cold semi-arid climate; at the same time, the average annual precipitation in the study area is around 320 mm, while the average annual temperature is almost 12.6 °C [81,82]. The city experiences humid and rainy weather in autumn, while it has a few snowy days during the winter season; at the same time, in spring, the city has a mild and fine climate, and during summers, the region can experience a semi-hot climate [81,82].

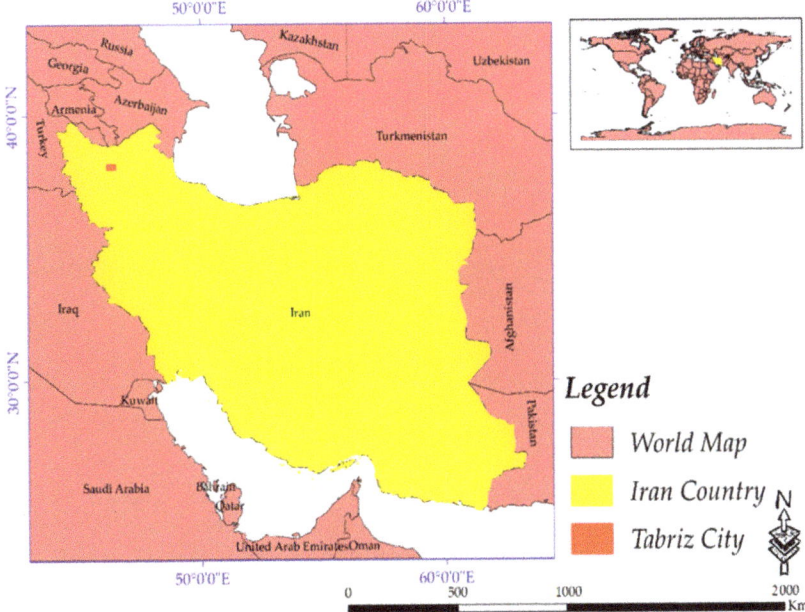

Figure 1. The geographical extent of the city.

3. Materials and Methods

3.1. Database and Data Acquisition

3.1.1. Optical Satellite Data

For the study area, we obtained four cloudless satellite data, from which a Landsat-5 image of the year 1989 was selected as the base image for the study. All the Landsat images were downloaded through the USGS portal. Based on the metadata, only the data with cloud coverage of less than 10% were selected. Then, we searched for and collected RS data between 1989 to 2020, and cloudless Landsat-5 data for 2005 and 2011 were found. Regarding data for the year 2020, an image of Landsat-8 operational land imager (OLI) was collected. The images include seven bands in the range of visible to thermal-infrared for Landsat-5 and nine bands for Landsat-8. The ground resolution of images in optical bands is 30 m, except for a panchromatic band of Landsat-8, which is 15 m. The technical characteristics of the Landsat data are clearly presented in Table 1.

Table 1. Characteristics of the optical data used.

Satellite Name	Sensor Mode	Resolution (m)	Path	Row	Date
Landsat-5	(Thematic Mapper) TM	30	168	34	30 June 1989
Landsat-5	TM	30	168	34	20 July 2005
Landsat-5	TM	30	168	34	5 July 2011
Landsat-8	OLI	30	168	34	11 June 2020

3.1.2. Synthetic Aperture Radar data

The study was focused on the Landsat missions to carry out the objectives, but the SAR data of Sentinel-1 (SLC and GRD) were also employed for extracting land cover, especially built-up areas, because built-up areas are associated more with seismic hazard assessment; therefore, the validity and reliability of its extraction should be taken into account. Table 2 shows the geometric attributes of SAR data.

Table 2. Geometric attributes of SAR data used.

Satellite Name	Platform	Product Type	Sensor Mode	Date
Sentinel-1	S1A	SLC	IW	11 June 2020
Sentinel-1	S1A	SLC	IW	23 June 2020
Sentinel-1	S1A	GRD	IW	11 June 2020

3.2. Data Preprocessing and Processing

First of all, all data of Landsat-5 and 8 were processed for atmospheric, radiometric corrections and the spatial resolution of them was enhanced to 15 m using a panchromatic band of Landsat-8. It is worth mentioning that in pan-sharpening, spectral information will remain unchanged, while the spatial resolution of higher pixel size images will be assigned to the lower one. Here, the 30-m spatial resolution of the Landsat-5 data was enhanced to just 15-m using the panchromatic band of the Landsat-8. These data were imported into TerrSet Software for classification, change detection and prediction using the Markov chain algorithm for the years 2011 and 2030. At the first stage, the regions of interest (ROIs) were extracted carefully; then, by the maximum likelihood algorithm, changes in land cover from Landsat satellite images were detected, classified and mapped. Furthermore, all ancillary data were processed and applied together with classified maps for the prediction steps using the Markov chain model. The overall approach for the current research includes three key procedures: (1) geometric correction of data; (2) classification of optical satellite data and prediction of the future potential landscape of the city; and (3) municipal performance evaluation and seismic hazard assessment (Figure 2). Ancillary reference data were collected from the Municipality of Tabriz and open street map (OSM) and were applied for training the Markov chain algorithm. These data included a digital elevation model (DEM), buildings, land use, places, railway, roads, green space, waterway and welfare services shapefile of the city (Table 3). Additionally, those data which were collected from the municipality of Tabriz were up to date and were based on the latest changes that occurred in the city.

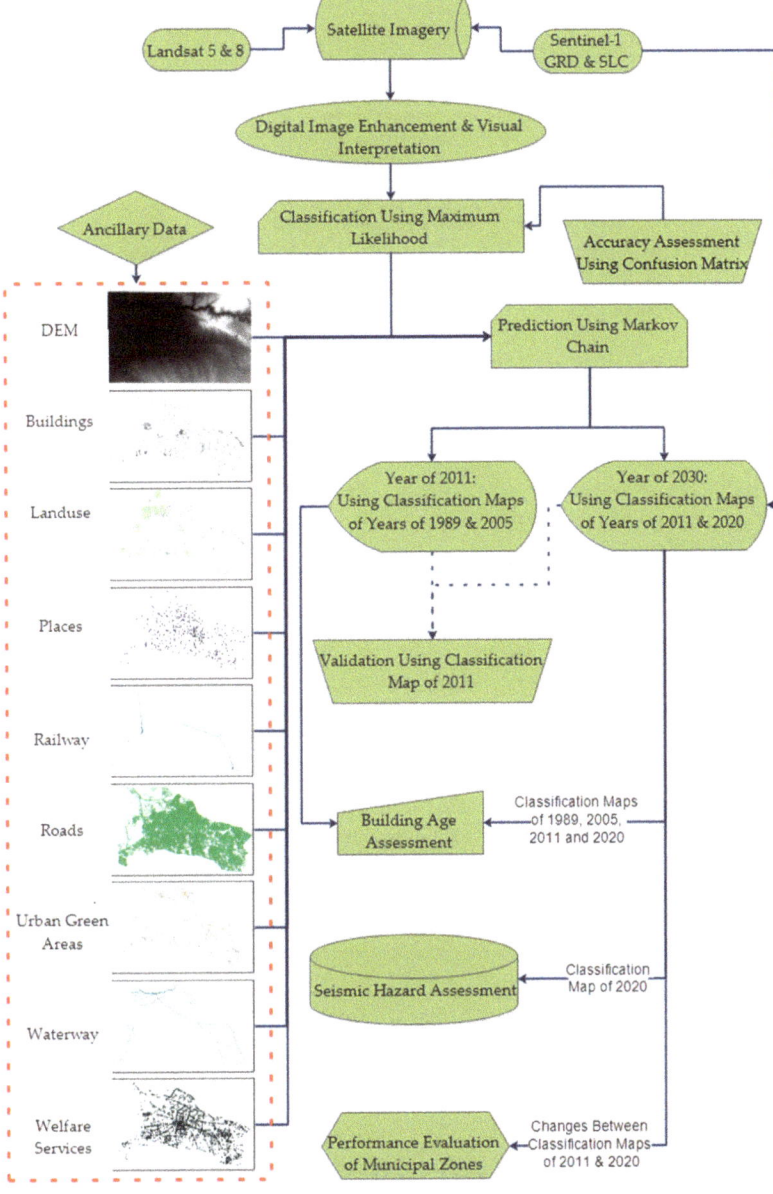

Figure 2. The methodology of the research.

Table 3. Information on ancillary reference data.

Ancillary Data	Description
DEM	30 m SRTM DEM downloaded from USGS website
Buildings	Polygon shapefile of buildings collected from the municipality of Tabriz
Land Use	Information on recently modified land use of a few place across the city, collected from the municipality of Tabriz
Places	Towns, crossroads and squares
Railway	Railway shapefile obtained from OSM
Roads	Roads shapefile obtained from OSM
Green Space	Information of recent made green areas over the city, collected from the municipality of Tabriz
Waterway	Waterway shapefile acquired from OSM
Welfare Services	Masques, hotels, educational institutes, public parks, sports centers and gyms, banks, petrol stations, hospitals, drug stores, markets and recreational facilities

3.3. Model Used in the Study

3.3.1. Maximum Likelihood

A maximum likelihood classifier was applied to extract surface information from RS data. This defined the statistical values with a normal distribution for each class in image's bands. On the other hand, the algorithm estimated the probability that one pixel would fall into the defined classes. This procedure was continued for all the pixels and the pixels were assigned for those classes that produced the highest probability as follows [83]:

$$g_i(x) - 1n\ p(\infty_i) - \frac{1}{2}1n|\Sigma_i| - \frac{1}{2}(x - m_i)\sum_i -1\ (x - m_i) \quad (1)$$

where the number of classes is defined by $g_i(x)$, which represents the number of imageries' bands, $p(\infty_i)$ describes the probability of class, which occurs in the images, the covariance matrix is defined by $|\Sigma_i|$; additionally, the inverse matrix is $\Sigma_i -1$, and m_i is the mean vector [83].

3.3.2. Markov Chain

Markov chain is a model from which the future potential landscape can be relatively detected, so that, based on the extracted information from the past data, it detects the future pattern of a land cover [25–27]. In this countable sequence, the chain moves state at discrete time steps [84,85]. It is worth mentioning that this sequence of time process is called a sequence-time Markov chain [84,86]. Markov chains have many applications in different fields. Overall, this model for land cover prediction produces promising findings [4,35]. The model calculated the following formulas [25]:

$$p_{ij} = \frac{n_{ij}}{n_i} \quad (2)$$

$$\sum_{j=1}^{k} p_{ij} = 1 \quad (3)$$

where transition probability is defined by p_{ij}, i and j describe two types of land cover, the total of pixels of each class is shown by n_i and n_{ij}, which represents the number of transformed pixels from class i to class j, and finally, k defines the number of land cover classes.

3.3.3. Logistic Regression Model

A good deterministic seismic hazard assessment is generally associated with effective and well-approved models [16]. In recent years, the LR model has been widely employed to analyze binary variables and, as a result, it has been introduced as an promising approach in environmental studies [19,42]. Therefore, the LR classification model was adopted in this study. It deals with independent and dependent parameters, where this relationship is nonlinear and can be calculated as Equations (4) and (5) [87,88]:

$$P = \frac{1}{1+e^{-Z}} \quad (4)$$

where P is an earthquake's probability occurrence ($0 \leq P \leq 1$), and e^{-Z} is a linear logistic factor ($-\infty \leq P \leq +\infty$) that is calculated based on Equation (5) [87]:

$$Z = \log it\ (p) = \ln(\frac{p}{1-p}) = b_0 + b_1 x_1 + \cdots + b_n x_n \quad (5)$$

where Z is a linear logistic factor, p is an earthquake's probability occurrence, n is the number of conditioning variables, and b_0 is the constant coefficient.

Factors Used for Seismic Hazard Assessment

Hazard studies give valuable information about human environments, which, if the results of these studies are taken into account, may protect them from such events in the future. Complex natural hazards such as landslides or flood mapping need considerable data collection and analysis [19,89]. However, in this study, a susceptibility map of potential sites of earthquakes is produced, in which the most important conditioning parameters for it are soil type, proximity to fault lines and lithology condition (Figure 3). A simple probabilistic seismic hazard analysis (PSHA) model was used, which is a useful algorithm, especially when in situ seismic data are not widely available. In the study area, the distribution of the faults is not complex and the Tabriz fault's orientation is straightforward. However, it must be noted that, due to the lack of actual seismic data, this model cannot address uncertainties well [74]. The occurrence probability and intensity of risk assessment depend on selected conditioning factors [17].

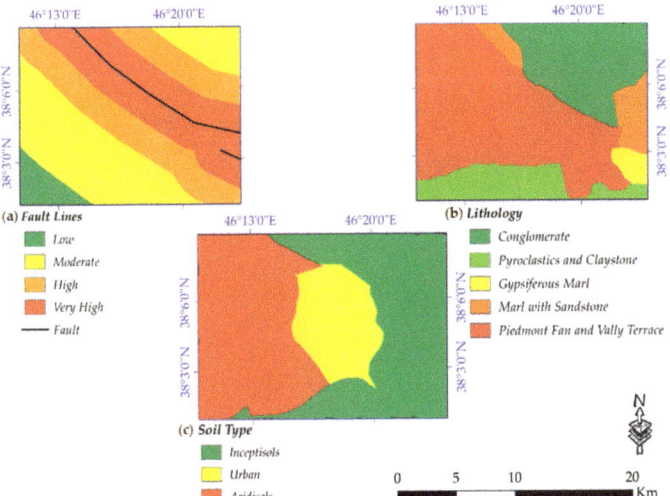

Figure 3. Conditioning parameters used for susceptibility mapping (a) fault lines; (b) lithology; and (c) soil type.

3.4. Accuracy Assessment and Validation

3.4.1. Confusion Matrix for the Classified Maps

Using Google Earth (GE) images, thirty (altogether ninety) ground control points (GCPs) were randomly extracted for each land cover (Figure 4) and then converted into ROIs for accuracy assessment. The overall accuracy and Kappa coefficient were employed in this research. These two models are measured based on Equations (6) and (7), respectively [90,91]:

$$OA = \frac{1}{N} \sum p_{ii} \qquad (6)$$

where OA defines the total accuracy of the model, test pixels are described by N, and $\sum p_{ii}$ represents the total number of correctly classified pixels.

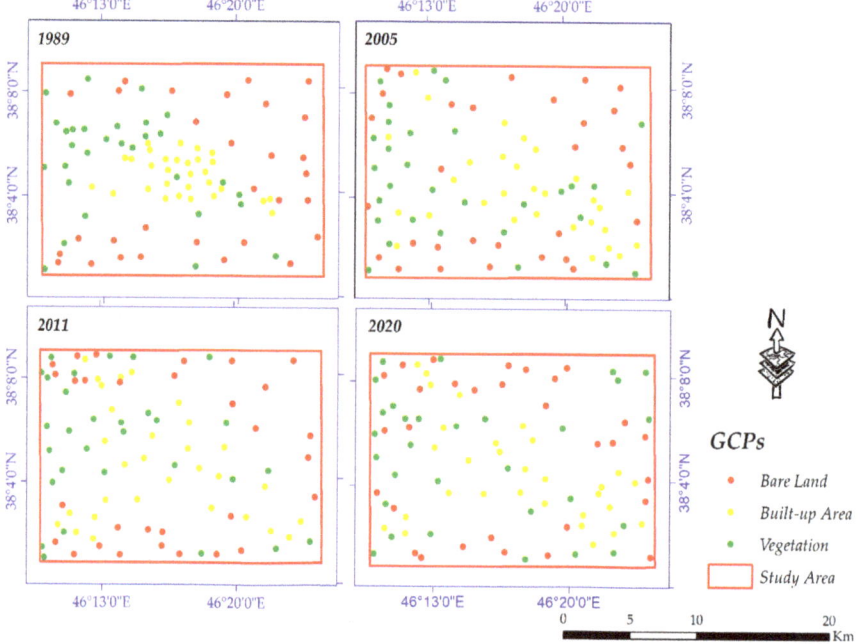

Figure 4. GCPs for all years.

Kappa coefficient is a statistical model that is employed to measure the reliability of the qualitative items [91]. It is universally accepted that this model is a more robust method than the simple calculations. The Kappa coefficient provides reliable and valuable information for the findings obtained. OA must be calculated first in order to measure it.

$$K = (OA - \frac{1}{q})(1 - \frac{1}{q}) \qquad (7)$$

where OA defines the total accuracy of the model; k and q are Kappa coefficient and unclassified pixels, respectively.

3.4.2. Validation of the Predicted Map of the Year 2011

The predicted land cover map using the Markov chain model for 2011 was validated by the generated land cover map of the same year. This was only conducted to determine the reliability and accuracy rate of the model that will be used for the prediction of the future landscape of the year 2030.

3.4.3. Validation of Extracted Land Cover Using SAR Data

GRD and SLC products of Sentinel-1 SAR data were used for the validation of mapped land cover. For this reason, a pair of SLC products for two close dates was preprocessed in sentinel application platform (SNAP) software and used for the extraction of land cover using RGB creation in the GIS environment. At the same time, the GRD product was also applied for this matter in order to ensure the complete reliability of the mapped land cover; the reliability of the land cover was essential because it was used to create many maps for the study area.

4. Results

4.1. Land Cover Classification

Reliable detection of landscape change using remote sensing data must strike a balance between affordability and product accuracy [78,92,93]. By applying the maximum likelihood method, vegetation, built-up and bare land surfaces in 1989, 2005, 2011 and 2020 were extracted for the study areas. This information was utilized to create a few maps of buildings by age, municipality performance and seismic hazard assessment. ROI extraction is one the most important steps in land cover classification, from which exact land cover can be extracted, and it also affects the overall accuracy [70,71,94]. Tables 4 and 5 detail the number of pixels and ROI separation characteristics, respectively. The mean pixel count of the extracted ROIs was used for obtaining the spectral signatures of the land cover. Therefore, an image-derived technique was applied for the extraction of the spectral signatures.

Table 4. Number of pixels used for ROIs.

ROI Summary	Pixel Count: 1989	Pixel Count: 2005	Pixel Count: 2011	Pixel Count: 2020
Vegetation	10,311	11,434	10,120	9312
Built Area	18,280	17,356	15,670	21,098
Bare Land	23,649	25,780	22,456	27,809

Table 5. ROI pair separation.

Years	1989	2005	2011	2020
Vegetation and Built Area	1.99575610	1.99296923	1.99971843	1.99938326
Vegetation and Bare Land	1.99787821	1.99999926	1.99999977	1.99999849
Built Area and Bare Land	1.99888996	1.99999990	1.99899203	1.99832448

The land cover maps for years 1989, 2005, 2011 and 2020 were generated using the maximum likelihood algorithm. Most of the new built-up areas occurred at the edges of the existing urbanized regions, which are displayed in orange color. Figure 5 details the spatial patterns of classified land cover from 1989 to 2005. The vegetation extent on the map is presented in green, built areas in orange and bare land in light yellow pixels. Figure 6 quantifies the changes which occurred from 1989 to 2005. For the sake of distribution clarity, two column charts of gain/losses and net changes were created for changes that occurred from 1989 to 2005. Moreover, vegetation lost around 20 km^2 and gained 18 km^2 from 1989 until 2005 (net change −2.63 km^2). At the same time, 49.47 km^2 was added to the built-up area, while only 4.07 km^2 was removed from it (net change +45.39 km^2). Finally, compared to built-up areas, losses for bare land are considerable, so that it lost around 55.35 km^2 from its areas and approximately 12 km^2 was added to bare land (net change −42.76 km^2).

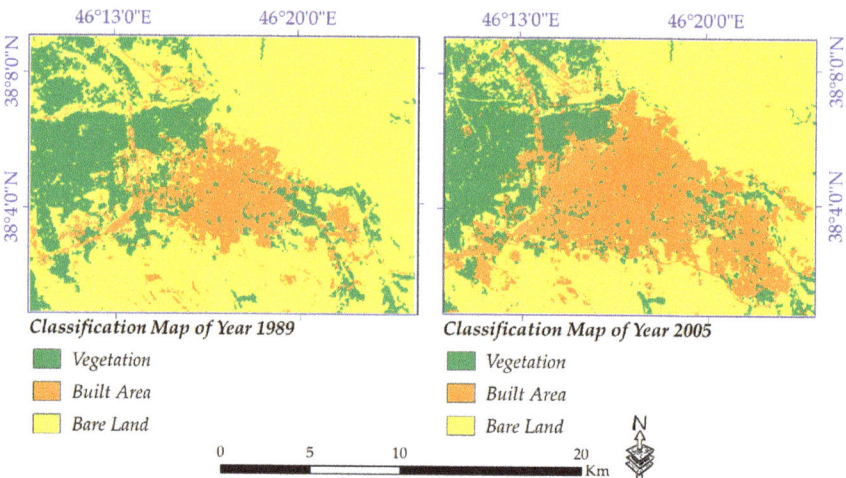

Figure 5. Spatial distribution of changes from 1989 to 2005.

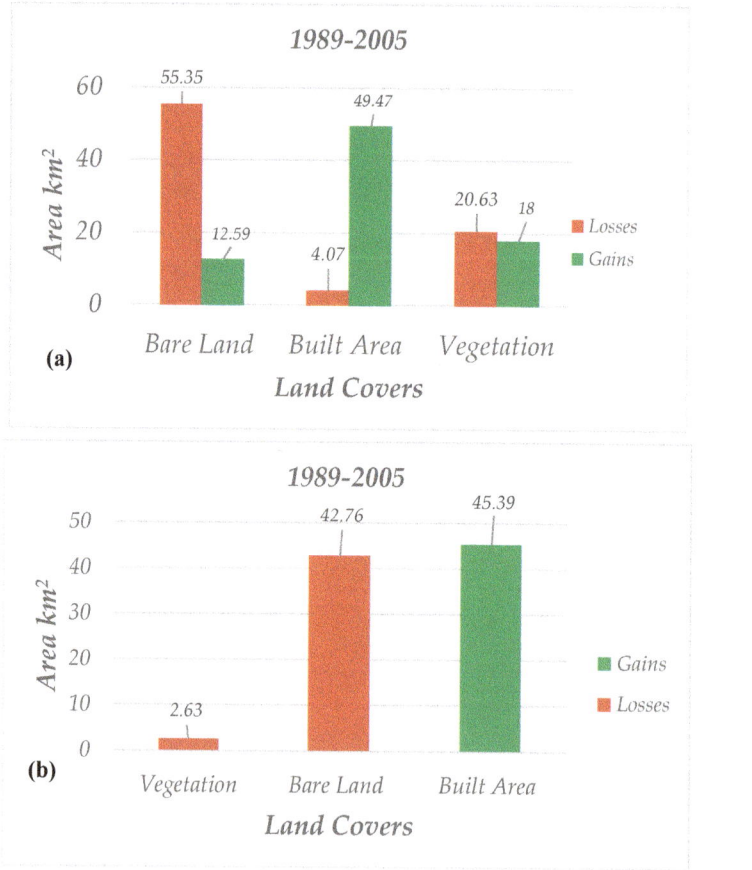

Figure 6. Statistical attributes for changes which occurred from 1989 to 2005; (**a**) gain and loss (**b**) net changes.

After visual inspection, general information over the city was gathered; the only drawback of our classification using the maximum likelihood algorithm was considering the airport band of Tabriz City as a built-up area. This bias may be because of the similarity of backscatters for roads inside the built-up areas with the airport band; however, this is a negligible area and can be addressed using simple editing using GIS.

Figure 7 represents the spatial trends of land cover from 2011 to 2020, from which increasing vegetation coverage inside the city and beyond is considerable. Two charts of gain/losses and net changes were also provided for changes which occurred from 2011 to 2020 (Figure 8); from these, it can be concluded that from 2011 onwards, only around 4 km^2 was added to the built-up areas. For many readers, this should be of great concern, but, based on an interview with the municipality of Tabriz (the interview was performed with a public affairs officer of the municipality on 18/08/2020 through phone call), from 2011 onwards, the city's buildings were constructed and grew vertically, meaning that old buildings with one or two floors were replaced by buildings with more than three floors. A summary of statistical reports for land cover changes from 2011 to 2020 is as follows: (1) net change for vegetation coverage was +20.56 km^2, meaning that approximately 33.83 km^2 was added to it and 13.27 km^2 was subtracted from it; (2) 18.64 km^2 was subtracted from built-up areas, while 22.73 km^2 was added to it; and (3) not surprisingly, bare land lost 46.14 km^2 and gained 21.68 km^2 so that the net change for it can be −24.47 km^2.

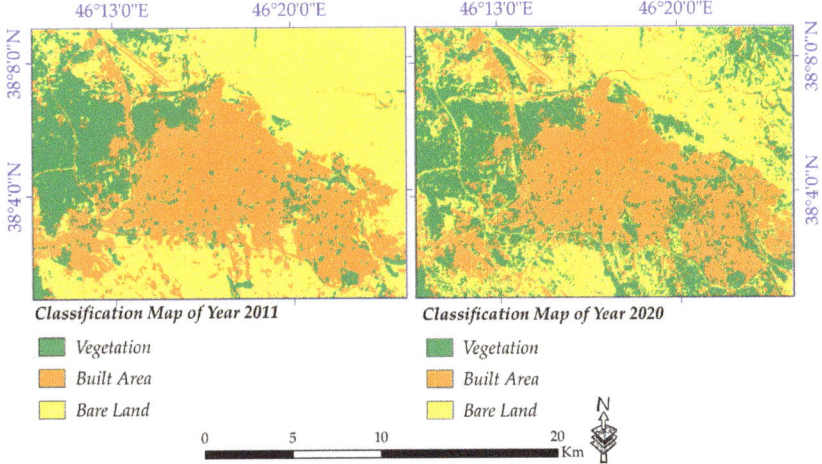

Figure 7. Spatial distribution of changes from 2011 to 2020.

Figure 9 clearly shows cross changes from one land cover to the other. For example, those areas that were once built-up areas but were replaced with vegetation coverage are displayed in light green. Dark green represents areas that once were bare land that have become vegetation. The areas indicated by the red color are areas that were originally vegetation but were replaced with built-up areas. At the same time, changes from bare land to built-up areas are indicated by the dark red color. Furthermore, the light yellow color shows areas that were replaced by bare land from vegetation. Additionally, the dark yellow color highlights areas that were built-up areas but then changed to bare grounds. This kind of map is important in showing changes among land cover between two specific years.

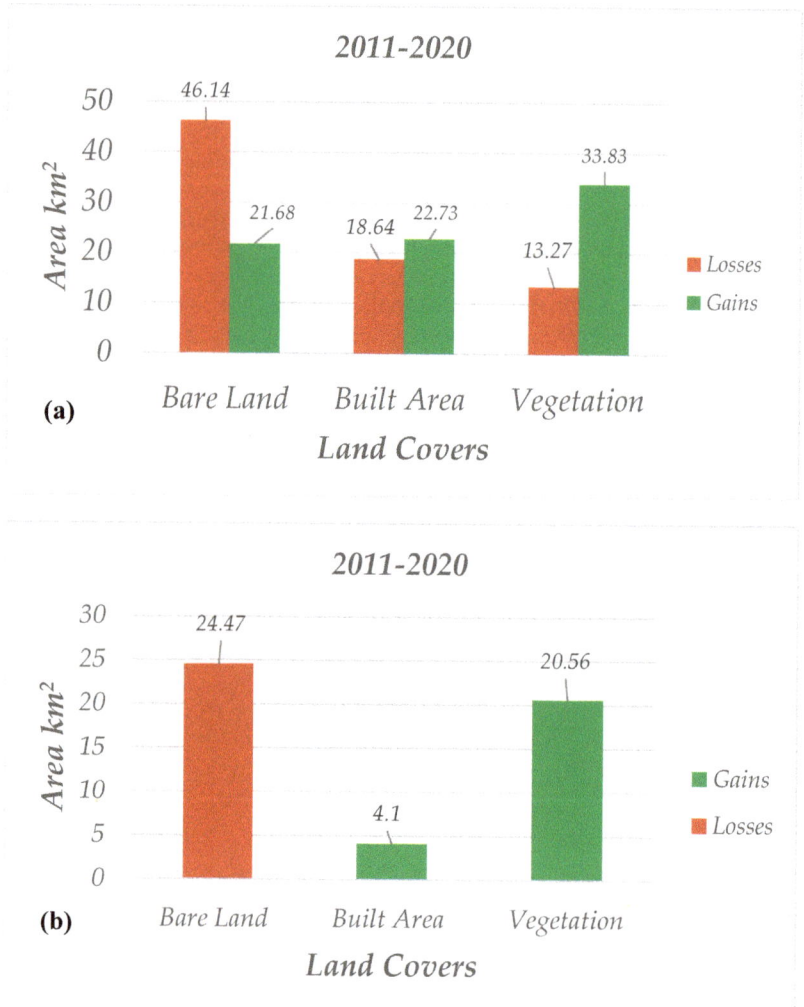

Figure 8. Statistical attributes for changes which occurred from 2011 to 2020; (**a**) gain and loss (**b**) net changes.

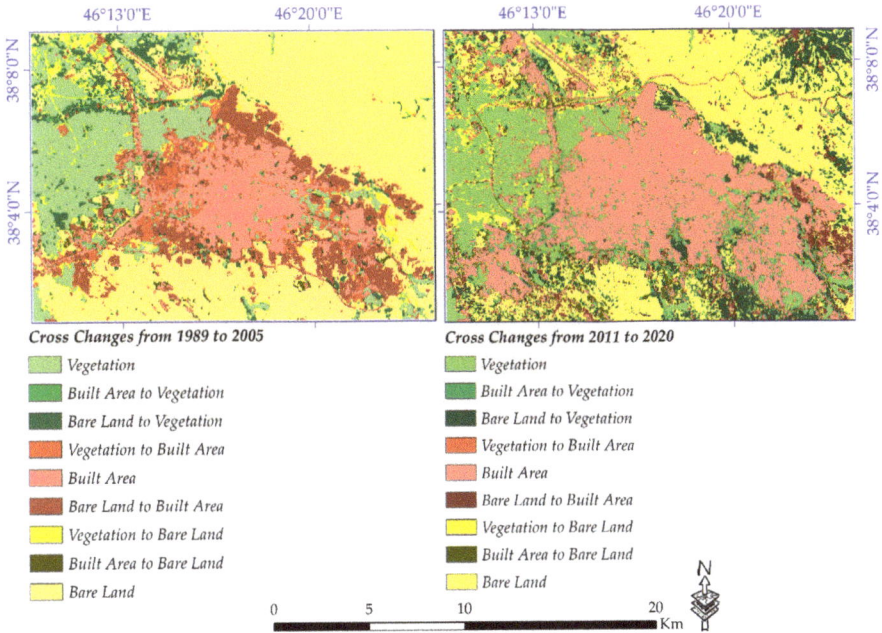

Figure 9. Cross change map of land cover.

4.2. Future Potential Landscape of Tabriz Using Markov Chain Model

The land cover potential pattern of the year 2030 was mapped for the study area. To understand the level of reliability of the model used (because there was no information for the year 2030), the land cover map of the year 2011 was also estimated using land cover maps of the years 1989 and 2005. Considering the predicted land cover map of 2011, around 86% of vegetation coverage was forecasted to remain unchanged (which is a high percentage), and changes from built-up area to vegetation were predicted by approximately 1%, while this rate was roughly 4% from bare land to vegetation. It was forecasted that 96% of the built-up areas would remain unchanged, which is also quite high and shows that the model works well, so the prediction for the year 2030 can be reliable to a great extent. The change prediction rates of vegetation to built-up areas and also the bare lands to built-up areas were estimated to be almost 6% and 8%, respectively. Like vegetation, around 86% of bare land was predicted to remain bare land by the year 2011 (which also represents a good prediction rate). Figure 10 and Table 6 detail the spatial pattern and statistical changes in land cover by the year 2011 (using land cover maps of years 1989 and 2005), respectively.

Table 6. Probability of land cover changes in 2011 predicted from maps of the years 1989 and 2005.

	Vegetation %	Built Area %	Bare Land %
Vegetation	0.8616	0.0618	0.0767
Built Area	0.0138	0.9604	0.0258
Bare Land	0.0440	0.0872	0.8688

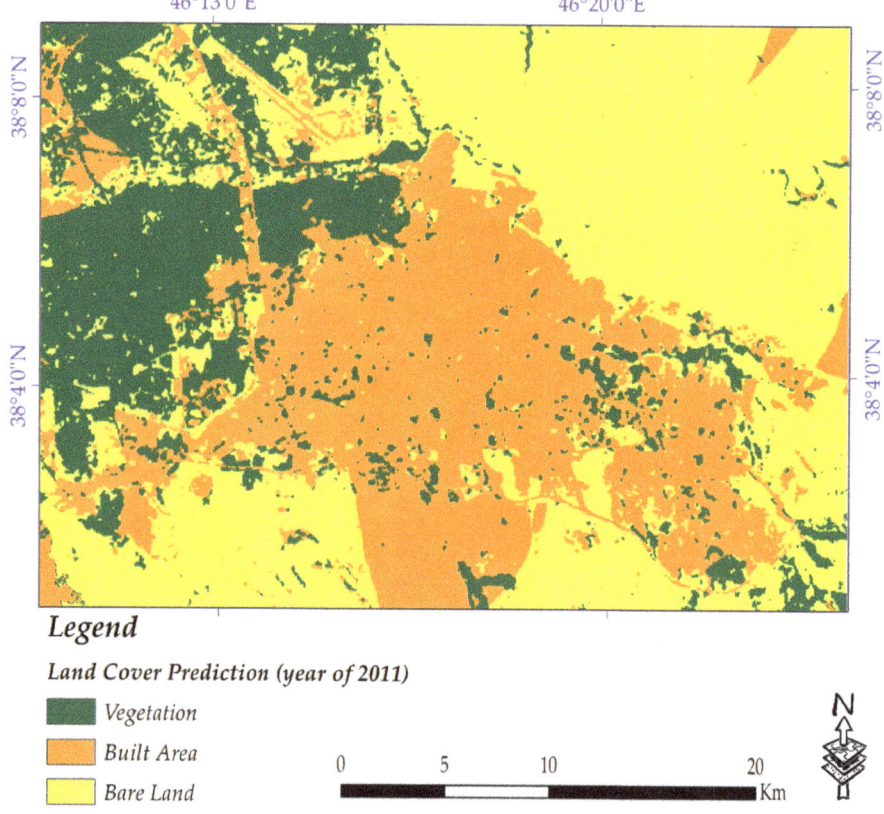

Figure 10. Predicted map of 2011.

Figure 11 and Table 7 highlight the spatial trends and statistical findings of land cover changes by the year 2030, respectively. Considerable findings were extracted regarding the predicted land cover from the map of 2030. Around 74% of vegetation coverage was forecasted to remain unchanged, meaning that almost 26% is likely to be replaced by other types of land cover. Changes from built-up area to vegetation are predicted by approximately 7%, while this rate is quite high for bare land to vegetation, at roughly 26%. It was estimated that 79% of the built-up areas would remain stable as themselves. The change prediction rates of vegetation and bare land to built-up areas are high, at an estimated rate of almost 9% and 11%, respectively. Approximately only 61% of bare land is likely to remain bare land by the year 2030, meaning that based on the changes which occurred until the year of 2020, the municipality plans to change bare land to other types of land cover. One of the considerable results in this regard could be the probability of changes from built-up areas to bare land, at approximately 13% (which is a quite high figure); based on the interview with the municipality, this is maybe because of the reconstruction of buildings over the city that occurred and was recorded by the satellite images used in this study. However, these are only predictions based on changes from the year 2011 to the year 2020.

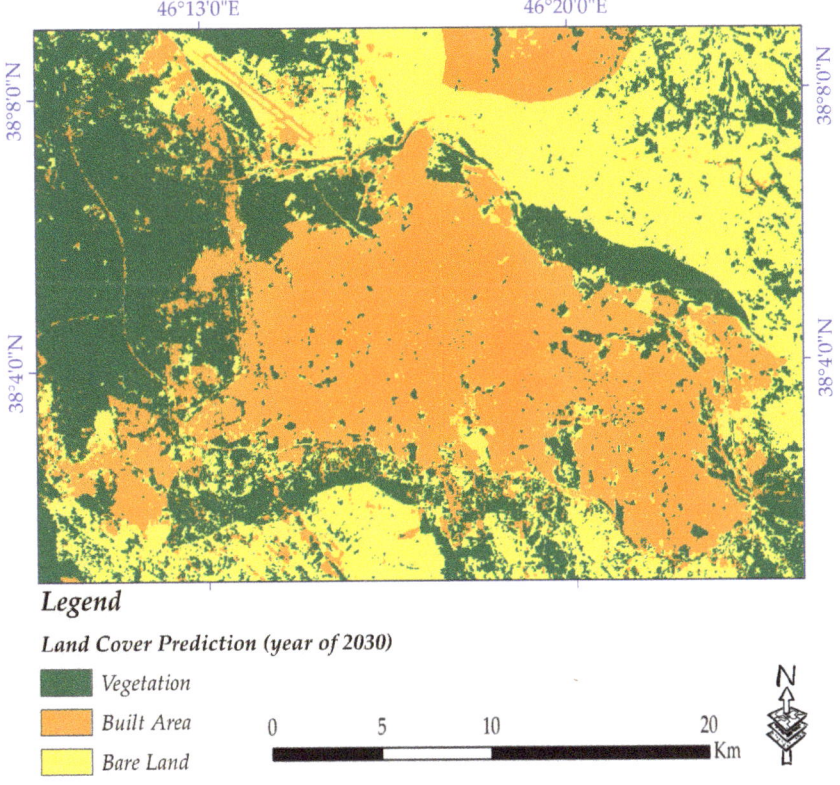

Figure 11. Predicted map of 2030.

Table 7. Probability of land cover changes in 2030 predicted from maps of years 2011 and 2020.

	Vegetation %	Built Area %	Bare Land %
Vegetation	0.7495	0.0915	0.1590
Built Area	0.0769	0.7927	0.1304
Bare Land	0.2637	0.1198	0.6164

4.3. Building Age Map of the Study Area since 1989

The built-up areas' distribution by age for the years of 1989, 2005, 2011 and 2020 was extracted (Figure 12 and Table 8). Pixels in light pink color are classified as built-up areas until 1989 and the pink color represents built-up areas that developed after the year of 2005. Areas indicated in red color are defined as newer urban areas that were constructed from 2005 until 2011. The newest built-up areas that have been constructed since 2011 are shown in the dark red color. Most of Tabriz City was constructed before 2011, but relatively new urban areas in and around the study area can be seen, which indicates that urban development has been gradually taking place. The proportion of built-up areas is presented in Table 8, which shows that the total built-up area constructed before 1989 is around 45 km^2. In the year 2005, the built-up area doubled to approximately 90 km^2. Almost 21 km^2 was added to built-up areas by the year 2011. Not surprisingly, only approximately 4 km^2 has been added to the built-up areas since 2011 (this does not mean that urbanization has stopped since then); this is because buildings have been reconstructed vertically (a few floors) instead of containing only one or two floors. Our interview with the municipality also confirmed that the old buildings with one or two floors are being reconstructed and replaced with buildings with three or more floors. This has good advantages, such as providing more land with the municipality for establishing other projects, while there is sufficient housing for citizens as well.

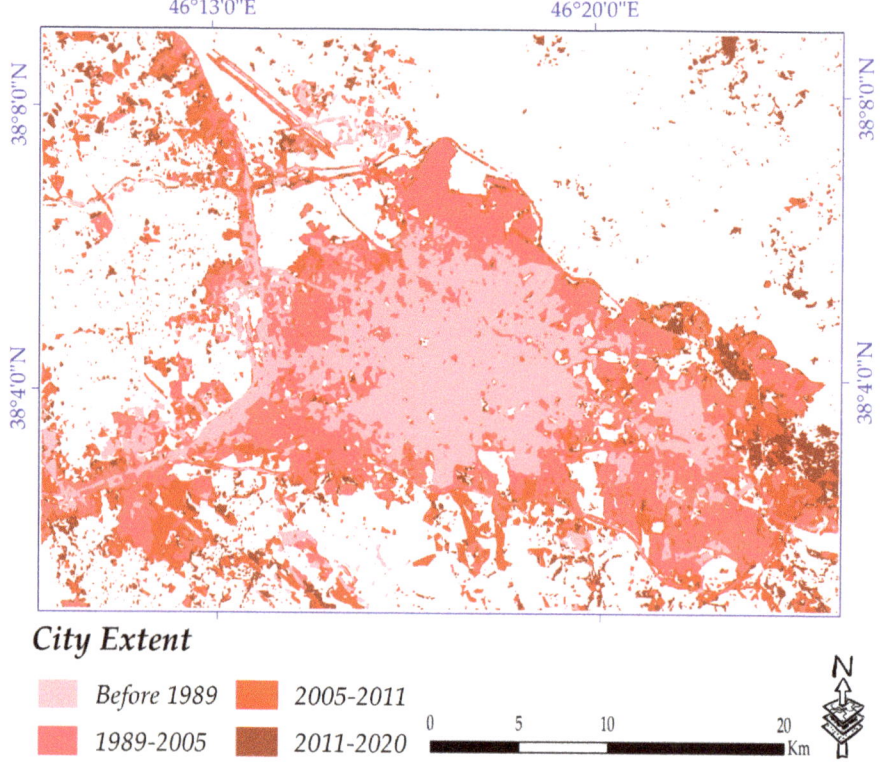

Figure 12. Spatial distribution of building age.

Table 8. Area of constructed regions by four different years.

Year	1989	2005	2011	2020
Area of built-up regions (km^2)	45.30	90.98	112.70	116.81

Urbanization Rate

Based on the built-up area extracted from the classified maps in this study, urbanization rate (UR) was calculated using the following user-defined equation:

$$UR = \frac{A}{T} \quad (8)$$

where UR is the urbanization rate, A is an extended area of built-up areas for each period, and T contributes to the time passed for urban growth. According to the equation, UR for a different period related to the current study is as follows: it is worth mentioning that the results are km^2 per year.

$$UR(2005) = \frac{45.68}{16} = 2.85$$

$$UR(2011) = \frac{21.72}{6} = 3.62$$

$$UR(2020) = \frac{4.11}{9} = 0.45$$

4.4. Municipal Performance Evaluation for 10 Municipal Zones of Tabriz City

Based on the changes that occurred from the year 2011 to 2020, the performance of municipal zones of Tabriz city was evaluated (considering the changes towards more built-up areas and green space along with less bare land), meaning that when more bare land for a municipal zone was converted to built-up and vegetation areas, it was considered that the zone worked well. Following our evaluation, the best and worst municipalities are municipal numbers one and four, respectively. Figure 13 and Table 9 show spatial and statistical changes in different types of land cover for each municipality, respectively.

Table 9. Quantitative results for municipal performance evaluation since 2011.

No. of Municipal Region	Land Cover	The Year 2011 (km²)	The Year 2020 (km²)	Percent of Changes	Rate of Change (km²)	Performance Rank
1	Vegetation	2.37	11.76	396.2	9.39	1
	Built Area	12.26	13.59	10.8	1.33	
	Bare Land	43.25	32.55	−24.7	−10.7	
2	Vegetation	2.41	8.12	236.9	5.70	3
	Built Area	13.55	14.31	5.6	0.7	
	Bare Land	9.28	2.81	−69.7	−6.46	
3	Vegetation	1.70	6.64	290.5	4.9	4
	Built Area	16.17	14.90	−7.8	−1.2	
	Bare Land	17.80	14.15	−20.5	−3.6	
4	Vegetation	7.28	6.58	−9.6	−0.7	10
	Built Area	17.25	16.82	−2.4	−0.4	
	Bare Land	0.83	1.96	136.1	1.13	
5	Vegetation	1.21	4.72	290	3.51	2
	Built Area	7.51	10.42	38.7	2.91	
	Bare Land	22.88	16.49	−27.9	−6.39	
6	Vegetation	25.07	23.67	−5.5	−1.3	9
	Built Area	14.82	15.74	6.2	0.9	
	Bare Land	26.47	26.94	1.7	0.4	
7	Vegetation	19.18	21.70	13.1	2.5	7
	Built Area	19.38	17.38	−10.3	−2	
	Bare Land	17.81	17.30	−2.8	−0.5	
8	Vegetation	0.03	0.07	133.3	0.04	8
	Built Area	3.85	3.76	−2.3	−0.09	
	Bare Land	----	0.05	----	----	
9	Vegetation	0.59	1.52	157.6	0.93	5
	Built Area	0.78	1.15	47.4	0.36	
	Bare Land	2.31	1.02	−55.8	−1.29	
10	Vegetation	0.38	1.31	244.7	0.93	6
	Built Area	8.18	8.19	0.1	0.009	
	Bare Land	2.31	1.37	−40.6	−0.94	

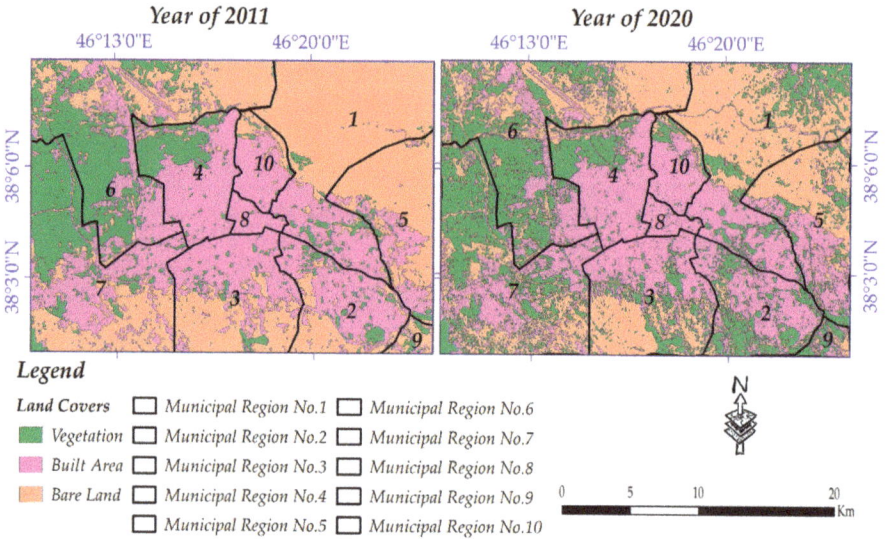

Figure 13. Municipal performance evaluation map.

4.5. Seismic Hazard Assessment

Four susceptible zones were finally reclassified. Figure 14 clearly shows the spatial patterns of areas to susceptible to earthquakes concerning the municipal zones of the city. Most of the municipal zone numbers 1, 9 and 10 are located in the low susceptibility zone, while the entire municipal zone number 8 and most of the municipal zone numbers 3, 4, 6 and 7 are located in the very high susceptibility zone.

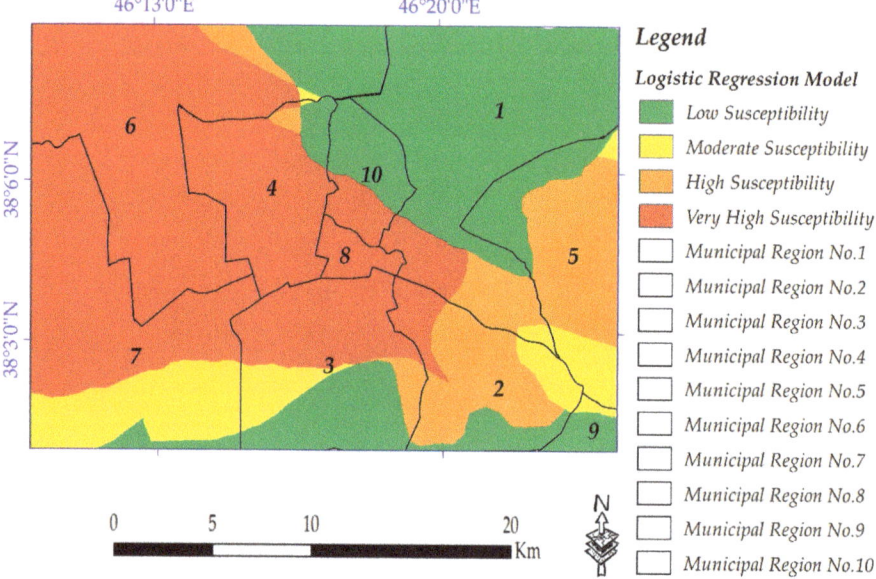

Figure 14. Earthquake susceptibility mapping.

4.6. Accuracy Assessment and Validation for This Study

4.6.1. Confusion Matrix for the Classified Maps

The land cover classification mapping using RS data is a relatively easy effort, while the accuracy and inaccuracy of it depends on proper data and models used [2,61]. An accuracy assessment using the confusion matrix was accomplished for all four classification maps. The overall accuracy for the classified maps of 1989, 2005, 2011 and 2020 was around 96% 96%, 93% and 94%, respectively. Meanwhile, these values for the Kappa coefficient were almost 93%, 92%, 85% and 88%, respectively (Table 10).

Table 10. Statistical results using the confusion matrix for all classified maps.

Statistical Parameters Year of Classified Maps	Overall Accuracy (%)	Kappa Coefficient (%)
1989	96.1354	0.9363
2005	96.0151	0.9222
2011	93.6413	0.8556
2020	94.0700	0.8873

4.6.2. Validation of the Predicted Map of the Year 2011

Using the land cover map of 2011, the predicted map of the same year was validated. Since there was not any information for the year 2030, the validation was not possible. However, fortunately, the prediction map of the year 2011 was well validated (meaning that it only predicted a few areas incorrectly; most were predicted well); therefore, it can be concluded that the prediction map of the year 2030 can be also correct to a great extent. These land cover products were then used to validate urban extent extraction, which confirmed that land cover extraction was done successfully. Validation interpretation was based on these three data: (1) initial land cover was the classified map of the year 2005 (2); predicted land cover for the year 2011; and (3) validation land cover (classified map of the year 2011). Figure 15 represents the validation of the predicted map of the year 2011. However, for interpretation of the image, two examples are presented here: (1) 1/1/1 means that these areas in all aforementioned images were vegetation, and (2) 2/3/3 means that these areas were bare land originally but were predicted as built-up areas.

4.6.3. Validation of the Extracted Land Cover for the Year 2020 as a Basis for Seismic Hazard Assessment

The magnitude of errors using conventional methods is a complex issue from which the extracted land cover from them cannot be directly applied for understanding the changes which occurred [3,52]. Additionally, uncertainties are inherent aspects of remotely sensed studies [3]; to minimize uncertainties, SAR data were also applied. This attempt ensured that urban land cover was not missed using optical images. To successfully validate the extracted built-up areas which were utilized for seismic hazard assessment of the city, SAR data were also used. A few small areas were marked by a few geometric symbols (square-shaped) in each set of satellite data (Figure 16). These areas were then enlarged and displayed with different shapes for each type of land cover, comparing the built-up surface from two SAR data of GRD and SLC products, ensuring that the built-up areas in both were successfully matched. After preprocessing and processing of SAR data in the SNAP environment, both bands of VV and VH were employed for RGB creation in the GIS environment. The VV band of the slave imagery was used for R and the VH band of the master one was employed for G and B windows.

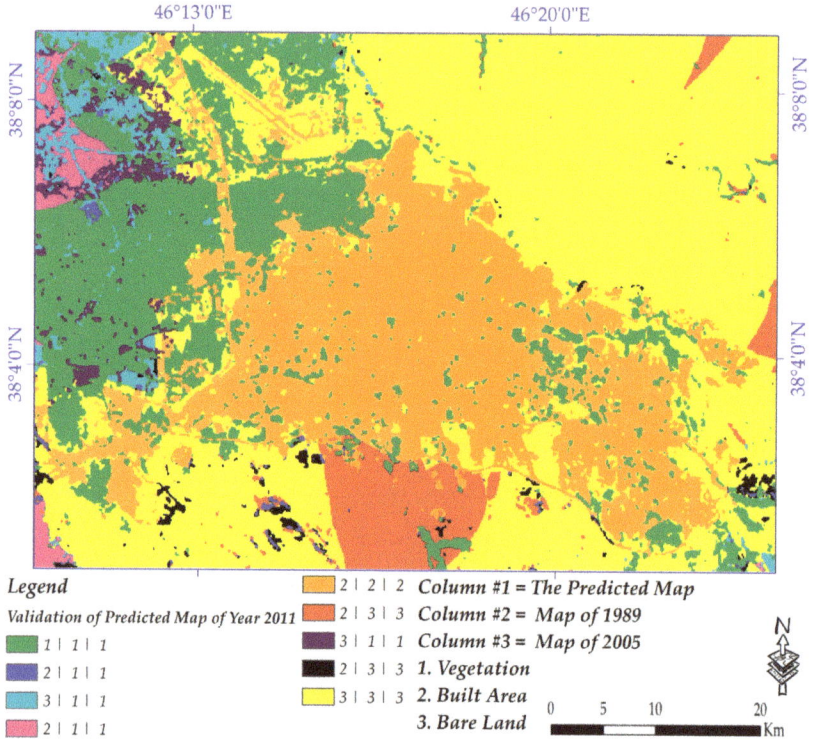

Figure 15. Validation of the predicted map of the year 2011.

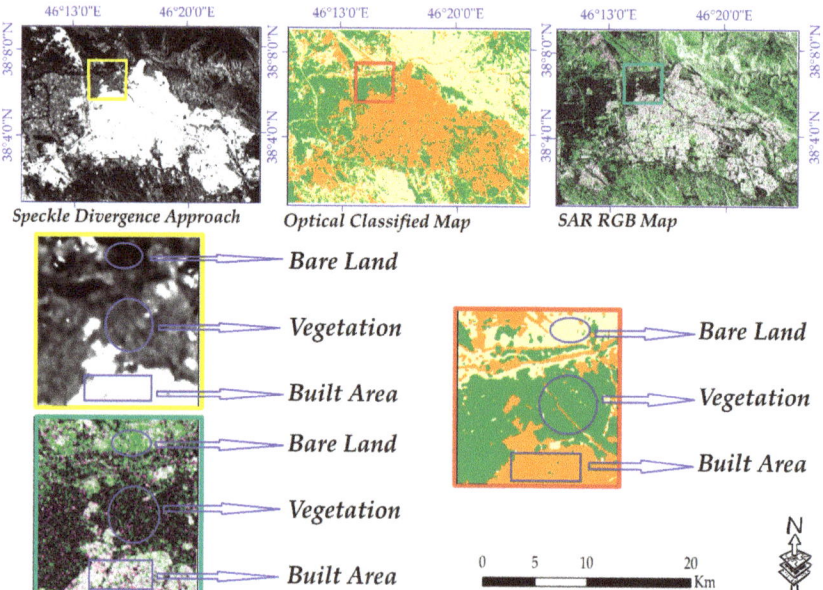

Figure 16. Land cover maps using both optical and SAR data.

5. Discussion

Generally, a few diverse factors may affect the results of change detection and prediction [9,84]. The land cover's spatiotemporal pattern and characteristics in this study were completely different from those which have been measured before, in which a comprehensive study including land cover change detection, future potential landscape, distribution of buildings by age, municipal performance evaluation and building damage assessment was carried out for the metropolitan area of Tabriz, especially for the municipality of Tabriz that has suffered from a lack of such studies. Previous studies were mainly focused on simple methods for measuring the land cover, seismic hazard and building vulnerability for the study area [80,95,96]. During the first period (1989–2005), the built-up area increased as a result of population growth and migration to the city, which was mainly based on destroying vegetation and bare land. During the second period (2011–2020), bare land was replaced by other types of land cover and, in this period of time, one of the most considerable findings was increased vegetation across the city, which, based on the findings, reflects the efforts of the municipal regions to increase the vegetation to a satisfactory level. Urban growth was mainly observed in the bare land and the vegetated areas, far from the economical areas.

The dynamics of the land cover are correlated [77]. Here, built-up areas grew considerably in the first period (1989 to 2005), while in the second period (2011 to 2020), this growth was negligible. For the second period, we found that vegetation areas experienced more positive changes than the first period among land cover. The built-up areas in both periods (1989 to 2005 and 2011 to 2020) showed the largest degree of change among all land covers, which could be linked to the rapid urbanization in Tabriz. The change in speed of bare land was relatively fast in both periods, in which it underwent the most significant land cover changes for the entire period. In addition, vegetation was directly linked to the civil projects of the municipality that turned bare land and old built-up areas into green lands. This analysis suggests that the land cover in Tabriz has considerably changed during the last three decades.

According to the findings for the year 2030, the general trend of change is toward more vegetation and built-up areas as well as less bare land, meaning that the municipality plans to convert more bare land to other types of land cover. Specifically, in the year of 2030, the vegetation and built-up areas will preserve most of their areas and will be larger because of the change from bare land to these types of land cover. Besides this, almost 60 percent of the area of bare land will be preserved, because always bare land will be used for the development of new projects. This implies that the landscape pattern of Tabriz has a tendency to be more optimized in this period.

Based on the results of the urbanization rate, from the year 1989 until 2005, the city of Tabriz experienced growth in urban areas of around 2.85 km^2 per year. At the same time, this rate was 3.62 km^2 per year from 2005 until 2011. Additionally, from 2011 to the year 2020, this rate was only 0.45 km^2 annually.

Seismic hazard assessment is always an essential part of sustainable development projects for urban areas [7,15]. Seismic hazard assessment produces a greater cost efficiency when focusing only on urban areas (rather than the entire area including vegetation and bare land), which has the greatest impact on people when they are destroyed. The application of seismic hazard assessment has not yet been conducted for a city like Tabriz, which is growing fast based on the population rate and its situation as a hub to the other cities of the region. Therefore, a seismic hazard assessment (even a simple one) could enable the local authorities and the policy-makers to direct urbanization to those areas with low or moderate susceptibility. Concerning the earthquake risk for the city, policy-makers should take this into account for future urban sprawl, meaning that they can design stricter policies for new buildings that are constructed or reconstructed in more susceptible areas.

Model validation is important to assess the level of the models' reliability and validity. Validation of the created maps was performed at three stages. In the first, the classified outputs of the years 1989, 2005, 2011 and 2020 were validated using a confusion matrix. In the second step, and since the land cover map of the year 2020 was selected for damage assessment, two other maps from Sentinel-1 SAR data were also employed. Finally, a classified map of the year 2011 was used to validate the prediction map of the year 2011 (only to assess whether the model could predict the future potential landscape or not, so that its prediction for the year 2020 could also be considered correct).

Based on the findings of this study, for arid and semi-arid regions (like the current study area), the maintenance of the existing vegetated areas rather than planting more grasses and trees in less suitable areas is recommended. Although this study, as well as many previous works, has demonstrated that the remotely sensed data and techniques can be well applied for monitoring changes in cities, we recommend that more high-resolution satellite imageries be used to gain further insights into such changes. Future research should focus on the deep learning techniques for change detection and the prediction of land cover; more details on seismic and risk assessments can also be obtained using deep learning algorithms for the study area. The only major limitation of this study was encountered when obtaining ancillary data from the municipality of Tabriz.

6. Conclusions

Cities need comprehensive and innovative plans in order to ensure progress based on sustainable development. Although it is very difficult to obtain absolute results from remotely sensed data, relative findings can be captured, which can be effective for any future planning. This study has emphasized changes in land cover and the future landscape in Tabriz City. Other important issues that the current research was focused on include information on building age, municipal performance evaluation and building damage assessment, which contributes to earthquake damage estimation. This study has also compared the results of optical satellite imagery with SAR data to extract the spatial distribution of buildings for the year 2020, which was the base map to evaluate municipal zone performances and seismic hazard assessment. The main findings of the current study are as follows: Landsat images for the years 1989, 2005, 2011 and 2020 were used to quantify the land cover changes from 1989 until 2020 and the results using the confusion matrix were promising. At the same time, by using and comparing SAR data, the accuracy of built-up areas for the year 2020 was well validated and verified. Referring to the assessment of the distribution of built-up areas by age for Tabriz City, we found that most of the built-up areas had been developed before 2011, and from then onwards, the city has been progressing vertically. Seismic hazard assessment for the future of the city was conducted by using a logistic regression model, from which results indicated that municipal zones 1 and 9 are located inside low susceptibility areas, while municipal zones 4, 6, 7, 8, and also most of zones 3 and 10, are located in highly susceptible regions. Further findings revealed that land cover prediction by using the Markov chain model provided a good opportunity to identify the future potential landscape of the city. Finally, based on the land cover maps of 2011 and 2020, the performances of the municipal zones were evaluated, from which results showed that municipal zone 1 followed by zone 5 have the best performances among all. Besides this, the performance of municipal zone 4 is negligible, as is much of municipal zone 6.

Author Contributions: A.M., S.K., K.V.K. and M.M. collected data and finalized the study; A.M., S.K. and M.M. worked on the methodology; A.M. and S.K. performed software analysis and validated the research findings; A.M. and S.K. finalized the formal analysis; A.M., S.K. and M.M. wrote the original draft; S.K., K.V.K. and M.M. reviewed and edited the original version; S.K., K.V.K. and M.M. funded the research. All authors have read and agreed to the published version of the manuscript.

Funding: The presented work is supported by National Elites Foundation of Iran (No. 102/1670) and the Japan Society for the Promotion of Science (JSPS), KAKENHI number 20H02411.

Conflicts of Interest: The authors declare no conflict of interest.

References

1. Wang, S.W.; Gebru, B.M.; Lamchin, M.; Kayastha, R.B.; Lee, W.-K. Land Use and Land Cover Change Detection and Prediction in the Kathmandu District of Nepal Using Remote Sensing and GIS. *Sustainability* **2020**, *12*, 3925. [CrossRef]
2. Fu, P.; Weng, Q. A time series analysis of urbanization induced land use and land cover change and its impact on land surface temperature with Landsat imagery. *Remote Sens. Environ.* **2016**, *175*, 205–214. [CrossRef]
3. Xian, G.; Homer, C. Updating the 2001 National Land Cover Database impervious surface products to 2006 using Landsat imagery change detection methods. *Remote Sens. Environ.* **2010**, *114*, 1676–1686. [CrossRef]
4. Yagoub, M.; Al Bizreh, A.A. Prediction of land cover change using Markov and cellular automata models: Case of Al-Ain, UAE, 1992–2030. *J. Indian Soc. Remote Sens.* **2014**, *42*, 665–671. [CrossRef]
5. Hasan, M.E.; Nath, B.; Sarker, A.; Wang, Z.; Zhang, L.; Yang, X.; Nobi, M.N.; Røskaft, E.; Chivers, D.J.; Suza, M. Applying Multi-Temporal Landsat Satellite Data and Markov-Cellular Automata to Predict Forest Cover Change and Forest Degradation of Sundarban Reserve Forest, Bangladesh. *Forests* **2020**, *11*, 1016. [CrossRef]
6. Zhu, Z.; Woodcock, C.E. Continuous change detection and classification of land cover using all available Landsat data. *Remote Sens. Environ.* **2014**, *144*, 152–171. [CrossRef]
7. Matsuoka, M.; Miura, H.; Midorikawa, S.; Estrada, M. Extraction of urban information for seismic hazard and risk assessment in Lima, Peru using satellite imagery. *J. Disaster Res.* **2013**, *8*, 328–345. [CrossRef]
8. Zhang, C.; Wei, S.; Ji, S.; Lu, M. Detecting large-scale urban land cover changes from very high resolution remote sensing images using cnn-based classification. *ISPRS Int. J. Geo-Inf.* **2019**, *8*, 189. [CrossRef]
9. El Jazouli, A.; Barakat, A.; Khellouk, R.; Rais, J.; El Baghdadi, M. Remote sensing and GIS techniques for prediction of land use land cover change effects on soil erosion in the high basin of the Oum Er Rbia River (Morocco). *Remote Sens. Appl. Soc. Environ.* **2019**, *13*, 361–374. [CrossRef]
10. Yirsaw, E.; Wu, W.; Shi, X.; Temesgen, H.; Bekele, B. Land use/land cover change modeling and the prediction of subsequent changes in ecosystem service values in a coastal area of China, the Su-Xi-Chang Region. *Sustainability* **2017**, *9*, 1204. [CrossRef]
11. Nhu, V.-H.; Mohammadi, A.; Shahabi, H.; Shirzadi, A.; Al-Ansari, N.; Ahmad, B.B.; Chen, W.; Ahmadi, M.; Khosravi, K.; Jaafari, A. Monitoring and Assessment of Water Level Fluctuations of the Lake Urmia and Its Environmental Consequences Using Multitemporal Landsat 7 ETM+ Images. *Int. J. Environ. Res. Public Health* **2020**, *17*, 4210. [CrossRef] [PubMed]
12. Alizadeh, M.; Alizadeh, E.; Asadollahpour Kotenaee, S.; Shahabi, H.; Beiranvand Pour, A.; Panahi, M.; Bin Ahmad, B.; Saro, L. Social vulnerability assessment using artificial neural network (ANN) model for earthquake hazard in Tabriz city, Iran. *Sustainability* **2018**, *10*, 3376. [CrossRef]
13. Tien Bui, D.; Shahabi, H.; Mohammadi, A.; Bin Ahmad, B.; Bin Jamal, M.; Ahmad, A. Land Cover Change Mapping Using a Combination of Sentinel-1 Data and Multispectral Satellite Imagery: A Case Study of Sanandaj County, Kurdistan, Iran. *Appl. Ecol. Environ. Res.* **2019**, *17*, 5449–5463. [CrossRef]
14. Gokon, H.; Koshimura, S.; Matsuoka, M. Object-based method for estimating tsunami-induced damage using TerraSAR-X data. *J. Disaster Res.* **2016**, *11*, 225–235. [CrossRef]
15. Karimzadeh, S.; Feizizadeh, B.; Matsuoka, M. From a GIS-based hybrid site condition map to an earthquake damage assessment in Iran: Methods and trends. *Int. J. Disaster Risk Reduct.* **2017**, *22*, 23–36. [CrossRef]
16. Sarkar, S.; Parihar, S.M.; Dutta, A. Fuzzy risk assessment modelling of East Kolkata Wetland Area: A remote sensing and GIS based approach. *Environ. Model. Softw.* **2016**, *75*, 105–118. [CrossRef]
17. Nobre, R.; Rotunno Filho, O.; Mansur, W.; Nobre, M.; Cosenza, C. Groundwater vulnerability and risk mapping using GIS, modeling and a fuzzy logic tool. *J. Contam. Hydrol.* **2007**, *94*, 277–292. [CrossRef]
18. Pourghasemi, H.R.; Beheshtirad, M.; Pradhan, B. A comparative assessment of prediction capabilities of modified analytical hierarchy process (M-AHP) and Mamdani fuzzy logic models using Netcad-GIS for forest fire susceptibility mapping. *Geomat. Nat. Hazards Risk* **2016**, *7*, 861–885. [CrossRef]
19. Pradhan, B. Landslide susceptibility mapping of a catchment area using frequency ratio, fuzzy logic and multivariate logistic regression approaches. *J. Indian Soc. Remote Sens.* **2010**, *38*, 301–320. [CrossRef]
20. Metternicht, G. Assessing temporal and spatial changes of salinity using fuzzy logic, remote sensing and GIS. Foundations of an expert system. *Ecol. Model.* **2001**, *144*, 163–179. [CrossRef]

21. Karnieli, A.; Gilad, U.; Ponzet, M.; Svoray, T.; Mirzadinov, R.; Fedorina, O. Assessing land-cover change and degradation in the Central Asian deserts using satellite image processing and geostatistical methods. *J. Arid Environ.* **2008**, *72*, 2093–2105. [CrossRef]
22. Kaliraj, S.; Meenakshi, S.M.; Malar, V. Application of Remote Sensing in Detection of Forest Cover Changes Using Geo-Statistical Change Detection Matrices- A Case Study of Devanampatti Reserve Forest, Tamilnadu, India. *Nat. Environ. Pollut. Technol.* **2012**, *11*, 261–269.
23. Shrestha, D. Image transformation and geo-statistical techniques to assess sedimentation in southern Nepal. *Asian J. Geoinform.* **2005**, *5*, 24–31.
24. Shetty, A.; Jayappa, K.; Ramakrishnan, R.; Rajawat, A. Shoreline Dynamics and Vulnerability Assessment Along the Karnataka Coast, India: A Geo-Statistical Approach. *J. Indian Soc. Remote Sens.* **2019**, *47*, 1223–1234. [CrossRef]
25. Halmy, M.W.A.; Gessler, P.E.; Hicke, J.A.; Salem, B.B. Land use/land cover change detection and prediction in the north-western coastal desert of Egypt using Markov-CA. *Appl. Geogr.* **2015**, *63*, 101–112. [CrossRef]
26. Mondal, M.S.; Sharma, N.; Garg, P.; Kappas, M. Statistical independence test and validation of CA Markov land use land cover (LULC) prediction results. *Egypt. J. Remote Sens. Space Sci.* **2016**, *19*, 259–272. [CrossRef]
27. Nath, B.; Wang, Z.; Ge, Y.; Islam, K.; Singh, R.P.; Niu, Z. Land Use and Land Cover Change Modeling and Future Potential Landscape Risk Assessment Using Markov-CA Model and Analytical Hierarchy Process. *ISPRS Int. J. Geo-Inf.* **2020**, *9*, 134. [CrossRef]
28. Li, K.; Feng, M.; Biswas, A.; Su, H.; Niu, Y.; Cao, J. Driving Factors and Future Prediction of Land Use and Cover Change Based on Satellite Remote Sensing Data by the LCM Model: A Case Study from Gansu Province, China. *Sensors* **2020**, *20*, 2757. [CrossRef]
29. Lan, H.; Stewart, K. Gap filling in large-area and long-term land use/land cover change time series using cloud-based Markov-Cellular Automata. In Proceedings of the AGU Fall Meeting 2019, San Francisco, CA, USA, 9–13 December 2019.
30. Balogun, I.; Ishola, K. Projection of future changes in landuse/landcover using cellular automata/markov model over Akure city, Nigeria. *J. Remote Sens. Technol.* **2017**, *5*, 22–31. [CrossRef]
31. Hua, A. Application of Ca-Markov model and land use/land cover changes in Malacca River Watershed, Malaysia. *Appl. Ecol. Environ. Res.* **2017**, *15*, 605–622. [CrossRef]
32. Ejikeme, J.; Igbokwe, J.; Igbokwe, E.; Paul, C. Application of Knowledge-Based Image Classification and Ca-Markov Chain Prediction Model for Landuse/Landcover Change Analysis of Onitsha and Environs, Anambra State. *Int. J. Adv. Res. Publ.* **2020**, *4*, 22–28.
33. Xu, X.; Du, Z.; Zhang, H. Integrating the system dynamic and cellular automata models to predict land use and land cover change. *Int. J. Appl. Earth Obs. Geoinf.* **2016**, *52*, 568–579. [CrossRef]
34. Deep, S.; Saklani, A. Urban sprawl modeling using cellular automata. *Egypt J. Remote Sens. Space Sci.* **2014**, *17*, 179–187. [CrossRef]
35. Rimal, B.; Zhang, L.; Keshtkar, H.; Haack, B.N.; Rijal, S.; Zhang, P. Land use/land cover dynamics and modeling of urban land expansion by the integration of cellular automata and markov chain. *ISPRS Int. J. Geo-Inf.* **2018**, *7*, 154. [CrossRef]
36. Rimal, B.; Zhang, L.; Keshtkar, H.; Wang, N.; Lin, Y. Monitoring and modeling of spatiotemporal urban expansion and land-use/land-cover change using integrated Markov chain cellular automata model. *ISPRS Int. J. Geo-Inf.* **2017**, *6*, 288. [CrossRef]
37. Ferchichi, A.; Boulila, W.; Farah, I.R. Propagating aleatory and epistemic uncertainty in land cover change prediction process. *Ecol. Inform.* **2017**, *37*, 24–37. [CrossRef]
38. Convertino, M.; Welle, P.; Muñoz-Carpena, R.; Kiker, G.A.; Chu-Agor, M.L.; Fischer, R.A.; Linkov, I. Epistemic uncertainty in predicting shorebird biogeography affected by sea-level rise. *Ecol. Model.* **2012**, *240*, 1–15. [CrossRef]
39. e Silva, L.P.; Xavier, A.P.C.; da Silva, R.M.; Santos, C.A.G. Modeling land cover change based on an artificial neural network for a semiarid river basin in northeastern Brazil. *Glob. Ecol. Conserv.* **2020**, *21*, e00811. [CrossRef]
40. Saputra, M.H.; Lee, H.S. Prediction of land use and land cover changes for north sumatra, indonesia, using an artificial-neural-network-based cellular automaton. *Sustainability* **2019**, *11*, 3024. [CrossRef]
41. Liu, X.; Lathrop, R., Jr. Urban change detection based on an artificial neural network. *Int. J. Remote Sens.* **2002**, *23*, 2513–2518. [CrossRef]

42. Choi, J.; Oh, H.-J.; Lee, H.-J.; Lee, C.; Lee, S. Combining landslide susceptibility maps obtained from frequency ratio, logistic regression, and artificial neural network models using ASTER images and GIS. *Eng. Geol.* **2012**, *124*, 12–23. [CrossRef]
43. Baroud, S.; Chokri, S.; Belhaous, S.; Hidila, Z.; Mestari, M. An Artificial Neural Network Combined to Object Oriented Method for Land Cover Classification of High Resolution RGB Remote Sensing Images. In Proceedings of the International Conference on Smart Applications and Data Analysis, Marrakesh, Morocco, 25–26 June 2020; pp. 221–232.
44. Wang, Q.; Shi, W.; Atkinson, P.M.; Li, Z. Land cover change detection at subpixel resolution with a Hopfield neural network. *IEEE J. Sel. Top. Appl. Earth Obs. Remote Sens.* **2014**, *8*, 1339–1352. [CrossRef]
45. Tatem, A.J.; Lewis, H.G.; Atkinson, P.M.; Nixon, M.S. Super-resolution land cover pattern prediction using a Hopfield neural network. *Remote Sens. Environ.* **2002**, *79*, 1–14. [CrossRef]
46. Tatem, A.J.; Lewis, H.G.; Atkinson, P.M.; Nixon, M.S. Multiple-class land-cover mapping at the sub-pixel scale using a Hopfield neural network. *Int. J. Appl. Earth Obs. Geoinf.* **2001**, *3*, 184–190. [CrossRef]
47. Tatem, A.J.; Lewis, H.G.; Atkinson, P.M.; Nixon, M.S. Increasing the spatial resolution of agricultural land cover maps using a Hopfield neural network. *Int. J. Geogr. Inf. Sci.* **2003**, *17*, 647–672. [CrossRef]
48. Li, X.; Ling, F.; Du, Y.; Feng, Q.; Zhang, Y. A spatial–temporal Hopfield neural network approach for super-resolution land cover mapping with multi-temporal different resolution remotely sensed images. *ISPRS J. Photogramm. Remote Sens.* **2014**, *93*, 76–87. [CrossRef]
49. Wu, K.; Zhong, Y.; Wang, X.; Sun, W. A novel approach to subpixel land-cover change detection based on a supervised back-propagation neural network for remotely sensed images with different resolutions. *IEEE Geosci. Remote Sens. Lett.* **2017**, *14*, 1750–1754. [CrossRef]
50. Zhang, S.L.; Chang, T.C. A study of image classification of remote sensing based on back-propagation neural network with extended delta bar delta. *Math. Probl. Eng.* **2015**, *2015*, 178598. [CrossRef]
51. Mu, L.; Wang, L.; Wang, Y.; Chen, X.; Han, W. Urban Land Use and Land Cover Change Prediction via Self-Adaptive Cellular Based Deep Learning With Multisourced Data. *IEEE J. Sel. Top. Appl. Earth Obs. Remote Sens.* **2019**, *12*, 5233–5247. [CrossRef]
52. Zhang, X.; Han, L.; Han, L.; Zhu, L. How well do deep learning-based methods for land cover classification and object detection perform on high resolution remote sensing imagery? *Remote Sens.* **2020**, *12*, 417. [CrossRef]
53. Kussul, N.; Lavreniuk, M.; Skakun, S.; Shelestov, A. Deep learning classification of land cover and crop types using remote sensing data. *IEEE Geosci. Remote Sens. Lett.* **2017**, *14*, 778–782. [CrossRef]
54. Shendryk, Y.; Rist, Y.; Ticehurst, C.; Thorburn, P. Deep learning for multi-modal classification of cloud, shadow and land cover scenes in PlanetScope and Sentinel-2 imagery. *ISPRS J. Photogramm. Remote Sens.* **2019**, *157*, 124–136. [CrossRef]
55. Haidara, I.; Tahri, M.; Maanan, M.; Hakdaoui, M. Efficiency of Fuzzy Analytic Hierarchy Process to detect soil erosion vulnerability. *Geoderma* **2019**, *354*, 113853. [CrossRef]
56. Kundu, S.; Khare, D.; Mondal, A. Landuse change impact on sub-watersheds prioritization by analytical hierarchy process (AHP). *Ecol. Inform.* **2017**, *42*, 100–113. [CrossRef]
57. Duro, D.; Franklin, S.; Dubé, M. Hybrid object-based change detection and hierarchical image segmentation for thematic map updating. *Photogramm. Eng. Remote Sens.* **2013**, *79*, 259–268. [CrossRef]
58. Redo, D.J.; Millington, A.C. A hybrid approach to mapping land-use modification and land-cover transition from MODIS time-series data: A case study from the Bolivian seasonal tropics. *Remote Sens. Environ.* **2011**, *115*, 353–372. [CrossRef]
59. Lyu, H.; Lu, H.; Mou, L. Learning a transferable change rule from a recurrent neural network for land cover change detection. *Remote Sens.* **2016**, *8*, 506. [CrossRef]
60. Sharma, A.; Liu, X.; Yang, X. Land cover classification from multi-temporal, multi-spectral remotely sensed imagery using patch-based recurrent neural networks. *Neural Netw.* **2018**, *105*, 346–355. [CrossRef] [PubMed]
61. Ienco, D.; Gaetano, R.; Dupaquier, C.; Maurel, P. Land cover classification via multitemporal spatial data by deep recurrent neural networks. *IEEE Geosci. Remote Sens. Lett.* **2017**, *14*, 1685–1689. [CrossRef]
62. Ndikumana, E.; Ho Tong Minh, D.; Baghdadi, N.; Courault, D.; Hossard, L. Deep recurrent neural network for agricultural classification using multitemporal SAR Sentinel-1 for Camargue, France. *Remote Sens.* **2018**, *10*, 1217. [CrossRef]

63. Polykretis, C.; Grillakis, M.G.; Alexakis, D.D. Exploring the impact of various spectral indices on land cover change detection using change vector analysis: A case study of Crete Island, Greece. *Remote Sens.* **2020**, *12*, 319. [CrossRef]
64. Lambin, E.F.; Strahlers, A.H. Change-vector analysis in multitemporal space: A tool to detect and categorize land-cover change processes using high temporal-resolution satellite data. *Remote Sens. Environ.* **1994**, *48*, 231–244. [CrossRef]
65. Johnson, R.D.; Kasischke, E. Change vector analysis: A technique for the multispectral monitoring of land cover and condition. *Int. J. Remote Sens.* **1998**, *19*, 411–426. [CrossRef]
66. Chen, J.; Chen, X.; Cui, X.; Chen, J. Change vector analysis in posterior probability space: A new method for land cover change detection. *IEEE Geosci. Remote Sens. Lett.* **2010**, *8*, 317–321. [CrossRef]
67. He, C.; Wei, A.; Shi, P.; Zhang, Q.; Zhao, Y. Detecting land-use/land-cover change in rural–urban fringe areas using extended change-vector analysis. *Int. J. Appl. Earth Obs. Geoinf.* **2011**, *13*, 572–585. [CrossRef]
68. Liu, H.; Zhou, Q. Developing urban growth predictions from spatial indicators based on multi-temporal images. *Comput. Environ. Urban Syst.* **2005**, *29*, 580–594. [CrossRef]
69. Hegazy, I.R.; Kaloop, M.R. Monitoring urban growth and land use change detection with GIS and remote sensing techniques in Daqahlia governorate Egypt. *Int. J. Sustain. Built Environ.* **2015**, *4*, 117–124. [CrossRef]
70. Mohammadi, A.; Baharin, B.; Shahabi, H. Land Cover Mapping Using a Novel Combination Model of Satellite Imageries: Case Study of a Part of the Cameron Highlands, Pahang, Malaysia. *Appl. Ecol. Environ. Res.* **2019**, *17*, 1835–1848. [CrossRef]
71. Mohammadi, A.; Shahabi, H.; Bin Ahmad, B. Land-Cover Change Detection in a Part of Cameron Highlands, Malaysia Using ETM+ Satellite Imagery and Support Vector Machine (SVM) Algorithm. *EnvironmentAsia* **2019**, *12*, 145–154. [CrossRef]
72. Waske, B.; Braun, M. Classifier ensembles for land cover mapping using multitemporal SAR imagery. *ISPRS J. Photogramm. Remote Sens.* **2009**, *64*, 450–457. [CrossRef]
73. Esch, T.; Schenk, A.; Ullmann, T.; Thiel, M.; Roth, A.; Dech, S. Characterization of land cover types in TerraSAR-X images by combined analysis of speckle statistics and intensity information. *IEEE Trans. Geosci. Remote Sens.* **2011**, *49*, 1911–1925. [CrossRef]
74. Abdikan, S.; Sanli, F.B.; Ustuner, M.; Calò, F. Land cover mapping using sentinel-1 SAR data. *Int. Arch. Photogramm. Remote Sens. Spat. Inf. Sci.* **2016**, *41*, 757. [CrossRef]
75. Gašparović, M.; Dobrinić, D. Comparative assessment of machine learning methods for urban vegetation mapping using multitemporal sentinel-1 imagery. *Remote Sens.* **2020**, *12*, 1952. [CrossRef]
76. Zhang, H.; Xu, R. Exploring the optimal integration levels between SAR and optical data for better urban land cover mapping in the Pearl River Delta. *Int. J. Appl. Earth Obs. Geoinf.* **2018**, *64*, 87–95. [CrossRef]
77. Lensch, G.; Schmidt, K.; Davoudzadeh, M. Introduction to the geology of Iran. *Neues Jahrb. Für Geol. Und Paläontologie-Abh.* **1984**, 155–164. [CrossRef]
78. Ghorbani, M. A summary of geology of Iran. In *The Economic Geology of Iran*; Springer: Berlin/Heidelberg, Germany, 2013; pp. 45–64.
79. Ghalamghash, J.; Mousavi, S.; Hassanzadeh, J.; Schmitt, A. Geology, zircon geochronology, and petrogenesis of Sabalan volcano (northwestern Iran). *J. Volcanol. Geotherm. Res.* **2016**, *327*, 192–207. [CrossRef]
80. Karimzadeh, S.; Miyajima, M.; Hassanzadeh, R.; Amiraslanzadeh, R.; Kamel, B. A GIS-based seismic hazard, building vulnerability and human loss assessment for the earthquake scenario in Tabriz. *Soil Dyn. Earthq. Eng.* **2014**, *66*, 263–280. [CrossRef]
81. Zarghami, M.; Abdi, A.; Babaeian, I.; Hassanzadeh, Y.; Kanani, R. Impacts of climate change on runoffs in East Azerbaijan, Iran. *Glob. Planet. Chang.* **2011**, *78*, 137–146. [CrossRef]
82. Ghorbani, M. Nature of Iran and its climate. In *The Economic Geology of Iran*; Springer: Berlin/Heidelberg, Germany, 2013; pp. 1–44.
83. Norouzi, M.; Bengio, S.; Jaitly, N.; Schuster, M.; Wu, Y.; Schuurmans, D. Reward augmented maximum likelihood for neural structured prediction. *Adv. Neural Inf. Process. Syst.* **2019**, *29*, 1723–1731.
84. Gagniuc, P.A. *Markov Chains: From Theory to Implementation and Experimentation*; John Wiley & Sons: Hoboken, NJ, USA, 2017.
85. Anderson, T.W.; Goodman, L.A. Statistical inference about Markov chains. *Ann. Math. Stat.* **1957**, 89–110. [CrossRef]

86. Hermanns, H. Interactive markov chains. In *Interactive Markov Chains*; Springer: Berlin/Heidelberg, Germany, 2002; pp. 57–88.
87. Chen, W.; Pourghasemi, H.R.; Zhao, Z. A GIS-based comparative study of Dempster-Shafer, logistic regression and artificial neural network models for landslide susceptibility mapping. *Geocarto Int.* **2017**, *32*, 367–385. [CrossRef]
88. Das, I.; Sahoo, S.; van Westen, C.; Stein, A.; Hack, R. Landslide susceptibility assessment using logistic regression and its comparison with a rock mass classification system, along a road section in the northern Himalayas (India). *Geomorphology* **2010**, *114*, 627–637. [CrossRef]
89. Mohammadi, A.; Shahabi, H.; Bin Ahmad, B. Integration of insartechnique, google earth images and extensive field survey for landslide inventory in a part of Cameron highlands, Pahang, Malaysia. *Appl. Ecol. Environ. Res.* **2018**, *16*, 8075–8091. [CrossRef]
90. Visa, S.; Ramsay, B.; Ralescu, A.L.; Van Der Knaap, E. Confusion Matrix-based Feature Selection. *MAICS* **2011**, *710*, 120–127.
91. McHugh, M.L. Interrater reliability: The kappa statistic. *Biochem. Med. Biochem. Med.* **2012**, *22*, 276–282. [CrossRef]
92. Ling, F.; Li, W.; Du, Y.; Li, X. Land cover change mapping at the subpixel scale with different spatial-resolution remotely sensed imagery. *IEEE Geosci. Remote Sens. Lett.* **2010**, *8*, 182–186. [CrossRef]
93. Torahi, A.A.; Rai, S. Modeling for prediction of land cover changes based on bio-physical and human factors in Zagros Mountains, Iran. *J. Indian Soc. Remote Sens.* **2013**, *41*, 845–854. [CrossRef]
94. Seibert, J.; McDonnell, J.J. Land-cover impacts on streamflow: A change-detection modelling approach that incorporates parameter uncertainty. *Hydrol. Sci. J. J. Des Sci. Hydrol.* **2010**, *55*, 316–332. [CrossRef]
95. Amiri, R.; Weng, Q.; Alimohammadi, A.; Alavipanah, S.K. Spatial–temporal dynamics of land surface temperature in relation to fractional vegetation cover and land use/cover in the Tabriz urban area, Iran. *Remote Sens. Environ.* **2009**, *113*, 2606–2617. [CrossRef]
96. Feizizadeh, B.; Blaschke, T. Land suitability analysis for Tabriz County, Iran: A multi-criteria evaluation approach using GIS. *J. Environ. Plan. Manag.* **2013**, *56*, 1–23. [CrossRef]

Publisher's Note: MDPI stays neutral with regard to jurisdictional claims in published maps and institutional affiliations.

© 2020 by the authors. Licensee MDPI, Basel, Switzerland. This article is an open access article distributed under the terms and conditions of the Creative Commons Attribution (CC BY) license (http://creativecommons.org/licenses/by/4.0/).

Article

Artificial Intelligence Applied to a Robotic Dairy Farm to Model Milk Productivity and Quality based on Cow Data and Daily Environmental Parameters

Sigfredo Fuentes [1,*], Claudia Gonzalez Viejo [1], Brendan Cullen [2], Eden Tongson [1], Surinder S. Chauhan [2] and Frank R. Dunshea [2]

1. Digital Agriculture, Food, and Wine Group, Faculty of Veterinary and Agricultural Sciences, The University of Melbourne, Parkville, VIC 3010, Australia; cgonzalez2@unimelb.edu.au (C.G.V.); eden.tongson@unimelb.edu.au (E.T.)
2. Agricultural Production System Modelling Group, Faculty of Veterinary and Agricultural Sciences, The University of Melbourne, Parkville, VIC 3010, Australia; bcullen@unimelb.edu.au (B.C.); ss.chauhan@unimelb.edu.au (S.S.C.); fdunshea@unimelb.edu.au (F.R.D.)
* Correspondence: sfuentes@unimelb.edu.au

Received: 27 April 2020; Accepted: 22 May 2020; Published: 24 May 2020

Abstract: Increased global temperatures and climatic anomalies, such as heatwaves, as a product of climate change, are impacting the heat stress levels of farm animals. These impacts could have detrimental effects on the milk quality and productivity of dairy cows. This research used four years of data from a robotic dairy farm from 36 cows with similar heat tolerance (Model 1), and all 312 cows from the farm (Model 2). These data consisted of programmed concentrate feed and weight combined with weather parameters to develop supervised machine learning fitting models to predict milk yield, fat and protein content, and actual cow concentrate feed intake. Results showed highly accurate models, which were developed for cows with a similar genetic heat tolerance (Model 1: n = 116, 456; R = 0.87; slope = 0.76) and for all cows (Model 2: n = 665, 836; R = 0.86; slope = 0.74). Furthermore, an artificial intelligence (AI) system was proposed to increase or maintain a targeted level of milk quality by reducing heat stress that could be applied to a conventional dairy farm with minimal technology addition.

Keywords: machine learning; heat stress; animal welfare; climate change; automation

1. Introduction

Robotic dairy farms or Automated Milking Systems (AMS) are the result of the implementation of state of the art technology related to robotics to increase milk yield through increased efficiency and automation [1,2]. These technologies are developed in response to the increasing market opportunities for the dairy industry globally, which is projected to grow by 35% by 2030 [3]. However, global demands will also be accompanied by 14 million traditional dairy farms shutting down production due to increased competitiveness and requirements for guaranteed milk quality and animal welfare [4]. The latter is considered a growing concern for consumers, which is achieved by AMS since it is based on the "milking when they like" system increasing wellbeing and welfare of cows [5]. Further potential advances to AMS technologies have been researched in recent years through the implementation of biometrics monitoring of animals to assess physiological changes in production systems [6]. Some of these technologies are noninvasive using visible (RGB) imagery/video, and infrared thermal imagery for heart rate, respiration rate, and body temperature assessments. These technologies could result in improvements in the monitoring of heat stress in farm animals.

Modeling heat stress in AMS has concentrated recently on the rumination and milking performance [7], identifying specific thresholds with production factors [8] and thermal comfort indices [9], mainly through the calculation of the temperature-humidity index (THI) using several models [10]. According to a study by Nascimento et al. [11], who compared nine different models to calculate THI, the equation from Berman et al. [12] (Equation (10) below) was significantly correlated with physiological data of cows such as respiration rate, heart rate, rectal, and skin temperatures. However, all previous methods use deterministic mathematical equations with minimal animal information in the analysis, and the noncontact biometric analysis could be cost-prohibitive for the near-future application to conventional dairy farms.

Artificial intelligence (AI) applied to Digital Agriculture deals with the implementation and integration of digital data, sensors, and tools on agricultural applications from the farm to consumers [13]. These technologies can include big data, sensor technology, sensor networks, remote sensing, robotics, and unmanned aerial vehicles (UAV). Data processing is performed using new and emerging technologies, such as computer vision, machine learning, and AI, among others. The implementation of AI not only should benefit high technological systems, such as AMS, but also conventional dairy farms to increase their competitiveness in the future.

This research was based on machine learning modeling using ubiquitous environmental data obtained from automatic meteorological stations and cow information available by all dairy farms as inputs. Target information related to important parameters related to milk productivity, milk quality, and actual feed of dairy cows was obtained from an AMS belonging to The University of Melbourne, Australia. High accurate machine learning (ML) models that can be applied to any dairy farm from AMS to conventional were obtained. Furthermore, this paper proposes an AI system model to be implemented in any dairy farm to automatically assess and ameliorate heat stress by implementing ML models developed with an automated sorting and gate system.

2. Materials and Methods

2.1. Site, Robotic Dairy Farm, and Data Acquisition

The study was conducted in a dairy farm located at The University of Melbourne Dookie College, Victoria, Australia (36°22′48″ S, 145°42′36″ E). This region had an average annual rainfall of 537 mm (monthly extremes: 30.5–57.6 mm) and mean daily solar exposure of 17 MJ m^{-2} (extremes: 7.3–27.3 m$^{2-1}$) from 1991–2019; data obtained from the Bureau of Meteorology (BoM) Dookie Agricultural College station 081013. The farm consists of 43 ha of irrigated pastures based on perennial ryegrass (Lolium perenne) and annual ryegrass (Lolium multiflorum). The herd in this site consists of Holstein-Friesian cows. The farm contains three Lely Astronaut robotic milking machines (Lely Holding S.à.r.l., Maassluis, The Netherlands), with a capacity of 60 cows per machine (maximum capacity of 180 cows) that move voluntarily for milking. As described by Dunshea et al. [14], cows wear an identification transponder neck collar (Lely Holding S.à.r.l., Maassluis, The Netherlands), which records the cows' activity. The robotic milking system can automatically record parameters such as lactation days counted from day 0 at calving up to the time of next calving including the dry cow period, lactation number, milking frequency per day, milk yield (kg day^{-1}), milk protein (%), milk fat (%) and somatic cells, programmed concentrate feed (kg day^{-1}), concentrate feed intake (kg day^{-1}), and liveweight (kg). Records of these data from June 2016 to March 2019 were used for this study.

Weather data were obtained from the meteorological station (Adcon Telemetry GmbH, Klosterneuburg, Austria), located at the Dookie Agricultural College, which provides data every 15 min for each day of the year. Parameters obtained were (i) temperature (T; °C), (ii) relative humidity (RH; %), (iii) rainfall as daily running total (mm), (iv) wind speed (km h^{-1}), and (v) wind direction (°). Based on these data, other variables such as dewpoint temperature (T_{dp}; °C; Equation (1); [15]), wet bulb temperature (T_{wet}; °C), and THI were calculated. The latter was calculated using the following nine different equations (Equations (2)–(10); [11]):

$$T_{dp} = \frac{243.5 \left(\frac{17.67 \times T}{243.5 + T} + \ln \frac{RH}{100} \right)}{17.67 - \left(\frac{17.67 \times T}{243.5 + T} + \ln \frac{RH}{100} \right)} \quad (1)$$

$$THI_1 = 0.4 \times (T + T_{wet}) \times 1.8 + 32 + 15 \quad (2)$$

$$THI_2 = (0.15 \times T + 0.85 \times T_{wet}) \times 1.8 + 32 \quad (3)$$

$$THI_3 = (T \times 0.35 + T_{wet} \times 0.65) \times 1.8 + 32 \quad (4)$$

$$THI_4 = 0.72 \times (T + T_{wet}) + 40.6 \quad (5)$$

$$THI_5 = (1.8 \times T + 32) - [(0.55 - 0.0055 \times RH) \times (1.8 + T - 26)] \quad (6)$$

$$THI_6 = \left(0.55 \times T + 0.2 \times T_{dp}\right) \times 1.8 + 32 + 17.5 \quad (7)$$

$$THI_7 = T + \left(0.36 \times T_{dp}\right) + 41.2 \quad (8)$$

$$THI_8 = (0.8 \times T) + \left(\frac{RH}{100}\right) \times (T - 14.4) + 46.4 \quad (9)$$

$$THI_9 = 3.43 + 1.058 \times T - 0.293 \times RH + 0.0164 \times T \times RH + 35.7 \quad (10)$$

where T_{wet} was calculated in batch using a customized code written in MATLAB®R2020a (Mathworks Inc., Natick, MA, USA; [16]), calculations were based on T, T_{dp}, and surface pressure and the bisection search method.

2.2. Statistical Data and Machine Learning Modeling

Mean values of *THI* calculated with Equation (10) along with milk yield, milk protein, and fat content, and concentrate feed intake were obtained and plotted to visualize the effects of the different seasons on each parameter. Statistical data obtained from the inputs and targets consisted of minimum, maximum, and mean values of each parameter.

Two ML models were developed based on artificial neural networks (ANN) using the Bayesian Regularization training algorithm. The latter was chosen as it showed the best accuracy and performance as well as no over or underfitting [17] after testing 17 different algorithms using a customized code written in MATLAB®R2020a. The inputs for the models (Figure 1) were based on the maximum values per day of the weather data (i) T, (ii) RH, (iii) rainfall, (iv) wind speed, (v) wind direction, (vi) T_{dp}, (vii) T_{wet}, (viii–xvi) *THI* calculated with the nine equations, and some data obtained from the robotic milking system, (xvii) programmed concentrate feed, (xviii) lactation days, (xix) lactation number, (xx) milking frequency, and (xxi) liveweight. The targets were also obtained from the robotic milking system. They consisted of (i) milk yield, (ii) milk protein, (iii) milk fat, and (iv) concentrate feed intake (i.e., cereal grain-based pellets fed to cows during milking, making up approximately 40% of cows diet). All data were normalized from −1 to 1. Model 1 was constructed using the data of cows with a similar heat tolerance (N = 36; heat tolerance range: 93–112) determined by estimation of Australian genomic breeding values for heat tolerance [18] following genotyping of each cow using hair follicle samples as per the commercial procedure (CLARIFIDE for dairy, Zoetis Australia Pty Ltd, Banyo, QLD, Australia). The genotyping experiment was approved by the University of Melbourne Faculty of Veterinary and Agricultural Science (FVAS) Animal Ethics Committee (AEC ID 1814645.1). In general, for heat stress, cows with Australian breeding values < 100 are less tolerant to hot, humid conditions than the average, while the cows with values > 100 are more tolerant than the average. Specifically, cows with breeding values of 93 will be 7% less heat tolerant than an average cow, and a cow with heat tolerance breeding values of 110 would be 10% more heat tolerant as compared to an average cow. In contrast, Model 2 was developed using data from all cows (N = 312) independent of their heat tolerance to create a general model. Samples were divided randomly as 70% for training and 30% for testing using a default derivative function. Ten neurons were chosen as the best number giving the highest accuracy and best performance based on the means squared error (MSE).

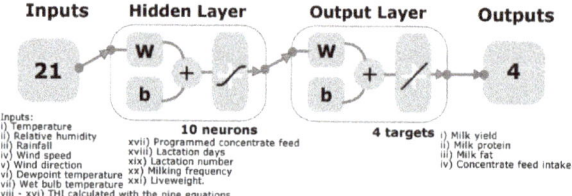

Figure 1. Diagram of the two-layer feedforward regression models with a tan-sigmoid function in the hidden layer and linear transfer function in the output layer. Abbreviations: THI: Temperature-humidity index; W: Weights; b: Bias.

3. Results

Figure 2 shows the mean values per season of each year for THI_9 and the four parameters used as targets in the ML models to represent the effect of different weather patterns on those variables. As expected, the highest *THI* were obtained in the summer seasons of all years (77.5–79.7) and the lowest in winter of all years (47.6–49.1). The highest average milk yield per cow was observed in winter (33.4 kg day^{-1}) and spring 2017 (33.5 kg day^{-1}) with the lowest yield in summer 2018–2019 (23.4 kg day^{-1}). The latter season also presented the lowest protein content in milk (3.1%) and concentrate feed intake (4.3 kg day^{-1}). Spring 2018 and autumn 2018 had the lowest (3.9%) and the highest milk fat content (4.6%), respectively.

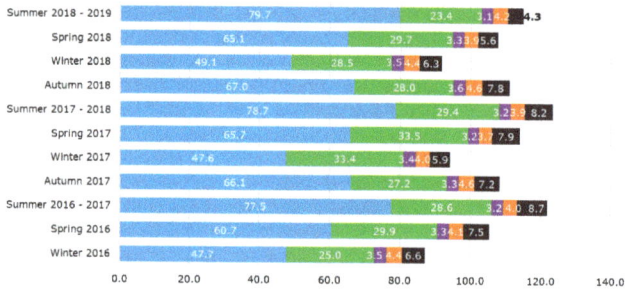

Figure 2. Mean values per season of each year for temperature-humidity index (THI_9) and the four parameters used as targets in the machine learning (ML) models to represent the effect of different weather patterns on potential heat stress, milk productivity, and quality.

Table 1 shows the minimum, maximum, and mean values per year of each parameter used as inputs to construct the ML models. The lowest mean temperature (19.3 °C) was observed during 2016, which, at the same time, presented the lowest mean THI_1–THI_9 (58.1–72.0), highest mean *RH* (95.6%), and daily rainfall (3.9 mm). On the contrary, 2019 had the highest maximum temperature (44.9 °C) and, until March, the lowest mean *RH* (69.2%), and daily rainfall (0.3 mm), as well as the highest mean THI_1–THI_9 (68.6–82.8). Data for lactation days = 0 are the day the calf was born, and milk production commenced. Due to the voluntary milking system on the farm, there are some days when cows are not milked (i.e., milking frequency = 0). Furthermore, there were cows on the farm with extended lactations (>600 days). These were 'carryover' cows that were in an extended lactation because they failed to get pregnant in a timely manner.

Table 1. Minimum, maximum, and mean values of the parameters used as inputs to develop the machine learning models.

Parameter/Year	2016 *				2017				2018				2019 *			
	Min	Max	Mean	SD	Min	Max	Mean	SD	Min	Max	Mean	SD	Min	Max	Mean	SD
T (°C)	7.9	37.8	19.3	7.15	8.3	42.0	22.2	8.27	8.9	43.3	22.6	7.90	16.3	44.9	31.7	6.07
RH (%)	66.0	100	95.6	6.20	56.2	100	92.3	9.27	44.2	100	87.9	11.67	39.2	92.6	69.2	12.13
T_{dp} (°C)	3.8	24.9	11.6	3.40	2.1	22.8	11.5	3.97	1.6	22.1	10.2	3.75	4.3	21.2	13.5	3.79
T_{wet} (°C)	6.6	25.3	13.8	3.56	6.8	25.3	14.7	4.20	5.9	24.5	14.1	3.97	11.3	24.1	18.6	3.11
Rainfall (mm day^{-1})	0.0	34.0	3.9	6.76	0.0	31.6	1.86	4.55	0.0	37.8	1.4	4.07	0.0	5.0	0.3	0.99
Wind speed (km h^{-1})	5.2	38.3	15.1	5.22	5.8	34.3	15.3	5.19	5.1	38.0	16.2	5.82	9.4	39.8	19.5	6.14
Wind direction (°)	127.7	360.0	344.0	26.54	247.2	360.0	345.1	23.76	112.2	360.0	341.8	32.81	241.7	360.0	338.9	30.62
THI_1	57.2	89.6	70.6	7.37	58.4	94.6	73.3	8.72	58.7	92.9	73.3	8.34	67.3	96.3	82.8	6.25
THI_2	44.5	78.6	58.1	7.08	45.4	81.2	60.2	8.42	44.5	79.2	59.6	8.01	54.9	80.5	68.6	6.06
THI_3	44.7	81.1	60.0	8.24	46.0	86.8	62.9	9.79	46.3	83.8	62.6	9.34	56.3	87.8	73.2	6.99
THI_4	50.8	83.2	64.2	7.37	52.0	88.2	66.9	8.72	52.3	86.5	66.9	8.34	60.9	89.9	76.4	6.25
THI_5	47.1	82.2	63.4	8.15	47.5	86.5	66.5	9.13	49.0	84.3	66.9	8.48	60.4	87.5	76.2	5.55
THI_6	59.1	91.5	72.0	7.33	60.2	97.0	74.7	8.75	60.5	94.4	74.6	8.39	68.7	98.7	84.3	6.56
THI_7	50.8	83.6	63.9	7.40	51.9	89.1	66.6	8.84	52.2	86.5	66.5	8.47	60.6	90.9	76.3	6.62
THI_8	47.1	82.0	63.4	8.09	47.5	86.3	66.5	9.07	49.0	84.0	66.8	8.41	60.4	87.2	76.0	5.49
THI_9	33.4	86.6	58.8	12.45	33.4	92.6	63.7	13.73	36.5	89.4	64.4	12.61	55.4	93.7	78.1	7.83
Programmed concentrate feed (kg day^{-1})	0.0	15.0	8.9	3.03	0.0	23.0	8.5	3.11	0.0	15.7	7.8	3.15	0.0	8.0	5.1	2.30
Lactation number	1.0	6.0	2.7	0.97	1.0	7.0	3.0	1.24	1.0	7.0	2.3	1.61	1.0	8.0	3.0	1.75
Lactation days	0.0	736.0	225	158.17	0.0	668.0	198.1	139.28	0.0	705.0	228.3	142.26	0.0	755.0	227.5	144.01
Milking frequency (per day)	0.0	5.0	2.4	0.71	0.0	6.0	2.5	0.75	0.0	6.0	2.4	0.84	0.0	5.0	1.9	0.81
Liveweight (kg)	373.0	938.0	677.7	82.85	428.0	951.0	668.2	78.25	335.0	959.0	655.4	84.57	410.0	896.0	629.5	71.86

* Values from 2016 cover from June to December and 2019 cover from January to March. Abbreviations: Min: Minimum; Max: Maximum; T: Temperature; RH: Relative humidity; T_{dp}: Dewpoint temperature; T_{wet}: Wet-bulb temperature; THI: Temperature-humidity index; SD: Standard deviation.

Table 2 shows the minimum, maximum, and mean values of the parameters used as targets for the ML models. It can be observed that 2017 presented the highest milk yield per cow on average (30.7 kg day^{-1}), although 2016 had the highest maximum milk yield per cow (65.4 kg day^{-1}). Likewise, for milk protein, 2017 had the highest maximum value (6.1%), while 2018 presented the highest mean value (3.4%). Regarding milk fat content, 2019 had the highest maximum and mean values (10.9% and 4.3%, respectively). In 2019, the lowest average concentrate feed intake (4.0 kg day^{-1}) was observed, while 2017 presented the highest mean (7.4 kg day^{-1}) and the highest maximum value (24.3 kg day^{-1}).

Table 2. Minimum, maximum, and mean values of the parameters used as targets to develop the machine learning models.

Parameter/Year	2016 *			2017			2018			2019 *		
	Min	Max	Mean	Min	Max	Mean	Min	Max	Mean	Min	Max	Mean
Milk yield (kg day^{-1})	0.0	65.4	28.1	0.0	60.2	30.7	0.0	61.2	28.8	0.0	52.1	21.2
Milk protein (%)	1.8	5.8	3.3	1.8	6.1	3.2	2.2	5.8	3.4	0.9	4.9	3.1
Milk fat (%)	1.0	10.7	4.2	0.8	10.2	4.0	0.7	10.3	4.2	0.7	10.9	4.3
Concentrate feed intake (kg day^{-1})	0.0	19.5	7.3	0.0	24.3	7.4	0.0	18.8	6.7	0.0	10.6	4.0

* Values from 2016 cover from June to December and 2019 cover from January to March. Abbreviations: Min: Minimum; Max: Maximum.

Table 3 shows the statistical results of both models to predict milk yield, milk fat, and protein content, and concentrate feed intake. It can be observed that both models presented similar results with high overall correlation coefficients (Model 1: R = 0.87; Model 2: 0.86; Figure 3). None of the models showed any signs of overfitting as the correlation coefficient of all stages was the same, and the performance of training (Model 1: MSE = 0.0186; Model 2: MSE = 0.0154) was lower than the testing stage (Model 1: MSE = 0.0189; Model 2: MSE = 0.0157). According to the 95% confidence bounds, Model 1 presented 3.88% outliers (4513 out of 116,456) and Model 2 presented 3.60% (23,998 out of 665,836).

Table 3. Statistical results of each stage of the machine learning models.

Stage	Samples (Cows x Days)	Observations (Samples x Targets)	R	b	Performance (MSE)
Model 1					
Training	20,380	81,520	0.87	0.76	0.0186
Testing	8734	34,936	0.86	0.76	0.0189
Overall	29,114	116,456	0.87	0.76	-
Model 2					
Training	116,521	466,084	0.86	0.74	0.0154
Testing	49,938	199,752	0.86	0.74	0.0157
Overall	166,459	665,836	0.86	0.74	-

Abbreviations: R: Correlation coefficient; b: Slope; MSE: Means squared error.

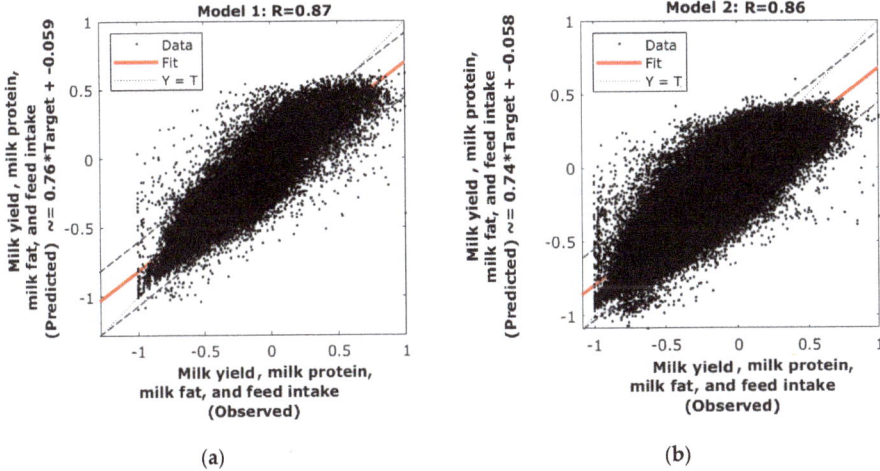

Figure 3. Overall regression graphs of (**a**) Model 1: Using the 36 cows with similar heat tolerance (93–112), and (**b**) Model 2: Using data from 312 cows.

4. Discussion

4.1. Seasonality and Milk Yield

During the four years included in this study (2016–2019), there was a clear variation within seasons reflected by environmental parameters (*THI*) and milk productivity parameters (Figure 2). Higher heat stress risks for cows were observed in the summer of 2018–2019. Even though the *THI* parameter had a higher tendency, it was not significantly greater compared to the *THI* of summers belonging to 2017 and 2016 (*THI* = 79.7 compared to 78.7 and 77.5, respectively). However, milk yield and quality parameters were lower for 2018 compared with previous years. The high variability among all parameters shown through the years considered for this study can be considered as an advantage for ML modeling. These differences can be further supported by the data presented in Tables 1 and 2 with more specific data per year. Prolonged periods of high temperature and relative humidity have shown to be detrimental to dairy cows performance due to heat stress [19]. This makes more critical the development of cost-effective methodologies to measure and alleviate heat stress during these periods of high *THI* [20].

4.2. Machine Learning Models

By investigating thermotolerance in cows from a genetic point of view, it could help to decrease economic losses associated with lower milk productivity, quality, and animal welfare [21,22]. Other methods have been based on the physical modification of the environment, such as shade and shelters, and dietary interventions to reduce heat stress effects, such as grape residue [23], açai [24], betaine [14,25], slowly fermentable grains [26], and other types of feed [27,28].

The ML models developed in this research (Model 1 and Model 2) do not differ much when considering 36 genetically similar cows for heat tolerance compared to a total of 320 cows. There is a slight difference in the slope for the general model considering all cows (Model 2; slope = 0.74) compared to Model 1 (slope = 0.76). Considering highly heat stress-tolerant cows helps to decrease underestimations made by Model 1 compared to Model 2. However, it can be considered that these differences are minimal when considering the number of cows deemed for Model 1 (n = 36) compared to Model 2 (n = 312). Furthermore, Model 1 presented a slightly higher percentage of outliers, considering them as outside the 95% confidence bounds, with 3.88% compared to 3.60% for Model 2 (Figure 3),

4.3. Artificial Intelligence to Manage Heat Stress and Milk Productivity

Physical modification of the environment to reduce ambient temperature or increase heat loss from the animal body, such as shading and fans, have been previously applied for lactating buffaloes with positive results [29], and in dairy cows using mixed-flow fans [30]. However, one of the most effective methods found is spraying water over animals using sprinkler systems [31–34]. This paper proposed the implementation of Model 2 with an automated system based on an individual cow assessment combined with environmental factors obtained from an automatic meteorological station (AME) (Figure 4). The AME can be easily connected to a processing unit (microprocessor or smartphone App) that can read the RFID from cows that are going to be milked to obtain cow information required by the model (Figure 1). The model outputs can be automatically set to specific thresholds for volume and milk quality that is desired by the dairy farm. The system can then automatically control gates to direct individual cows either to a cooling system with water sprinklers, the cows to reduce heat stress or to normal milking sections. The heat-stressed cows will be assessed the next day again, if they continue to be heat stressed, they will go to the sprinkler system and get milked to avoid mastitis.

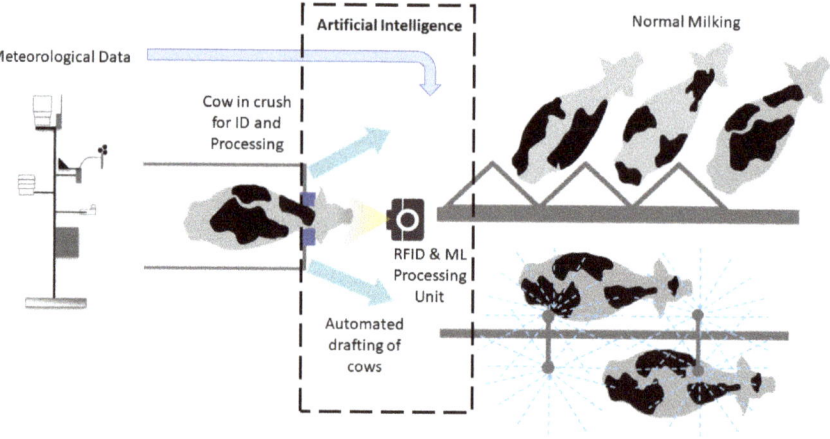

Figure 4. Proposed artificial intelligence (AI) application based on the automated processing of meteorological station and radio frequency identification system (RFID) for specific cow data input and machine learning (ML) processing. This system activates the gate system to draft cows to a cooling system or normal milking.

The technical advantages of the proposed system (Figure 4) are: (i) ML modeling is based on readily available environmental data by most of the dairy farms and from government services with meteorological stations close to the farms; (ii) the environmental data can be automatically extracted from government services, such as the Bureau of Meteorology (BoM, Australia) [35] or by direct connectivity of a nearby automatic meteorological station to the RFID & ML Processing Unit (Figure 4); (iii) the digital database per cow can be implemented as part of the system to incorporate data such as programmed concentrate feed, lactation days and number, milking frequency, and liveweight. This information will need to be updated by the dairy farm personnel; (iv) cows can be identified by the system with normal RFID systems to extract cow data automatically from databases, and (v) the system requires an automated gate system to draft cows to the heat stress sprinkler system or the normal milking facilities.

The managerial advantages that could be obtained by implementing the system proposed are: (i) milk volume and quality information available in real-time, per cow, and according to daily environmental conditions; (ii) prediction of actual concentrate feed intake per cow for feed monitoring management compared to programmed concentrate feed; (iii) real-time information to manage heat stress in a per cow basis to increase efficiency and maintain milk volumes and quality set as objectives, and (iv) data recorded from specific dairy farms can be incorporated in the model to increase the accuracy of target predictions.

With these considerations, an AI system for dairy farms can be implemented with reasonable investment affordable to small and medium dairy farmers. An alternative or complementary approach to an engineering solution may be to introduce dietary interventions such as betaine or antioxidants to cows likely to experience heat stress [14,28]. However, the time lag before the tissue concentrations of these nutrients are optimized could reduce the immediacy of this approach.

It should be noted that individual pasture intake could not be included in the model as the cows grazed as a single herd, so it was not measured. While this could no doubt add precision to the model, individual pasture intake cannot be measured under commercial grazing systems, and inclusion in the model would reduce its commercial utility.

5. Conclusions

The machine learning models developed in this research may be applied to assess automatically animal welfare, milk productivity, and quality. Based on the inputs of the models, this machine learning modeling technique can be applied to any dairy farm. Implementation of Artificial Intelligence in dairy farms and the ML models developed here will require minimal technological additions, automated gate, and cooling systems. This paper has shown a practical application of AI using detailed information from a robotic dairy farm for the benefit of small and medium dairy farms to increase competitiveness in an increasingly demanding international market.

Author Contributions: Conceptualization, S.F. and C.G.V.; data curation, B.C.; formal analysis, S.F. and C.G.V.; investigation, S.F., C.G.V., B.C., S.S.C., and F.R.D.; methodology, S.F. and C.G.V.; project administration, B.C. and F.R.D.; resources, B.C. and F.R.D.; software, S.F.; supervision, B.C. and F.R.D.; validation, S.F., C.G.V., B.C., S.S.C., and F.R.D.; visualization, S.F., C.G.V., and E.T.; writing—original draft, S.F. and C.G.V.; writing—review and editing, B.C., S.S.C., and F.R.D. All authors have read and agreed to the published version of the manuscript.

Funding: This research received no external funding.

Conflicts of Interest: The authors declare no conflict of interest.

References

1. Britt, J.; Cushman, R.; Dechow, C.D.; Dobson, H.; Humblot, P.; Hutjens, M.; Jones, G.; Ruegg, P.; Sheldon, I.; Stevenson, J. Invited review: Learning from the future—A vision for dairy farms and cows in 2067. *J. Dairy Sci.* **2018**, *101*, 3722–3741. [CrossRef] [PubMed]
2. Hostiou, N.; Fagon, J.; Chauvat, S.; Turlot, A.; Kling-Eveillard, F.; Boivin, X.; Allain, C. Impact of precision livestock farming on work and human-animal interactions on dairy farms. A review. *Biotechnol. Agron. Soc. Environ.* **2017**, *21*, 268–275.
3. McCullough, C. Global dairy outlook by 2030: Industry. *The Dairy Mail* **2019**, *26*, 12–15.
4. Bear, C.; Holloway, L. Beyond resistance: Geographies of divergent more-than-human conduct in robotic milking. *Geoforum* **2019**, *104*, 212–221. [CrossRef]
5. Heyden, T. The Cows That Queue up to Milk Themselves. Available online: https://robohub.org/the-cows-that-queue-up-to-milk-themselves-bbc-news/ (accessed on 3 May 2020).
6. Jorquera-Chavez, M.; Fuentes, S.; Dunshea, F.R.; Warner, R.D.; Poblete, T.; Jongman, E.C. Modelling and Validation of Computer Vision Techniques to Assess Heart Rate, Eye Temperature, Ear-Base Temperature and Respiration Rate in Cattle. *Animals* **2019**, *9*, 1089. [CrossRef] [PubMed]
7. Ji, B.; Banhazi, T.; Ghahramani, A.; Bowtell, L.; Wang, C.; Li, B. Modelling of heat stress in a robotic dairy farm. Part 3: Rumination and milking performance. *Biosyst. Eng.* **2020**. [CrossRef]

8. Ji, B.; Banhazi, T.; Ghahramani, A.; Bowtell, L.; Wang, C.; Li, B. Modelling of heat stress in a robotic dairy farm. Part 2: Identifying the specific thresholds with production factors. *Biosyst. Eng.* **2019**. [CrossRef]
9. Ji, B.; Banhazi, T.; Ghahramani, A.; Bowtell, L.; Wang, C.; Li, B. Modelling of heat stress in a robotic dairy farm. Part 1: Thermal comfort indices as the indicators of production loss. *Biosyst. Eng.* **2019**. [CrossRef]
10. Ekine-Dzivenu, C.; Mrode, R.A.; Ojango, J.M.; Okeyo Mwai, A. Evaluating the impact of heat stress as measured by temperature-humidity index (THI) on test-day milk yield of dairy cattle in Tanzania. In Proceedings of the Seventh All Africa conference on Animal Agriculture, Accra, Ghana, 29 July–2 August 2019.
11. Nascimento, F.G.d.O.; Aguiar, H.C.P.; Rodrigues, G.M.; Guimarães, E.C.; Nascimento, M.R.B.d.M. What is the best temperature-humidity index equation to indicate heat stress in crossbred dairy calves in a tropical environment? *Ciência Rural* **2019**, *49*, e20180132. [CrossRef]
12. Berman, A.; Horovitz, T.; Kaim, M.; Gacitua, H. A comparison of THI indices leads to a sensible heat-based heat stress index for shaded cattle that aligns temperature and humidity stress. *Int. J. Biometeorol.* **2016**, *60*, 1453–1462. [CrossRef]
13. Carolan, M. Automated agrifood futures: Robotics, labor and the distributive politics of digital agriculture. *J. Peasant Stud.* **2020**, *47*, 184–207. [CrossRef]
14. Dunshea, F.R.; Oluboyede, K.; DiGiacomo, K.; Leury, B.J.; Cottrell, J.J. Betaine improves milk yield in grazing dairy cows supplemented with concentrates at high temperatures. *Animals* **2019**, *9*, 57. [CrossRef] [PubMed]
15. Bolton, D. The computation of equivalent potential temperature. *Mon. Weather Rev.* **1980**, *108*, 1046–1053. [CrossRef]
16. Goodwin, R. Wet Bulb, Matlab Central File Exchange. Available online: https://www.mathworks.com/matlabcentral/fileexchange/50785-wet-bulb (accessed on 20 April 2020).
17. Gonzalez Viejo, C.; Torrico, D.; Dunshea, F.; Fuentes, S. Development of Artificial Neural Network Models to Assess Beer Acceptability Based on Sensory Properties Using a Robotic Pourer: A Comparative Model Approach to Achieve an Artificial Intelligence System. *Beverages* **2019**, *5*, 33. [CrossRef]
18. Nguyen, T.T.; Bowman, P.J.; Haile-Mariam, M.; Nieuwhof, G.J.; Hayes, B.J.; Pryce, J.E. Implementation of a breeding value for heat tolerance in Australian dairy cattle. *J. Dairy Sci.* **2017**, *100*, 7362–7367. [CrossRef]
19. Ouellet, V.; Cabrera, V.; Fadul-Pacheco, L.; Charbonneau, É. The relationship between the number of consecutive days with heat stress and milk production of Holstein dairy cows raised in a humid continental climate. *J. Dairy Sci.* **2019**, *102*, 8537–8545. [CrossRef]
20. Gunn, K.M.; Holly, M.A.; Veith, T.L.; Buda, A.R.; Prasad, R.; Rotz, C.A.; Soder, K.J.; Stoner, A.M. Projected heat stress challenges and abatement opportunities for US milk production. *PLoS ONE* **2019**, *14*. [CrossRef]
21. Summer, A.; Lora, I.; Formaggioni, P.; Gottardo, F. Impact of heat stress on milk and meat production. *Anim. Front.* **2019**, *9*, 39–46. [CrossRef]
22. Sigdel, A.; Abdollahi-Arpanahi, R.; Aguilar, I.; Peñagaricano, F. Whole Genome Mapping Reveals Novel Genes and Pathways Involved in Milk Production Under Heat Stress in US Holstein Cows. *Front. Genet.* **2019**, *10*, 928. [CrossRef]
23. Alba, D.F.; Campigotto, G.; Cazarotto, C.J.; dos Santos, D.S.; Gebert, R.R.; Reis, J.H.; Souza, C.F.; Baldissera, M.D.; Gindri, A.L.; Kempka, A.P. Use of grape residue flour in lactating dairy sheep in heat stress: Effects on health, milk production and quality. *J. Therm. Biol.* **2019**, *82*, 197–205. [CrossRef]
24. dos Santos, D.d.S.; Klauck, V.; Campigotto, G.; Alba, D.F.; Dos Reis, J.H.; Gebert, R.R.; Souza, C.F.; Baldissera, M.D.; Schogor, A.L.B.; Santos, I.D. Benefits of the inclusion of açai oil in the diet of dairy sheep in heat stress on health and milk production and quality. *J. Therm. Biol.* **2019**, *84*, 250–258. [CrossRef] [PubMed]
25. Hall, L.; Dunshea, F.; Allen, J.; Rungruang, S.; Collier, J.; Long, N.; Collier, R.J. Evaluation of dietary betaine in lactating Holstein cows subjected to heat stress. *J. Dairy Sci.* **2016**, *99*, 9745–9753. [CrossRef] [PubMed]
26. Gonzalez-Rivas, P.A.; Sullivan, M.; Cottrell, J.J.; Leury, B.J.; Gaughan, J.B.; Dunshea, F.R. Effect of feeding slowly fermentable grains on productive variables and amelioration of heat stress in lactating dairy cows in a sub-tropical summer. *Trop. Anim. Health Prod.* **2018**, *50*, 1763–1769. [CrossRef] [PubMed]
27. Coppock, C.E. Reduce Heat Stress in High-Producing Dairy Cows with Feeding and Nutrition Management. In *Dairy Science Handbook*; CRC Press: Boca Raton, FL, USA, 2019; pp. 315–324.
28. Dunshea, F.R.; Leury, B.J.; Fahri, F.; DiGiacomo, K.; Hung, A.; Chauhan, S.; Clarke, I.J.; Collier, R.; Little, S.; Baumgard, L. Amelioration of thermal stress impacts in dairy cows. *Anim. Prod. Sci.* **2013**, *53*, 965–975. [CrossRef]

29. Ahmad, M.; Bhatti, J.A.; Abdullah, M.; Ullah, R.; ul Ain, Q.; Hasni, M.S.; Ali, M.; Rashid, A.; Qaisar, I.; Rashid, G. Different ambient management intervention techniques and their effect on milk production and physiological parameters of lactating NiliRavi buffaloes during hot dry summer of subtropical region. *Trop. Anim. Health Prod.* **2019**, *51*, 911–918. [CrossRef]
30. Yao, C.; Shi, Z.; Zhao, Y.; Ding, T. Effect of Mixed-Flow Fans with a Newly Shaped Diffuser on Heat Stress of Dairy Cows Based on CFD. *Energies* **2019**, *12*, 4315. [CrossRef]
31. Tresoldi, G.; Schütz, K.E.; Tucker, C.B. Cooling cows with sprinklers: Effects of soaker flow rate and timing on behavioral and physiological responses to heat load and production. *J. Dairy Sci.* **2019**, *102*, 528–538. [CrossRef]
32. Sruthi, S.; Sasidharan, M.; Anil, K.; Harikumar, S.; Simon, S. Effect of automated intermittent wetting and forced ventilation on the physiological parameters and milk production of Murrah buffaloes in humid tropics. *Pharma Innov. J.* **2019**, *8*, 315–319.
33. Imbabi, T.; Hassan, T.; Radwan, A.; Soliman, A. Production, haematological and biochemical metabolites of Egyptian buffaloes (Bubalus bubalis) during the hot summer months in Egypt. *Slovak J. Anim. Sci.* **2019**, *52*, 152–159.
34. Pinto, S.; Hoffmann, G.; Ammon, C.; Heuwieser, W.; Levit, H.; Halachmi, I.; Amon, T. Effect of two cooling frequencies on respiration rate in lactating dairy cows under hot and humid climate conditions. *Annal. Anim. Sci.* **2019**, *19*, 821–834. [CrossRef]
35. Meteorology, B.O. BoM Automated Data Services. Available online: http://www.bom.gov.au/catalogue/data-feeds.shtml (accessed on 10 February 2020).

© 2020 by the authors. Licensee MDPI, Basel, Switzerland. This article is an open access article distributed under the terms and conditions of the Creative Commons Attribution (CC BY) license (http://creativecommons.org/licenses/by/4.0/).

Article

Non-Invasive Sheep Biometrics Obtained by Computer Vision Algorithms and Machine Learning Modeling Using Integrated Visible/Infrared Thermal Cameras

Sigfredo Fuentes [1,*], Claudia Gonzalez Viejo [1], Surinder S. Chauhan [2], Aleena Joy [2], Eden Tongson [1] and Frank R. Dunshea [2,3]

1. Digital Agriculture, Food and Wine Sciences Group, School of Agriculture and Food, Faculty of Veterinary and Agricultural Sciences, The University of Melbourne, Parkville, VIC 3010, Australia; cgonzalez2@unimelb.edu.au (C.G.V.); eden.tongson@unimelb.edu.au (E.T.)
2. Animal Nutrition and Physiology, Faculty of Veterinary and Agricultural Sciences, The University of Melbourne, Parkville 3010, Australia; ss.chauhan@unimelb.edu.au (S.S.C.); aleenajoyj@student.unimelb.edu.au (A.J.); fdunshea@unimelb.edu.au (F.R.D.)
3. Faculty of Biological Sciences, The University of Leeds, Leeds LS2 9JT, UK
* Correspondence: sfuentes@unimelb.edu.au

Received: 26 September 2020; Accepted: 4 November 2020; Published: 6 November 2020

Abstract: Live sheep export has become a public concern. This study aimed to test a non-contact biometric system based on artificial intelligence to assess heat stress of sheep to be potentially used as automated animal welfare assessment in farms and while in transport. Skin temperature (°C) from head features were extracted from infrared thermal videos (IRTV) using automated tracking algorithms. Two parameter engineering procedures from RGB videos were performed to assess Heart Rate (HR) in beats per minute (BPM) and respiration rate (RR) in breaths per minute (BrPM): (i) using changes in luminosity of the green (G) channel and (ii) changes in the green to red (a) from the CIELAB color scale. A supervised machine learning (ML) classification model was developed using raw RR parameters as inputs to classify cutoff frequencies for low, medium, and high respiration rate (Model 1). A supervised ML regression model was developed using raw HR and RR parameters from Model 1 (Model 2). Results showed that Models 1 and 2 were highly accurate in the estimation of RR frequency level with 96% overall accuracy (Model 1), and HR and RR with R = 0.94 and slope = 0.76 (Model 2) without statistical signs of overfitting

Keywords: animal welfare; skin temperature; artificial intelligence; heart rate; respiration rate

---

1. Introduction

Live animal exports have been lately under scrutiny by the public and animal welfare advocates [1], especially live export though shipping, related to welfare conditions and heat stress during long trips up to six weeks by sea, which in extreme cases can result in the death of animals in rates up to 2–3.8% [2]. Specifically, these mortality rates have been recently found in animal shipments from Australia through the Persian Gulf, which can reach temperatures of 36 °C with 95% relative humidity resulting in heat stress [3].

Heat stress events for animals are not only restricted to animal transport through sea or land, but it can also happen in farms due to increased ambient temperatures related to climate change, which can directly impact the health and welfare of animals [4–7]. There have been several types of research investigating the genetic resilience and adaptation of animals to heat stress [8–11] and mitigation strategies [12–14]. Many of these studies have based their assessment of heat stress on environmental indices, such as ambient temperature and relative humidity combined to form a temperature-humidity index (THI) [14–16]. The THI can be coupled with direct assessment of the effects of heat stress using physiological responses through manual monitoring [17,18], using sensors directly located on animals [19], behavioral assessments or including molecular, cellular and metabolic biomarkers [20–22]. These methods, though very reliable and robust, are intensive, requiring animal restraining, are labor-intensive, and time-consuming, also requiring specialized instrumentation and technical know-how from the personnel acquiring the data. Moreover, the use of intravaginal/rectal devices or contact sensors can be stressful for the animals [23].

Applications of artificial intelligence (AI) and machine learning modeling (ML) have been recently implemented to analyze environmental factors, such as THI, and its effects on heat stress of dairy cows and final productivity and quality of milk to maximize the utility of big data available from robotic dairy farms [24]. Further, AI and ML have been applied for processing and modeling remotely sensed information, which may offer a powerful tool to automatically extract critical physiological data from videos and infrared thermal imagery from animals and welfare analysis or the effects on quality of products [23,25–27].

Non-invasive methods to assess heat stress, based on remote sensing, have shown to be promising, as they avoid biases in the physiological data obtained from animals due to stresses imposed by wearable sensors, such as collars, polar sensors (for respiration and heart rate measurements), or intravaginal/rectal sensors for body temperature measurements [23,28]. Specifically, computer vision and infrared thermal remote sensing techniques have been recently applied to assess animal stress based on skin temperature and respiration rate [23,29] or the detection of heart rate and respiratory rates in pigs through luminosity changes from RGB videos of animals [29,30].

One of the main constraints in applying remote sensing techniques on sheep involves the thick fleece from unshorn animals, which presents a thick resistance layer from the skin. The advantage of utilizing these remote sensing techniques on pigs, especially hairless breeds, is that reflectance from visible and infrared thermal wavelengths are a direct representation of skin changes. Hence, non-invasive methods are required to be applied to body sections with less hair or wool in sheep and with outputs that can be representative, such as the head and face parts [23,30,31]. Specifically, these areas mainly correspond to the nose for respiration and heart rate and the whole head for skin temperature, especially focused on the eye section, since they are the only exposed internal organs to the environment, which may represent core body temperatures.

This study aimed to test a non-contact biometric system based on artificial intelligence to assess heat stress of sheep to be potentially used as automated animal welfare assessment in farm and while in transport. Specifically, it was focused on the automatic tracking of regions of interest (ROI) from sheep RGB videos and infrared thermal videos (IRTV) and the assessment of physiological information such as skin temperature, respiration rate (RR), and heart rate (HR) modeled using machine learning algorithms of sheep subjected to thermoneutral and controlled heat stress conditions. The system proposed was based on an affordable and integrated RGB 4K video camera and a high-resolution thermal infrared camera. It was further recommended an artificial intelligence approach to extract information automatically from sheep that could be coupled to blockchain [32,33] to have an independent assessment of animal welfare to be applied in the farm and transport or vessel environments. The latter could allow research on automated systems to ameliorate heat stress on farm animals or during transportation, such as mister or sprinklers, and will offer a blockchain system for control and certification of good practices on the farm or transportation to abattoirs or export markets.

2. Materials and Methods

2.1. Location, Animal Treatments, and Data Acquisition

This study was based on live animals and approved by the Faculty of Veterinary and Agricultural Sciences, University of Melbourne Animal Ethics Committee (AEC#1914872.1). It was conducted at

The University of Melbourne (UoM), Dookie Campus, Victoria, Australia (36°22′48″ S, 145°42′36″ E). Twelve sheep (Merino lambs 4–5 months old) were acclimatized to indoor facilities and housed in the individual pens for 3 days before starting measurements. They were fed a mixed ration (50% pellets, 25% oaten, and 25% Lucerne chaff) formulated to meet or exceed the National Research Council (NRC) [34] requirements, complemented with fresh water ad libitum. Room exhaust ventilation was performed using fans through the whole time of the experiments to simulate ventilation usually performed during live sheep export shipments. The latter mainly rely on mechanical ventilation using fans to remove heat and water vapor produced by animals, to ventilate moisture produced from manure pads, and to remove any possible build-up of noxious gases. After acclimatization, sheep were relocated to metabolic cages and housed in two temperature and relative humidity control rooms (Figure 1), conditioned to have two treatments with six sheep exposed to cyclic heat stress: (i) room at 28–40 °C and 40–60% relative humidity (RH), the cycles consisted of high temperatures of 36–40 °C every day from 8:00 to 16:00, and then reduced to 28–30 °C, and thermoneutral (control) conditions (ii) room at 18–21 °C and RH between 40 and 50%. The temperature (T) and RH were recorded every 30 min in each room using a universal serial bus (USB) temperature and humidity data logger (TechBrands; Electus Distribution, Rydalmere, NSW, Australia). These data were used to calculate the *THI* using the formula from Equation (1), which was specially developed for sheep [35].

$$THI = T - \lceil (0.31 - 0.31RH)(T - 14.4) \rceil \quad (1)$$

An integrated RGB video and infrared thermal video (IRTV) camera, FLIR® Duo Pro (FLIR Systems, Wilsonville, OR, USA) was fixed in each room using a small rack and tripod for stabilization (Figure 1A). This device has two cameras to record simultaneously RGB videos (Resolution: 4000 × 3000; Field of View: 56° × 45°) and IRTV (with a resolution of 336 × 256; Field of View: 35° × 27°; Thermal Sensitivity: <50 mK; Thermal Frame Rate: 9 Hz; Accuracy: ±5 °C). The camera has Bluetooth® connectivity (Bluetooth Special Interest Group, Kirkland, WA, USA) and, hence, can be controlled remotely using a FLIR smartphone/tablet personal computer (PC) application, FLIR® UAS 2 (FLIR Systems, Wilsonville, OR, USA). The RGB video and IRTV data were recorded three times daily (8:00; 12:00; 16:00) during 1 min each time for four weeks to have a wider range of physiological data.

Two kinds of measurements were conducted using: (i) traditional/manual techniques, and (ii) non-contact biometrics based on remote sensing (Figure 1B). The manual methods consisted of (i) heart rate (HR) using an elitecare® Sprague stethoscope (eNurse, Brisbane, QLD, Australia) and a timer, (ii) respiration rate (RR) visually with a chronometer assessing animal inhalations and exhalations, (iii) skin temperature from the right flank, below the wool in contact with the skin using a digital thermometer (Model: DT-K11A; Honsun, Shanghai, China), and (iv) rectal temperature using the same type of digital thermometer (Model: DT-K11A). The remotely sensed data (FLIR camera) were recorded using two FLIR® Duo integrated cameras for 1 min on each side of the room to capture all sheep three times a day, as previously mentioned. The thermal videos were used to assess skin temperature. In contrast, the RGB videos were recorded to evaluate HR and RR using computer vision analysis and customized ML modeling developed by the Digital Agriculture Food and Wine Group (DAFW) from UoM based on changes in luminosity within the RGB (HR) and Lab (RR) channels that have been developed for humans and animals based on the photoplethysmography (PPG) principle [23,30,31,36–38]. For the validation/calibration purposes of these newly developed ML models, only the most representative recordings were used; therefore, not all sheep were analyzed due to chamber size restrictions.

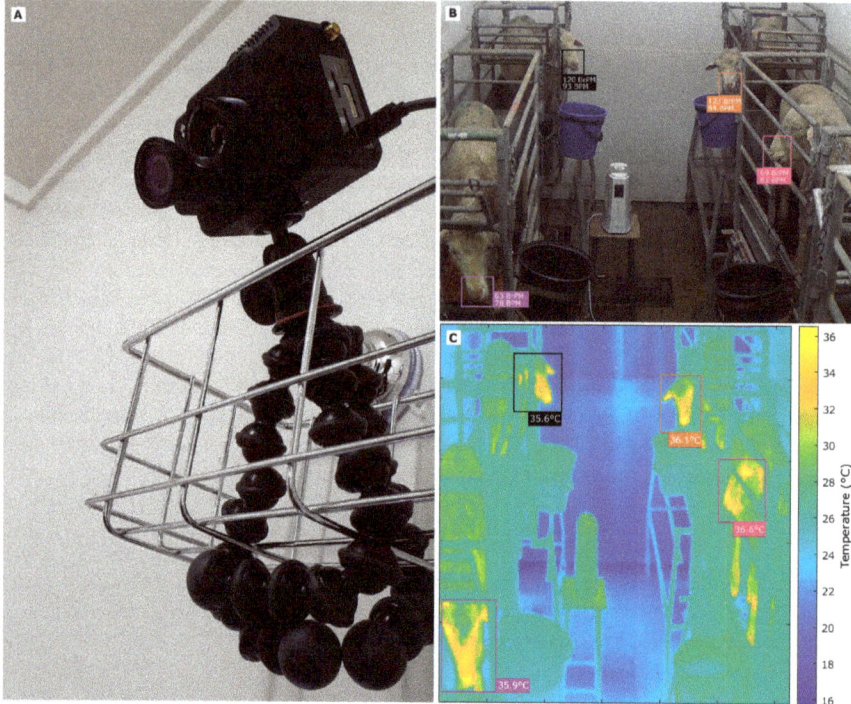

Figure 1. Images showing the experimental layout for thermoneutral (control) and heat stress chambers and implementation of feature tracking algorithms and machine learning (ML) Models 1 and 2 developed, with (**A**) the FLIR® Dup Pro camera setup, (**B**) shows the selected region of interest (ROI: nose) from each sheep visible and extraction of corresponding respiration rate and heart rate values from the video analysis using machine learning (Models 1 and 2), and (**C**) the selected region of interest (ROI: face) from each sheep and automatic tracking and extraction of temperature values (°C) from the infrared thermal video (IRTV) analysis. Abbreviations: BrPM: breaths per minute; BPM: beats per minute.

2.2. Computer Vision Analysis to Obtain Biometrics

The radiometric IRTVs were saved in sequence file extension (seq) and batch converted to Audio Video Interleaved (AVI) using the Sense Batch software (Sense Software, Warszawa, Mazowsze, Poland). The latter was also used to extract in batch and parallel the radiometric data from each frame from all thermal videos in comma-separated values (csv) files. The IRTV was imported to MATLAB® R2020a (MathWorks Inc., Natick, MA, USA) and the Video Labeler functions from the Computer Vision Toolbox™ 9.2 in MATLAB® R2020a were then used to select and track ROIs focusing on the head from each animal (automatic). Specifically, for sheep, the face was selected because the hottest visible spots are found in the eyes and nose (Figure 1C). Once the ROIs were tracked, labels were saved automatically, and a customized algorithm written in MATLAB® R2020a by the DAFW Group from UoM was used to obtain the maximum (Max), mode, and standard deviation (SD) of the temperatures from each frame from the selected ROI. Additionally, the mean, Max, mode, and SD from the Max temperatures from all frames were calculated.

For the analysis of raw signals related to HR and RR, the RGB videos acquired in QuickTime Movie (MOV) file-extension were used. These were analyzed using the Video Labeler functions from the Computer Vision Toolbox™ 9.2 in MATLAB® R2020a and the point tracker algorithm, which can detect features defined as a region of interest (ROI) and track one or more region of interest (ROI) based on the Kanade–Lucas–Tomasi (KLT) algorithm. For this specific study, the nose section was used as ROI for both HR and RR analysis (Figure 1B), as this is the area in which less wool may be found, as other areas may interfere with the readings creating biases in the data extracted. The ROI labels obtained were automatically exported and used to crop the RGB videos to get smaller videos only from the nose area from each sheep selected. These cropped videos were then automatically analyzed to obtain the signal changes from luminosity and different channels (RGB, CIELAB) using a modified version of the raw video analysis (RVA) algorithm developed as a function to measure HR in humans using the photoplethysmography (PPG) method [38] developed by the DAFW Group from UoM. This algorithm applies a fast Fourier transformation (FFT) for the transformation of the time signal to frequency and uses a second-order Butterworth filter with cutoff frequencies (Hz) for analysis. To assess raw signals related to RR, the RVA algorithm was modified (RVAm) to determine the luminosity changes in the "a" channel from the CIELAB color scale (green to red). The raw signals from computer vision analysis of cropped videos were evaluated within the cutoff frequency range 0.33–3.1 Hz for a ML classification model (Model 1, detailed in the machine learning modeling subsection), and the respiration cutoff frequency ranges used were according to the outputs of Model 1 described in detail below for low: 0.2–1.2 Hz; medium: 1.2–2.2 Hz, and high: 2.2–3.2 Hz. On the other hand, to assess HR, the luminosity changes in the green (G) channel of the RGB color scale were used within a frequency range of 0.83–3.00 Hz, since the normal and stressed HR for sheep had a lower spread in values compared to RR.

All the steps mentioned above were automated into a pipeline code using components as functions, which are represented in the diagram of Figure 2, in which the only supervised processes are the initial ROI selection for the IRTVs and RGB Videos.

2.3. Statistical Analysis and Machine Learning Modeling

Linear regression analysis for temperature data with intercept passing through the origin and $p \leq 0.05$ as criteria were used to compare the skin and rectal temperature measurements using the manual methods against each other and the non-invasive infrared thermal biometrics (IRTV) with XLSTAT ver. 2020.3.1 (Addinsoft, New York, NY, USA). Furthermore, linear regression analysis for RR and HR data measured manually and from videos using computer vision analysis with a single frequency range for RR and using frequency ranges for low, medium, and high, as previously mentioned, were performed. Statistical parameters, such as determination coefficient (R^2), p-value, and root means squared error (RMSE) were calculated to test the goodness of fits.

Based on a proposed parameter engineering procedure, raw RR-related parameters obtained from RGB Video analysis, using the RVAm algorithm and a single frequency range (0.33–3.1 Hz), of mean, minimum (Min), maximum (Max), and standard deviation (SD) of luminosity changes and mean, SD, frequency, and amplitude were used as inputs to develop an initial ML supervised pattern recognition model to classify the sheep cropped videos into low, medium, and high respiration frequencies (Model 1; Figure 2D). For this procedure, a customized MATLAB® code, developed by the authors, was used to test 17 artificial neural networks (ANN) training algorithms [39]. The Bayesian Regularization algorithm was selected as the best performing algorithm from this procedure based on the accuracy [correlation coefficient (R)] and best performance (means squared error (MSE)) with no signs of overfitting. This algorithm does not require a validation stage as it updates the weights and biases according to the optimization of the model, and is very effective on avoiding overfitting especially for small and/or noisy datasets [39–41]. Samples were divided randomly with 70% used for training ($n = 94$), and 30% for testing ($n = 40$). Figure 2D shows the model diagram with the two-layer feedforward network with a tan-sigmoid function in the hidden layer and Softmax function in the output layer. Ten neurons were selected as the best performance with no under- or over-fitting, which was obtained from a neuron trimming test (data not shown).

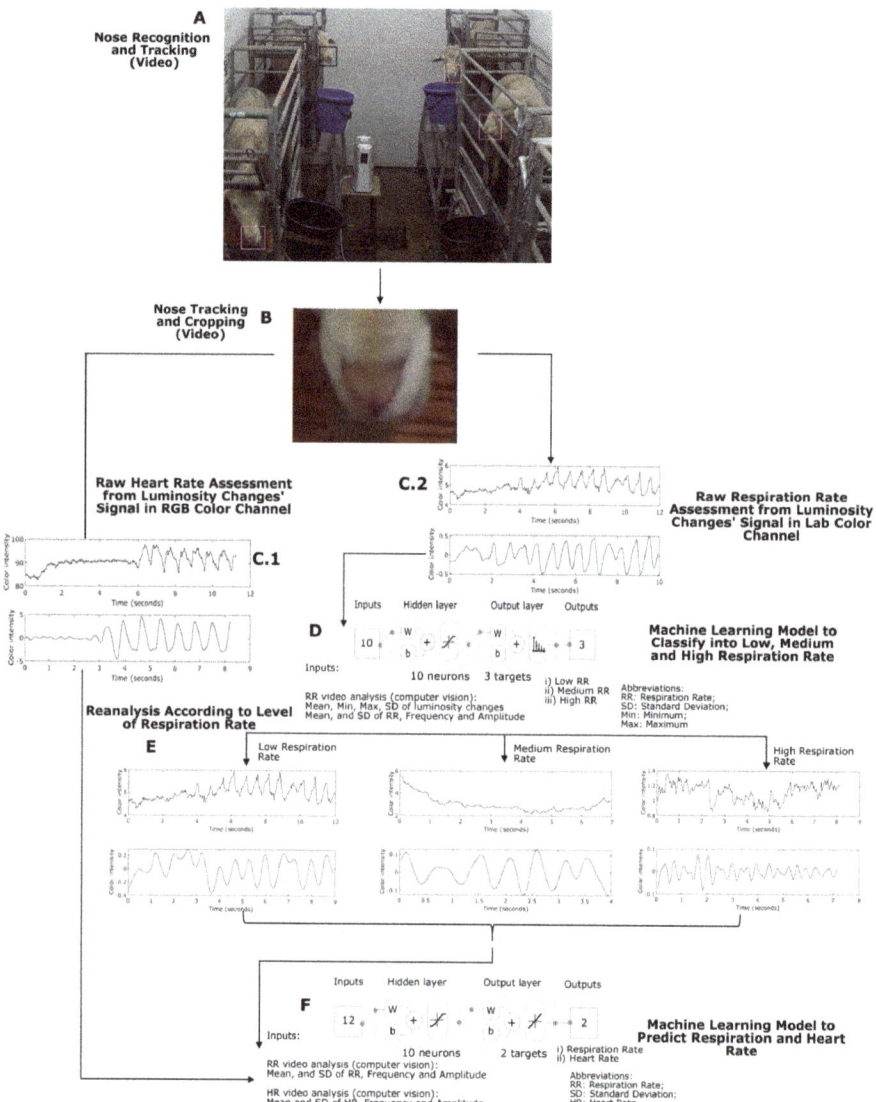

Figure 2. Diagram showing the algorithms pipeline for RGB video analysis process from (**A**) region of interest selection; (**B**) cropped videos from sheep feature to be analyzed; (**C1**) raw video analysis (RVA) for heart rate (HR) signals using the green channel (RGB); (**C2**) modified RVA (RVAm) for respiration rate (RR) analysis using the "a" channel (CIELAB) wide cutoff frequency range; (**D**) machine learning pattern recognition (Model 1) to obtain actual cutoff frequency range; (**E**) re-analysis of RR signals; (**F**) regression machine learning (Model 1) to obtain accurate HR an RR. Model diagram abbreviations: w: weights; b: bias.

Once the videos of sheep were classified automatically into low, medium, and high RR by cutoff frequency ML analysis, the videos are automatically reanalyzed using the corresponding frequency ranges, low: 0.2–1.2 Hz; medium: = 1.2–2.2 Hz; high: 2.2–3.2 Hz, by calling three separated functions. From this analysis, the outputs from raw RR and HR parameters were used as inputs to develop a fitting/regression model to predict the real values of RR and HR based on the manual measurements as targets (Model 2). Again, 17 different training algorithms [40] for artificial neural networks (ANN) were assessed in batch to find the best model based on output statistics. The Bayesian Regularization algorithm was selected as the best performing from this procedure. For modeling purposes, samples were divided randomly as follows: 70% (n = 94; observations (n × targets) = 188) for training and 30% (n = 40; observations (n × targets) = 80) for testing. Figure 2F depicts the model diagram showing the two-layer feedforward network with a tan-sigmoid function in the hidden layer and a linear transfer function in the output layer. Ten neurons were selected as the best performance with no under- or over-fitting, which was obtained from a neuron trimming test (data not shown).

Multivariate data analysis based on a biplot (variables and samples) of principal component analysis (PCA) was performed using XLSTAT to find relationships and patterns among the data between real physiological parameters and estimated using computer vision tools and models proposed. The cutoff point of 60% of data variability explained by the total of both PC1 and PC2 was considered to test significance [42]. The THI index calculated using Equation (1) was also included to compare data from sheep in control and heated chambers.

3. Results

Figure 3a shows the results from the linear regression of rectal and skin temperatures measured with the manual/traditional methods compared to those obtained from the IRTV analysis. There was a narrow distribution of temperatures from all sources (from around 35–40 °C) since the study was performed on live animals. The linear regression passing through the origin (0,0) was statistically significant ($p < 0.001$) and presented a very high correlation and determination coefficients (R = 0.99; R^2 = 0.99; RMSE = 0.66; slope = 0.97) between these two parameters with 3.6% of outliers (4 out of 110) based on the 95% confidence intervals. On the other hand, Figure 3b shows the results from the linear regression of observed skin temperature (manual/traditional methods) and the values obtained from the remote sensing analysis using the IRTVs. These relationships were also statistically significant ($p < 0.001$) with an R^2 = 0.99 (R = 0.99); RMSE = 1.66; slope = 1.02. Based on the 95% confidence intervals, it only had 2.73% of outliers (3 out of 110). Similarly, Figure 3c shows the results from the linear regression of observed rectal temperature (manual/traditional methods) and the values obtained from the remote sensing analysis using the IRTVs. The lineal model resulted with very high correlation and determination coefficients (R = 0.99; R^2 = 0.99) and was statistically significant ($p < 0.001$) with RMSE = 1.71; slope = 0.98 and 3.6% of outliers (4 out of 110) based on the 95% confidence intervals.

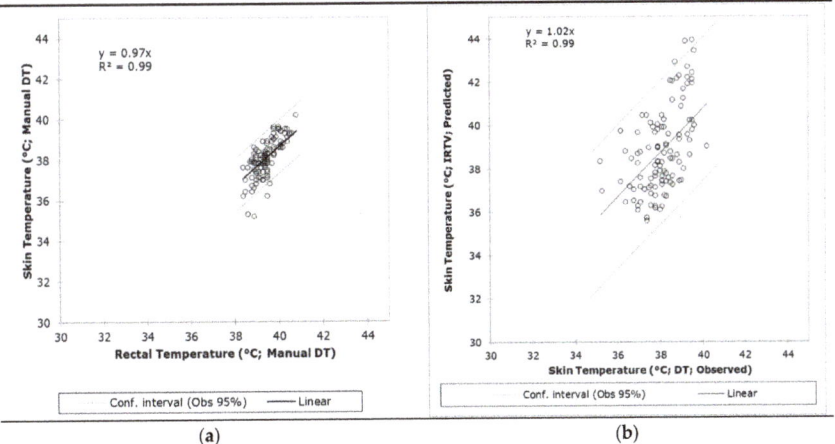

Figure 3. Cont.

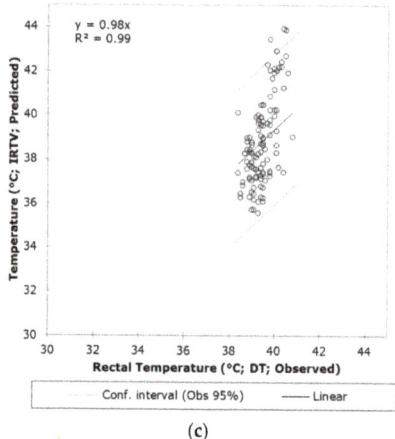

(c)

Figure 3. Linear regressions comparing results from (**a**) rectal vs. skin temperatures measured manually using a digital thermometer (DT), (**b**) observed skin temperature (manual) vs. temperature from the infrared thermal video analysis (IRTV), and (**c**) observed rectal temperature (manual) vs. temperature from the IRTV analysis. Abbreviations: Obs: observed, Conf: Confidence.

Figure 4A shows the linear regression between RR and HR measured manually and raw signal analysis related to HR and RR using computer vision analysis with a single cutoff frequency range for respiration rate (0.33–31. Hz) and HR (0.83–3.00 Hz). It can be observed that the correlation and determination coefficients were very low (R = 0.15; R^2 = 0.02; $p < 0.001$) with RMSE = 23.73 and slope = 0.09 mainly represented by the poor correlation found for the RR raw data. On the other hand, Figure 4B shows the same manually measured HR and RR rates against the raw computer vision analysis using different cutoff frequency ranges for RR according to low, medium, or high values (low: 0.2–1.2 Hz; medium: = 1.2–2.2 Hz; high: 2.2–3.2 Hz). It can be observed that the correlation increased significantly (R = 0.78; R^2 = 0.61; $p < 0.001$) compared to Figure 4A, with RMSE = 21.81 and slope = 0.67.

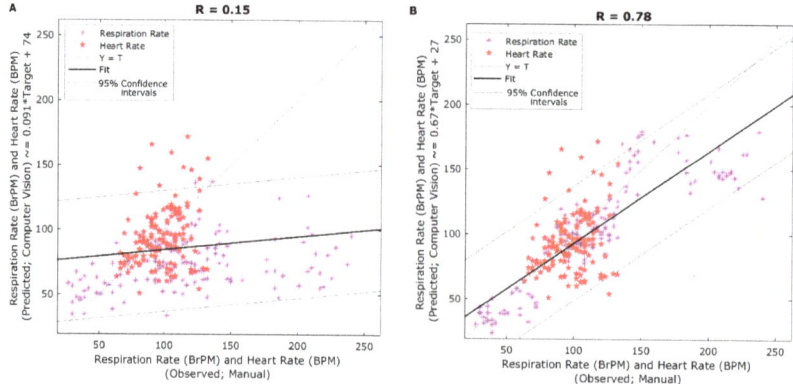

Figure 4. Linear regressions comparing results from (**A**) respiration rate (RR) and heart rate (HR) measured manually (x-axis) and using computer vision analysis with a single cutoff frequency range for respiration rate (0.33–3.1 Hz; y-axis), and (**B**) respiration rate and heart rate measured manually (x-axis) and using computer vision analysis with the corresponding cutoff frequency range for respiration rate according to low, medium or high respiration rate (low: 0.2–1.2 Hz; medium: = 1.2–2.2 Hz; high: 2.2–3.2 Hz; y-axis).

Table 1 shows the results from the ML pattern recognition model to classify cropped videos from sheep into low, medium, and high RR, according to the cutoff frequencies for these three levels. It can be observed that it presented a very high overall accuracy (96%) with no signs of overfitting as the MSE of the training stage (MSE < 0.01) was lower than the testing (MSE = 0.10). Figure 5 depicts the receiver operating characteristics (ROC) curve with the true positive (sensitivity) and false-positive (specificity) rates of the three categories, with all three categories within the true positive side of the curve; the high RR group presented the lowest sensitivity.

Table 1. Results of the artificial neural networks pattern recognition model (Model 1) showing the accuracy, error, and performance based on means squared error (MSE) for each stage for the selection of cutoff frequency related to low, medium, and high respiration rate signals from computer vision analysis.

Stage	Samples	Accuracy	Error	Performance (MSE)
Training	94	100%	0%	<0.01
Testing	40	85%	15%	0.10
Overall	134	96%	4%	-

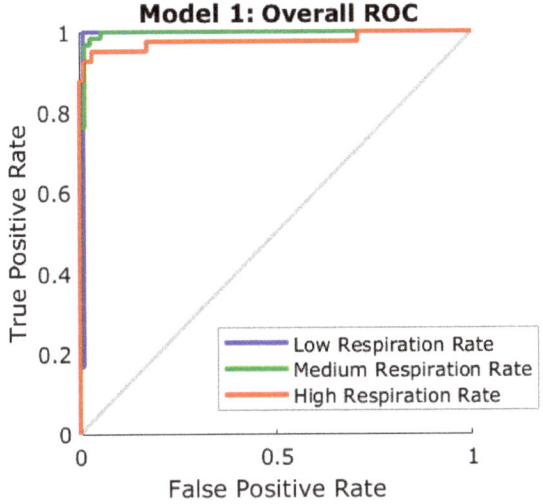

Figure 5. Receiver operating characteristics (ROC) curve of the pattern recognition (Model 1) depicting the true-positive and false-positive rates.

Table 2 shows the results of the ML model developed using the results from the remote sensing analysis proposed to obtain physiological parameters using RGB video, computer vision, and ML modeling (Figure 2) to extract parameters used as inputs to predict RR and RH. It can be observed that the overall model presented a high correlation (R = 0.94) and slope close to the unity (0.92) with no signs of overfitting as the performance MSE value of the training stage (MSE = 72) was lower than the testing (MSE = 512). Furthermore, the overall model presented 9.7% of outliers (13 out of 134) based on the 95 confidence intervals (Figure 6).

Table 2. Results of the artificial neural networks regression model (Model 2) showing statistical data such as correlation coefficient (R), slope, and performance based on means squared error (MSE) for each stage.

Stage	Samples	Observations	R	Slope	Performance (MSE)
Training	94	188	0.98	0.94	72
Testing	40	80	0.84	0.86	512
Overall	134	268	0.94	0.92	-

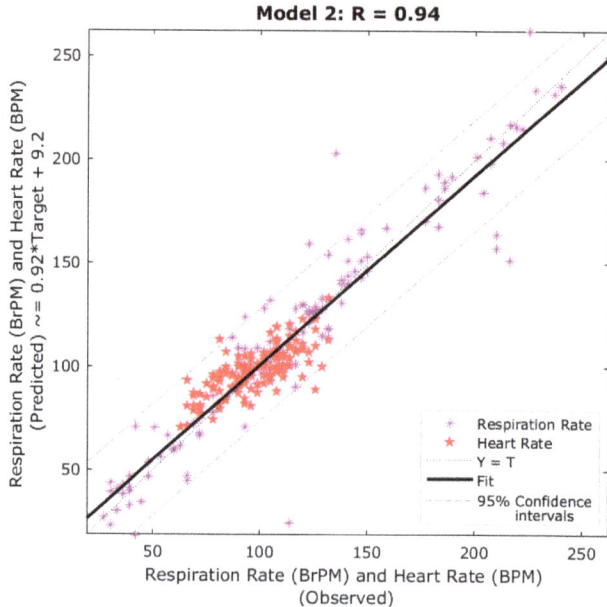

Figure 6. Artificial neural network overall fitting model showing the correlation coefficient (R), observed (x-axis), and predicted (y-axis) respiration rate and heart rate values. Abbreviations: BrPM: breaths per minute; BPM: beats per minute; T: Targets.

Figure 7 shows the PCA comparing the HR, RR, and skin temperature measured with manual techniques (HRreal, RRreal, and SkTreal) with those predicted using Model 2 (HRM2, and RRM2) and measured by computer vision algorithms (SkTcv), as well as the THI (Equation (1)) for sheep from both treatments (control and heat stress) in different days/times of measurements. The resulting PCA described a total of 82.92% of total data variability (PC1: 65.04%; PC2: 17.88%). It can be observed that the manual measurements and those assessed using the proposed methods were closely related. Furthermore, skin temperature was related to THI. As expected, there was a clear separation and clustering between the sheep physiology under control treatment (blue circles) compared to those under heat stress (red crosses), with the latter associated with higher RR, skin temperatures, and THI. PC1, which is the main responsible for the separation of the data according to the treatments, is more related to RR and skin temperature, with HR with lower variability related to PC2. The THI values obtained in this study ranged from 18 to 20 for control and between 27 and 36 for heat stress conditions (data not shown).

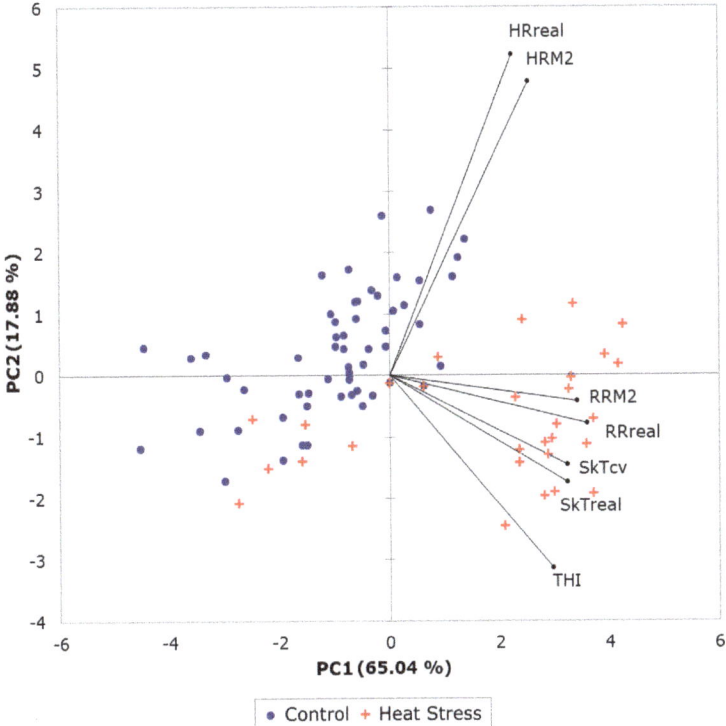

Figure 7. Principal components analysis of data measured with manual techniques (i) SkTreal: skin temperature real, (ii) HRreal: heart rate real, (iii) RRreal: respiration rate real, and those measured using computer vision analysis and predicted using the machine learning models (iv) SkTcv: skin temperature computer vision, (v) RRM2: respiration rate from Model 2, (vi) HRM2: heart rate from Model 2, and the temperature-humidity index (THI) from each day of the control and heat stress rooms.

4. Discussion

4.1. Selection of Critical Sheep ROIs, Features Tracking, and Automation

Due to constraints in the experimental chambers related to interference from the metabolic cages and sheep head movement through them, especially while feeding (Figure 8), it was not possible to use single co-registered ROIs for RGB videos and IRTV. The latter would have simplified the modeling procedure; however, results could have been only applicable to animals with visible features and with no obstructions through the whole video recordings, rather than those with obstructions like bars from the cage (Figure 8A,B). Hence, the methodology proposed has greater practical applications in penned and transported animals. By selecting the whole head of the sheep as ROI for IRTV analysis (Figure 8; red rectangles), with automated maximum temperature extraction, it gives a higher probability of extracting meaningful temperature information from the eyes, nose or mouth regions at any specific time from cropped videos even when the head moves across obstacles, such as bars from the cages (Figure 8). Furthermore, without obstructions and considering nose and mouth regions as ROI, a simplified skin temperature extraction could also have been easily implemented by signal analysis of peaks and valleys that represents mathematically the variability between temperatures related to inhalations and exhalations using a similar RVA analysis of the signal (Figure 2). However, considering obstacles (metabolism cage bars) and head movement through them (Figure 8), the temperature variability could have been biased and difficult to discriminate from those related to obstacles (Figure 8A), which would have rendered this potential simple procedure with higher errors in skin temperature estimation similar to Figure 4A. On the contrary, the method proposed in this research resulted in high accuracy compared to measured skin and rectal temperatures with closer relationship from the 1:1 line for skin temperatures as expected (Figure 3).

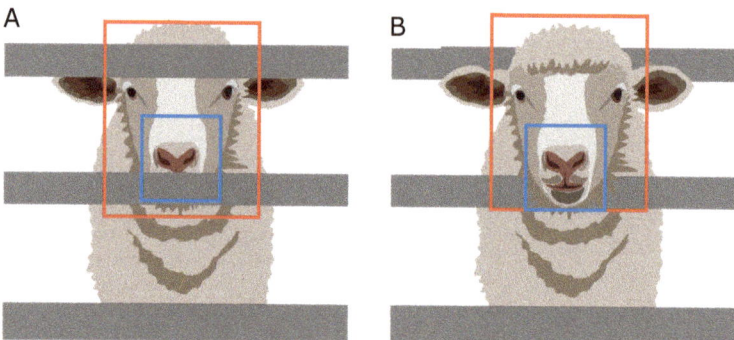

Figure 8. Examples of sheep with head and features tracked as region of interest (ROI) for the head (blue rectangles) for infrared thermal video (IRTV) analysis and nose/mouth regions (blue rectangles) for heart and respiration rates analysis from RGB video analysis behind bars (**A**) and through the bars (**B**) of cages.

In the case of HR and RR, ROIs were selected from the nose/mouth region of the animals since they have more hairless skin exposed (Figure 8; blue rectangles). The nose region is considered the best area to measure RR as it is where inhalation and exhalation occurs and the area in which a large number of blood vessels are found in sheep [43,44], which also allows measuring HR more accurately. Specifically, from these ROIs, changes in luminosity are related to the rushing in and out of the bloodstream and cooling down and warming up of skin surfaces, which can be related to HR and RR, respectively. This would happen obviously on surfaces of living and breathing organisms, making inanimate obstacle's luminosity unchanged and easy to discriminate either by signal analysis from RGB video or to be detected by ML modeling.

4.2. Computer Vision Analysis of Raw Signals Obtained from Videos

The sensitivity of the raw signal analysis extracted from computer vision algorithms related to raw HR and RR can be determined by the use of specific cutoff frequency ranges (Hz) as determined by the RVA and RVAm algorithms [38]. Since the range of RR observed for sheep under control and heated environments is very high (27–240 BrPM) compared to HR (63–132 BPM), the selection of cutoff frequency to extract RR raw data is critical. By selecting the whole range cutoff frequency range, results showed low sensitivity when compared to observed RR data (Figure 4A).

Accuracies for HR and respiration rates found through computer vision analysis with different cutoff frequencies for sheep to obtain raw HR and RR (Figure 4B) were in accordance to those using similar methodologies for cattle [31] and pigs [30]. In the case of HR, this study had a narrower range between 55 and 135 BPM, with the lower range consistent to the average HR reported for lambs without stress of 57 ± 5 [45].

For RR rate analysis, a review showed that by using cutoff frequencies between 0.20 and 0.40 Hz in the case of sheep and goats corresponded to RR of 12–24 BrPM. However, a breathing frequency study in ruminants recorded RR of 54 BrPM, equivalent to 0.9 Hz, which is within the reference range for adult sheep of 12–72 breaths per minute [46]. In this study, RR ranged from around 45–260 BrPM, which is consistent with ranges found in other sheep studies under normal and heat stress conditions, such as BrPM values between 31 and 247 BrPM [47]. Hence, higher RR corresponding to stressed sheep corresponded to three times higher than those associated with around 1 Hz frequency. By using maximum values of 3 Hz for higher RR values, resulting in more accurate raw RR obtained from computer vision algorithms (Figure 4B).

4.3. Machine Learning Modeling to Extract Further Sheep Biometrics

The ANN pattern recognition algorithm (Model 1) was able to pre-process the data to obtain these specific cutoff frequency ranges for RR analysis, which increased the accuracy and performance of Model 2 compared to lower accuracies and performances using a single frequency cutoff range for both HR and RR and computer vision analysis (Figure 4B). By using computer vision analysis for skin temperature extraction and ANN models 1 and 2 allowed full automation in the estimation of HR and RR from RGB video and IRTV. Furthermore, the integrated FLIR cameras used have direct connectivity drivers to be used within MATLAB® environments that can allow real-time extraction of sheep biometrics using the codes developed in this study as shown in Figure 1B, C. From Figure 2, the only supervised procedure is the initial ROIs selection for visible animals, making the rest of the process automatic through the pipeline of algorithms, functions and ML models proposed (Figure 2).

Pre-processing of videos using a wider cutoff frequency range for signal analysis of cropped videos from sheep, plus classification (Model 1) and re-analysis, it takes around 20 s for a 1-min video approximately, using parallel computing capabilities on a 4-core laptop PC. Hence, using higher computer capabilities, it was estimated for this time requirement to be between 3–5 s to allow signal stabilization and cut start to initiate real-time rendering of outputs as shown in Figure 1B, C. Specifically for HR in humans, the same pre-analysis periods can be found for commercial software, such as FaceReader (Noldus, Wageningen, The Netherlands) and the computer application Cardiio (Cardiio, Inc., Cambridge, MA, USA) for smartphones and tablet PCs.

4.4. Comparison between Non-Invasive Biometrics and Environmental Heat Stress Indices

In previous studies, sheep exhibit heat stress with THI $\geq$ 23 for Mediterranean dairy sheep and THI $\geq$ 27 in Comisana dairy sheep [48]. These THI values are consistent with the ranges for heat stress treatment applied in this study (THI: 26–36). Sheep regulate heat through panting mainly; hence, the RR is the main heat regulatory mechanism for these animals [47]. Very high RR values found for heat-stressed sheep (260 BrPM) were related to the highest THIs between 27 and 36 (Figure 7). From the same figure, vectors related to THI, skin temperature, and RR (for both observed and extracted through biometrics) were related as expected, which help to maintain a relatively constant maximum HR showing heat stress regulation. Higher THI is related to higher panting from sheep, which helps reducing skin and the internal temperature of animals. Some sheep from the heat stress treatments appear in the PCA graph close to the control cluster, which may correspond to more genetically resilient sheep to heat stress [49,50].

4.5. Artificial Intelligence System Proposed Based on Algorithms and Models Developed

Automatic ROI selection from sheep can be achieved through the training of deep learning algorithms to recognize specific sheep features, such as those for the head and nose/mouth regions. This has been achieved using convolutional networks for pigs [51], wildlife animals [52], marine animals [53], and chimpanzee faces [54], among others. Other methodologies based on ML, such as discriminant analysis, independent components Analysis, and ANN, among others, have been used for animal detection, classification, and tracking [55–59].

Previously, an AI system to reduce heat stress and increase milk production and quality has been proposed for dairy farms based on the analysis of THI data and cow management information through ML [24]. From the latter study, automatic drafting doors are controlled from ML outputs that transport the cows to the milking area or to a sprinkler-based system to reduce heat stress. A similar method can be implemented here using the algorithms and models developed in this study (Figure 9A). The system proposed has the advantage that there are no physical obstacles between the camera/analysis hub and the individual sheep monitored. In open environments, this AI system could be coupled with virtual fencing systems through collars and the internet of things (IoT) to automate the separation of heat-stressed sheep towards a sprinkler cooling system (Figure 9A). Virtual fencing has been successfully applied for automated cattle control systems [60,61], for sheep [62,63], with acceptable ethical frameworks assessed for their implementation [62,64].

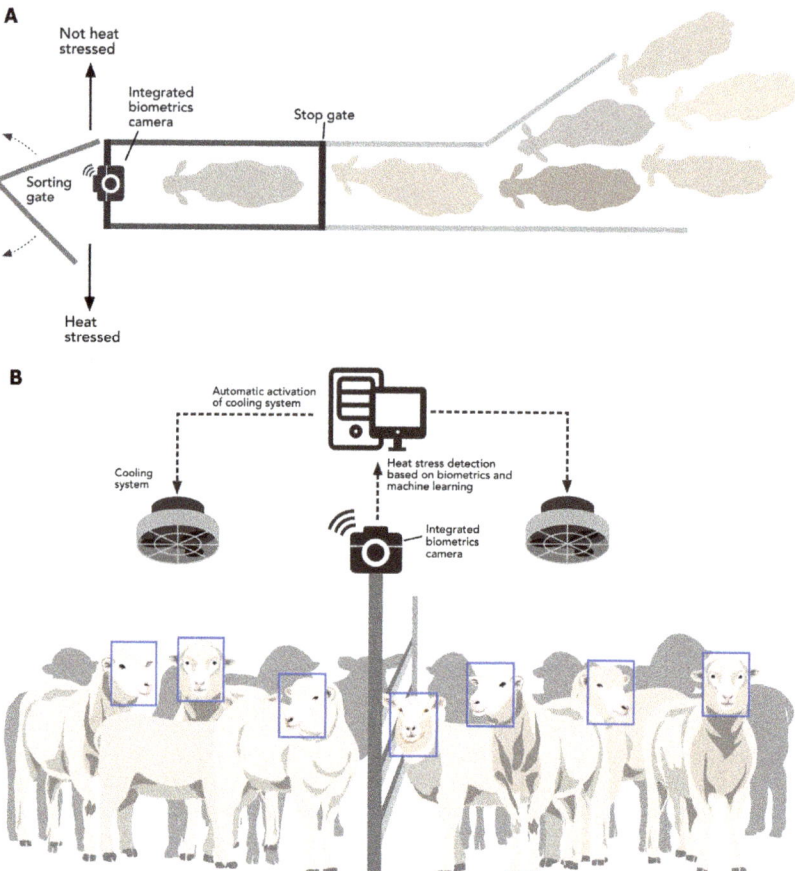

Figure 9. Representation of potential artificial intelligence applications implementing models developed in this study for (**A**) in farm detection of heat-stressed sheep and isolation towards a cooling or shaded area from non-stressed sheep by synchronization with an automated sorting gate, and (**B**) detection of heat stress for sheep in transport coupled with cooling fan systems.

The AI systems proposed can also be implemented in confined sheep in the preparation or during transport (Figure 9B), which could be coupled to blockchain [32,33] to have an unbiased control and independent assessment of animal welfare to be applied in the farm and transport or vessel environments.

5. Conclusions

This study proposed the implementation of automated computer vision algorithms and machine learning models to obtain critical biometrics from recorded RGB, and infrared thermal videos from sheep, to help in the automated assessment of heat stress. The implementation of the proposed system requires affordable hardware capabilities, such as the FLIR integrated cameras, which can include dedicated AI micro-processors and blockchain technology. The user-friendly AI system proposed would be able to analyze non-invasive biometrics from sheep in the farm automatically, and through their transport to secure animal welfare, through independent analysis of information incorporating blockchain technology for control purposes. Advances proposed in this paper could offer an AI-based system to monitor animal welfare in farms, and also as a tool to assess animal welfare in transport by land or sea independently, using blockchain. The latter not only could serve governments to audit live animal exports but also the industry in general, for more transparency to the public in their treatment of living animals for human consumption.

Author Contributions: Conceptualization, S.F., and C.G.V.; data curation, C.G.V.; formal analysis, C.G.V.; funding acquisition, S.S.C.; investigation, S.F., C.G.V., S.S.C., A.J., and F.R.D.; methodology, S.F., C.G.V., and A.J.; project administration, S.S.C.; resources, S.S.C., and F.R.D.; software, S.F., and C.G.V.; supervision, S.F., C.G.V., and F.R.D.; validation, S.F., and C.G.V.; visualization, S.F., C.G.V., and E.T.; writing—original draft, S.F., and C.G.V; writing—review and editing, S.F., C.G.V., S.S.C., A.J., E.T., and F.R.D. All authors have read and agreed to the published version of the manuscript.

Funding: This study was partially funded by The University of Melbourne Early Career Researcher Grant 2019.

Conflicts of Interest: The authors declare no conflict of interest.

References

1. Rice, M.; Hemsworth, L.M.; Hemsworth, P.H.; Coleman, G.J. The Impact of a Negative Media Event on Public Attitudes towards Animal Welfare in the Red Meat Industry. *Animals* **2020**, *10*, 619. [CrossRef] [PubMed]
2. Norman, G. Available online: https://www.mla.com.au/research-and-development/search-rd-reports/final-report-details/National-livestock-export-industry-sheep-cattle-and-goat-transportperformance-report2017/3852#:~{}:text=National%20livestock%20export%20industry%20sheep%2C%20cattle%20and%20goat%20transport%20performance%20report%202017&text=The%20overall%20mortality%20rate%20for,of%200.80%25%20observed%20in%202016 (accessed on 11 July 2020).
3. Davey, A.; Fisher, R. Available online: https://www.agriculture.gov.au/sites/default/files/documents/draft-ris-animals-australia-attachment-pegasus-report.pdf (accessed on 11 July 2020).
4. Das, R.; Sailo, L.; Verma, N.; Bharti, P.; Saikia, J. Impact of heat stress on health and performance of dairy animals: A review. *Vet. World* **2016**, *9*, 260. [CrossRef] [PubMed]
5. Rojas-Downing, M.M.; Nejadhashemi, A.P.; Harrigan, T.; Woznicki, S.A. Climate change and livestock: Impacts, adaptation, and mitigation. *Clim. Risk Manag.* **2017**, *16*, 145–163. [CrossRef]
6. Lacetera, N. Impact of climate change on animal health and welfare. *Anim. Front.* **2019**, *9*, 26–31. [CrossRef] [PubMed]
7. Gonzalez-Rivas, P.A.; Chauhan, S.S.; Ha, M.; Fegan, N.; Dunshea, F.R.; Warner, R.D. Effects of heat stress on animal physiology, metabolism, and meat quality: A review. *Meat Sci.* **2020**, *162*, 108025. [CrossRef] [PubMed]
8. Mallard, B.A.; Husseini, N.; Cartwright, S.; Livernois, A.; Hodgins, D.; Altvater-Hughes, T.; Beard, S.; Karrow, N.; Canovas, A.; Schmied, J. Resilience of High Immune Response (HIR) Genetics in the Context of Climate Change: Effects of Heat Stress on Cattle with Diverse Immune Response Genotypes. *J. Immunol.* **2020**, *204* (Suppl. 1), 92.4.
9. Mayorga, E.J.; Renaudeau, D.; Ramirez, B.C.; Ross, J.W.; Baumgard, L.H. Heat stress adaptations in pigs. *Anim. Front.* **2019**, *9*, 54–61. [CrossRef]
10. Joy, A.; Dunshea, F.R.; Leury, B.J.; Clarke, I.J.; DiGiacomo, K.; Chauhan, S.S. Resilience of Small Ruminants to Climate Change and Increased Environmental Temperature: A Review. *Animals* **2020**, *10*, 867. [CrossRef] [PubMed]
11. Osei-Amponsah, R.; Chauhan, S.S.; Leury, B.J.; Cheng, L.; Cullen, B.; Clarke, I.J.; Dunshea, F.R. Genetic selection for thermotolerance in ruminants. *Animals* **2019**, *9*, 948. [CrossRef]
12. Rashamol, V.P.; Sejian, V.; Bagath, M.; Krishnan, G.; Archana, P.R.; Bhatta, R. Physiological adaptability of livestock to heat stress: An updated review. *J. Anim. Behav. Biometeorol.* **2018**, *6*. [CrossRef]
13. Henry, B.; Eckard, R.; Beauchemin, K. Adaptation of ruminant livestock production systems to climate changes. *Animal* **2018**, *12*, s445–s456. [CrossRef] [PubMed]

14. Collier, R.J.; Baumgard, L.H.; Zimbelman, R.B.; Xiao, Y. Heat stress: Physiology of acclimation and adaptation. *Anim. Front.* **2019**, *9*, 12–19. [CrossRef]
15. Berman, A.; Horovitz, T.; Kaim, M.; Gacitua, H. A comparison of THI indices leads to a sensible heat-based heat stress index for shaded cattle that aligns temperature and humidity stress. *Int. J. Biometeorol.* **2016**, *60*, 1453–1462. [CrossRef]
16. Ekine-Dzivenu, C.; Mrode, R.A.; Ojango, J.M.; Okeyo Mwai, A. Evaluating the impact of heat stress as measured by temperature-humidity index (THI) on test-day milk yield of dairy cattle in Tanzania. *Livestock Sci.* **2019**, 104314.
17. Collier, R.J.; Renquist, B.J.; Xiao, Y. A 100-Year Review: Stress physiology including heat stress. *J. Dairy Sci.* **2017**, *100*, 10367–10380. [CrossRef]
18. Osei-Amponsah, R.; Dunshea, F.R.; Leury, B.J.; Cheng, L.; Cullen, B.; Joy, A.; Abhijith, A.; Zhang, M.H.; Chauhan, S.S. Heat Stress Impacts on Lactating Cows Grazing Australian Summer Pastures on an Automatic Robotic Dairy. *Animals* **2020**, *10*, 869. [CrossRef]
19. Peters, R. Sensor Based Measurements of Maximum Day Temperature Effects on Eating-, Ruminating-, Lying-, Inactive Time and Number of Steps in 6 Dutch Dairy Farms. Master's Thesis, Utrecht University, Utrecht, The Netherlands, 2019.
20. Belhadj Slimen, I.; Najar, T.; Ghram, A.; Abdrrabba, M. Heat stress effects on livestock: Molecular, cellular and metabolic aspects, a review. *J. Anim. Physiol. Anim. Nutr.* **2016**, *100*, 401–412. [CrossRef]
21. Sejian, V.; Bhatta, R.; Gaughan, J.; Dunshea, F.; Lacetera, N. Adaptation of animals to heat stress. *Animal* **2018**, *12*, s431–s444. [CrossRef]
22. Pragna, P.; Sejian, V.; Bagath, M.; Krishnan, G.; Archana, P.; Soren, N.; Beena, V.; Bhatta, R. Comparative assessment of growth performance of three different indigenous goat breeds exposed to summer heat stress. *J. Anim. Physiol. Anim. Nutr.* **2018**, *102*, 825–836. [CrossRef]
23. Jorquera-Chavez, M.; Fuentes, S.; Dunshea, F.R.; Jongman, E.C.; Warner, R.D. Computer vision and remote sensing to assess physiological responses of cattle to pre-slaughter stress, and its impact on beef quality: A review. *Meat Sci.* **2019**, *156*, 11–22. [CrossRef]
24. Fuentes, S.; Gonzalez Viejo, C.; Cullen, B.; Tongson, E.; Chauhan, S.S.; Dunshea, F.R. Artificial Intelligence Applied to a Robotic Dairy Farm to Model Milk Productivity and Quality based on Cow Data and Daily Environmental Parameters. *Sensors* **2020**, *20*, 2975. [CrossRef]
25. Gorczyca, M.T.; Milan, H.F.M.; Maia, A.S.C.; Gebremedhin, K.G. Machine learning algorithms to predict core, skin, and hair-coat temperatures of piglets. *Comput. Electron. Agric.* **2018**, *151*, 286–294. [CrossRef]
26. Ramesh, V. A Review on Application of Deep Learning in Thermography. *Int. J. Eng. Manag. Res.* **2017**, *7*, 489–493.
27. Laloë, D. Available online: http://dataia.eu/sites/default/files/Outils%20com/Livestock%20and%20AI%20-%20Denis%20Laloe.pdf (accessed on 13 July 2020).
28. Hoffmann, G.; Herbut, P.; Pinto, S.; Heinicke, J.; Kuhla, B.; Amon, T. Animal-related, non-invasive indicators for determining heat stress in dairy cows. *Biosyst. Eng.* **2019**. [CrossRef]
29. Barbosa Pereira, C.; Dohmeier, H.; Kunczik, J.; Hochhausen, N.; Tolba, R.; Czaplik, M. Contactless monitoring of heart and respiratory rate in anesthetized pigs using infrared thermography. *PLoS ONE* **2019**, *14*, e0224747. [CrossRef]
30. Jorquera-Chavez, M.; Fuentes, S.; Dunshea, F.R.; Warner, R.D.; Poblete, T.; Morrison, R.S.; Jongman, E.C. Remotely Sensed Imagery for Early Detection of Respiratory Disease in Pigs: A Pilot Study. *Animals* **2020**, *10*, 451. [CrossRef]
31. Jorquera-Chavez, M.; Fuentes, S.; Dunshea, F.R.; Warner, R.D.; Poblete, T.; Jongman, E.C. Modelling and Validation of Computer Vision Techniques to Assess Heart Rate, Eye Temperature, Ear-Base Temperature and Respiration Rate in Cattle. *Animals* **2019**, *9*, 1089. [CrossRef]
32. Subramanian, N.; Chaudhuri, A.; Kayikci, Y. Blockchain Applications in Food Supply Chain. In *Blockchain and Supply Chain Logistics*; Palgrave Pivot: Cham, Switzerland, 2020; pp. 21–29.
33. Kamilaris, A.; Fonts, A.; Prenafeta-Boldú, F.X. The rise of blockchain technology in agriculture and food supply chains. *Trends Food Sci. Technol.* **2019**, *91*, 640–652. [CrossRef]
34. National Research Council. *Nutrient Requirements of Small Ruminants: Sheep, Goats, Cervids, and New World Camelids*; The National Academies Press: Washington, DC, USA, 2007.

35. Marai, I.; El-Darawany, A.; Fadiel, A.; Abdel-Hafez, M. Physiological traits as affected by heat stress in sheep—A review. *Small Rumin. Res.* **2007**, *71*, 1–12. [CrossRef]
36. Aubakir, B.; Nurimbetov, B.; Tursynbek, I.; Varol, H.A. Vital sign monitoring utilizing Eulerian video magnification and thermography. In Proceedings of the 38th Annual International Conference of the IEEE Engineering in Medicine and Biology Society (EMBC), Orlando, FL, USA, 16–20 August 2016; pp. 3527–3530.
37. Alghoul, K.; Alharthi, S.; Al Osman, H.; El Saddik, A. Heart rate variability extraction from videos signals: ICA vs. EVM comparison. *IEEE Access* **2017**, *5*, 4711–4719. [CrossRef]
38. Gonzalez Viejo, C.; Fuentes, S.; Torrico, D.D.; Dunshea, F.R. Non-contact heart rate and blood pressure estimations from video analysis and machine learning modelling applied to food sensory responses: A case study for chocolate. *Sensors* **2018**, *18*, 1802. [CrossRef]
39. Gonzalez Viejo, C.; Torrico, D.; Dunshea, F.; Fuentes, S. Development of Artificial Neural Network Models to Assess Beer Acceptability Based on Sensory Properties Using a Robotic Pourer: A Comparative Model Approach to Achieve an Artificial Intelligence System. *Beverages* **2019**, *5*, 33. [CrossRef]
40. Amini, M.; Abbaspour, K.C.; Khademi, H.; Fathianpour, N.; Afyuni, M.; Schulin, R. Neural network models to predict cation exchange capacity in arid regions of Iran. *Eur. J. Soil Sci.* **2005**, *56*, 551–559. [CrossRef]
41. Markopoulos, A.P.; Georgiopoulos, S.; Manolakos, D.E. On the use of back propagation and radial basis function neural networks in surface roughness prediction. *J. Ind. Eng. Int.* **2016**, *12*, 389–400. [CrossRef]
42. Deep, K.; Jain, M.; Salhi, S. *Logistics, Supply Chain and Financial Predictive Analytics: Theory and Practices*; Springer: Singapore, 2019.
43. Dawes, J.; Prichard, M.M. Studies of the vascular arrangements of the nose. *J. Anat.* **1953**, *87*, 311.
44. Mitchell, J.; Thomalla, L.; Mitchell, G. Histological studies of the dorsal nasal, angularis oculi, and facial veins of sheep (Ovis aries). *J. Morphol.* **1998**, *237*, 275–281. [CrossRef]
45. Mir, S.; Nazki, A.; Raina, R. Comparative electrocardiographic studies, and differing effects of pentazocine on ECG, heart and respiratory rates in young sheep and goats. *Small Rumin. Res.* **2000**, *37*, 13–17. [CrossRef]
46. Konold, T.; Bone, G.E. Heart rate variability analysis in sheep affected by transmissible spongiform encephalopathies. *BMC Res. Notes* **2011**, *4*, 539. [CrossRef] [PubMed]
47. Lees, A.; Sullivan, M.; Olm, J.; Cawdell-Smith, A.; Gaughan, J. A panting score index for sheep. *Int. J. Biometeorol.* **2019**, *63*, 973–978. [CrossRef]
48. Sejian, V.; Bhatta, R.; Gaughan, J.; Malik, P.K.; Naqvi, S.; Lal, R. *Sheep Production Adapting to Climate Change*; Springer: Singapore, 2017.
49. Sejian, V.; Bagath, M.; Krishnan, G.; Rashamol, V.; Pragna, P.; Devaraj, C.; Bhatta, R. Genes for resilience to heat stress in small ruminants: A review. *Small Rumin. Res.* **2019**, *173*, 42–53. [CrossRef]
50. Mengistu, U.; Puchala, R.; Sahlu, T.; Gipson, T.; Dawson, L.; Goetsch, A. Conditions to evaluate differences among individual sheep and goats in resilience to high heat load index. *Small Rumin. Res.* **2017**, *147*, 89–95. [CrossRef]
51. Psota, E.T.; Mittek, M.; Pérez, L.C.; Schmidt, T.; Mote, B. Multi-pig part detection and association with a fully-convolutional network. *Sensors* **2019**, *19*, 852. [CrossRef] [PubMed]
52. Nguyen, H.; Maclagan, S.J.; Nguyen, T.D.; Nguyen, T.; Flemons, P.; Andrews, K.; Ritchie, E.G.; Phung, D. Animal recognition and identification with deep convolutional neural networks for automated wildlife monitoring. In Proceedings of the IEEE international conference on data science and advanced Analytics (DSAA), Tokyo, Japan, 19–21 October 2017; pp. 40–49.
53. Zhuang, P.; Xing, L.; Liu, Y.; Guo, S.; Qiao, Y. Marine Animal Detection and Recognition with Advanced Deep Learning Models. In Proceedings of the Working Notes of CLEF 2017—Conference and Labs of the Evaluation Forum, Dublin, Ireland, 11–14 September 2017.
54. Schofield, D.; Nagrani, A.; Zisserman, A.; Hayashi, M.; Matsuzawa, T.; Biro, D.; Carvalho, S. Chimpanzee face recognition from videos in the wild using deep learning. *Sci. Adv.* **2019**, *5*, eaaw0736. [CrossRef] [PubMed]
55. Tydén, A.; Olsson, S. Edge Machine Learning for Animal Detection, Classification, and Tracking. Master's Thesis, Automatic Control. Department of Electrical Engineering, Linköping University, Linköping, Sweden, 2020; p. 49.
56. Hossein, S.; Zargham, H.B.; Hamid, D.C. A new face recognition method using PCA, LDA and neural network. *Int. J. Comput. Sci. Eng.* **2008**, *2*, 218–223.
57. Corkery, G.; Gonzales-Barron, U.A.; Butler, F.; Mc Donnell, K.; Ward, S. A preliminary investigation on face recognition as a biometric identifier of sheep. *Trans. Asabe* **2007**, *50*, 313–320. [CrossRef]

58. Kumar, S.; Singh, S.K. Biometric recognition for pet animal. *J. Softw. Eng. Appl.* **2014**, *2014*, 1945–3124. [CrossRef]
59. Noviyanto, A.; Arymurthy, A.M. Automatic cattle identification based on muzzle photo using speed-up robust features approach. In Proceedings of the 3rd European Conference of Computer Science, Heraklion, Crete, Greece, 13–17 September 2007.
60. Bishop-Hurley, G.; Swain, D.L.; Anderson, D.; Sikka, P.; Crossman, C.; Corke, P. Virtual fencing applications: Implementing and testing an automated cattle control system. *Comput. Electron. Agric.* **2007**, *56*, 14–22. [CrossRef]
61. Campbell, D.L.; Ouzman, J.; Mowat, D.; Lea, J.M.; Lee, C.; Llewellyn, R.S. Virtual Fencing Technology Excludes Beef Cattle from an Environmentally Sensitive Area. *Animals* **2020**, *10*, 1069. [CrossRef]
62. Marini, D.; Meuleman, M.D.; Belson, S.; Rodenburg, T.B.; Llewellyn, R.; Lee, C. Developing an ethically acceptable virtual fencing system for sheep. *Animals* **2018**, *8*, 33. [CrossRef]
63. Brunberg, E.; Sørheim, K.; Bergslid, I.R.K. The ability of ewes with lambs to learn a virtual fencing system. *Animal* **2017**, *2017*, 1–6. [CrossRef]
64. Lee, C.; Colditz, I.G.; Campbell, D.L. A framework to assess the impact of new animal management technologies on welfare: A case study of virtual fencing. *Front. Vet. Sci.* **2018**, *5*, 187. [CrossRef] [PubMed]

Publisher's Note: MDPI stays neutral with regard to jurisdictional claims in published maps and institutional affiliations.

© 2020 by the authors. Licensee MDPI, Basel, Switzerland. This article is an open access article distributed under the terms and conditions of the Creative Commons Attribution (CC BY) license (http://creativecommons.org/licenses/by/4.0/).

Article

Assessment of Smoke Contamination in Grapevine Berries and Taint in Wines Due to Bushfires Using a Low-Cost E-Nose and an Artificial Intelligence Approach

Sigfredo Fuentes [1], Vasiliki Summerson [1], Claudia Gonzalez Viejo [1,*], Eden Tongson [1], Nir Lipovetzky [2], Kerry L. Wilkinson [3,4], Colleen Szeto [3,4] and Ranjith R. Unnithan [5]

1. Digital Agriculture, Food and Wine Sciences Group, School of Agriculture and Food, Faculty of Veterinary and Agricultural Sciences, The University of Melbourne, Parkville, VIC 3010, Australia; sfuentes@unimelb.edu.au (S.F.); vsummerson@student.unimelb.edu.au (V.S.); eden.tongson@unimelb.edu.au (E.T.)
2. School of Computing and Information Systems, Melbourne School of Engineering, The University of Melbourne, Parkville, VIC 3010, Australia; nir.lipovetzky@unimelb.edu.au
3. School of Agriculture, Food and Wine, The University of Adelaide, Waite Campus, PMB 1, Glen Osmond, SA 5064, Australia; kerry.wilkinson@adelaide.edu.au (K.L.W.); colleen.szeto@adelaide.edu.au (C.S.)
4. The Australian Research Council Training Centre for Innovative Wine Production, PMB 1, Glen Osmond, SA 5064, Australia
5. School of Engineering, Department of Electrical and Electronic Engineering, The University of Melbourne, Parkville, VIC 3010, Australia; r.ranjith@unimelb.edu.au
* Correspondence: cgonzalez2@unimelb.edu.au; Tel.: +61-412055704

Received: 24 August 2020; Accepted: 4 September 2020; Published: 8 September 2020

Abstract: Bushfires are increasing in number and intensity due to climate change. A newly developed low-cost electronic nose (e-nose) was tested on wines made from grapevines exposed to smoke in field trials. E-nose readings were obtained from wines from five experimental treatments: (i) low-density smoke exposure (LS), (ii) high-density smoke exposure (HS), (iii) high-density smoke exposure with in-canopy misting (HSM), and two controls: (iv) control (C; no smoke treatment) and (v) control with in-canopy misting (CM; no smoke treatment). These e-nose readings were used as inputs for machine learning algorithms to obtain a classification model, with treatments as targets and seven neurons, with 97% accuracy in the classification of 300 samples into treatments as targets (Model 1). Models 2 to 4 used 10 neurons, with 20 glycoconjugates and 10 volatile phenols as targets, measured: in berries one hour after smoke (Model 2; R = 0.98; R^2 = 0.95; b = 0.97); in berries at harvest (Model 3; R = 0.99; R^2 = 0.97; b = 0.96); in wines (Model 4; R = 0.99; R^2 = 0.98; b = 0.98). Model 5 was based on the intensity of 12 wine descriptors determined via a consumer sensory test (Model 5; R = 0.98; R^2 = 0.96; b = 0.97). These models could be used by winemakers to assess near real-time smoke contamination levels and to implement amelioration strategies to minimize smoke taint in wines following bushfires.

Keywords: climate change; machine learning; electronic nose; smoke taint; wine sensory

---

1. Introduction

When bushfires occur within the grape growing season, vineyards can be affected at critical stages (véraison to harvest) [1], which could result in different levels of smoke contamination in berries and smoke taint in wines [2,3]. The intensity, number, and severity of bushfires are increasing due to climate change as well as the window of opportunity [4].

The growing concerns in Australia regarding bushfire scale and frequency are shared by wine regions around the world, including the USA, Canada, South Africa, Portugal, Chile, and others [5]. To assess the potential risk of smoke taint, the industry typically relies on the analysis of grape samples by commercial laboratories to quantify smoke taint marker compounds (i.e., volatile phenols and their glycoconjugates), but this can be prohibitively expensive for some producers [6,7]. Alternatively, grapes can be harvested and vinified so that sensory analysis can be conducted in-house. However, depending on the timing of smoke exposure, these approaches may not inform decision-making within the time-constraints of vintage.

To date, there has been little research into the use of affordable in-field technology to assess grapevine smoke contamination. Recently, the authors' group published a study evaluating short-range remote sensing in the thermal and near-infrared spectrum, combined with machine learning, as a novel approach to assessing smoke contamination in grapevine leaves, berries, and wines, with high levels of accuracy [5]. These tools may support rapid decision-making, enabling the implementation of management strategies that reduce the risk of contamination carrying over into wine, as smoke taint.

Electronic noses (e-nose) are comprised of an array of metal oxide semiconductor sensors (MOS) sensitive to different gases that can measure a variety of volatiles in the environment [8]. Early developments of e-noses involve arrays of 5–8 tin-oxide type of MOS sensors, requiring the use of sealed chambers and/or a complete setup of different devices to heat the sample and obtain headspace to be injected in the e-nose chamber, which has made the e-noses non-portable as they require a laboratory setup [9,10]. Some studies have explored different signal extraction methods, such as the Lorentzian model, which has resulted in a powerful and rapid-response technique [11]. Ayhan et al. [12] explored the fluctuation-enhanced sensing method to detect and classify gases with improved accuracy when developing classification models using machine learning algorithms. Some applications include medical diagnostics [13], space shuttles and stations [14–16], crime and security [17], and food and beverages, such as rapeseed to detect volatile compounds in pressed oil [18], wine [19], and beer [20], among others. The latter study describes a low-cost e-nose developed with nine gas sensors to assess the aroma profile of beers coupled with machine learning modeling. Examples of the implementation of e-noses for food science can be found from early literature reviews [21] through the implementation of disease diagnostics [22], more recent applications to assess food quality [23], meat quality assessment [24], for food control [23], assessment of food authentication and adulteration [25], and for the wine industry [26–30]. However, the e-noses used in the past range in complexity, accessibility to users, and cost.

Low-cost e-noses can be used in the field to assess smoke contamination levels coupled with the internet of things (IoT) for data transmission and analysis from different locations or nodes within vineyards. However, a more efficient approach could be to mount e-noses to assess gases in different parts of vineyards and to generate geo-referenced maps of these gases on unmanned terrestrial vehicles (UTV), robots [31], or unmanned aerial vehicles (UAV) [32]. The levels of smoke-related contaminants could be modeled using machine learning algorithms to infer the levels of contaminants in berries, and therefore, the risk of smoke taint in the final wine. However, they could not be used to directly "sniff" these contaminants from bunches since smoke-derived volatile compounds are rapidly metabolized in berries, leading to the formation of glycoconjugates, which are odorless [2,5–7,33–35].

This study evaluated the potential for low-cost e-noses to be used to assess wines made from grapes exposed to different levels (densities) of smoke. The e-nose measurements were used as inputs in machine learning modeling strategies, and the concentrations of smoke taint marker compounds in berries and wines used as targets. Further, targets were obtained from a sensory analysis trial, during which consumers assessed the wines made from each treatment. In total, five machine learning models were created based on e-nose data to assess (i) the level of contamination in grapevines related to smoke exposure from wine samples using classification models (Model 1); (ii) to evaluate smoke-related compounds from wines, such as 20 glycoconjugates and 10 volatile phenols in berries after 1 h smoke (Model 2), (iii) smoke-related compounds in berries measured at harvest (Model 3), (iv) for wines made from treatments (Model 4), and (v) consumer sensory analysis using 12 wine descriptors (Model 5; Figure 1). The models obtained were of high accuracy, which could allow the implementation of this artificial intelligence (AI) technology in the winemaking process to assess the effect of ameliorating management techniques in the field (Model 1) through micro-vinifications, to assess the best timing for skin contact during fermentation for red wines, the addition of activated carbon to adsorb smoke-related compounds, wine filtration using membranes, reverse osmosis, and other commercial fining agents, among others [34,35].

Not only could the implementation of this technique help winemakers evaluate the different amelioration techniques mentioned above, but it could also monitor almost real-time changes in the aroma profiles of wine and assess which technology could best maintain a certain quality or style target.

This paper described how the e-nose was implemented for the different treatments and wine samples used and the specific machine learning algorithms used to develop five machine learning models with their respective analyses for accuracy and performance. A discussion on potential applications of the e-nose and models was also described for the wine industry to monitor and reduce smoke taint in wines.

2. Materials and Methods

2.1. Description of Treatments and Wine Samples

Field trials involving the application of smoke and/or in-canopy misting to Cabernet Sauvignon grapevines have been reported previously [3]. Briefly, three different smoke treatments were applied to vines (at approximately 7 days post-véraison): (i) low-density smoke exposure (LS), (ii) high-density smoke exposure (HS), and (iii) high-density smoke exposure, with in-canopy misting (HSM). Two controls were also included: (iv) a control without misting (C; no smoke treatment) and (v) a control with misting (CM; no smoke treatment). Treatments were applied to six adjacent vines, except for HSM, which was applied to five adjacent vines (i.e., one vine was missing). Smoke treatments involved exposure of grapevines to straw-derived smoke using a purpose-built tent for 1 h. At least one buffer vine separated treatments. The wine was subsequently produced on a small scale (i.e., ~5 kg per fermentation, performed in triplicate for each treatment), as described previously [3].

2.2. Electronic Nose

Wine samples were measured (in triplicate) using a portable, user-friendly, and low-cost e-nose, comprising nine different sensors, which were sensitive to different gases, as mentioned in Table 1, plus a humidity and temperature sensor (AM2320; Guangzhou Aosong Electronics Co., Ltd., Guangzhou, China). Sensor details have already been reported [20]. A total of 100 mL of wine was poured into a 500 mL beaker, and the e-nose was placed on top of the container for 1 min to capture the gases present in the sample. The e-nose was calibrated for 20–30 s before and after measuring each sample to reset the readings to baseline. Values from all sensors were automatically recorded in a comma-separated values (.csv) file to facilitate analysis.

Table 1. Sensors, attached to the electronic nose, and the gasses they are sensitive to.

Sensor Name	Gases	Manufacturer
MQ3	Ethanol	
MQ4	Methane	
MQ7	Carbon monoxide (CO)	
MQ8	Hydrogen	Henan Hanwei Electronics Co., Ltd., Henan, China
MQ135	Ammonia, alcohol, and benzene	
MQ136	Hydrogen sulfide	
MQ137	Ammonia	
MQ138	Benzene, alcohol, and ammonia	
MG811	Carbon dioxide (CO_2)	

2.3. Chemical Analysis of Glycoconjugates and Volatile Phenols

Volatile phenols (Table 2) were evaluated in wine samples using stable isotope dilution analysis (SIDA) methods, as previously described [15,17–19]. Isotopically labeled standards of d_3-guaiacol, d_3-4-methylguaiacol, d_7-o-cresol, and d_3-syringol were prepared in house by the Australian Wine Research Institute's (AWRI) Commercial Services Laboratory (Adelaide, Australia) using published methods [15,17,18]. Measurements were performed using an Agilent 6890 gas chromatography coupled to a 5973 mass-spectrometer (Agilent Technologies, Forest Hill, VIC, Australia). The limit of quantitation for volatile phenols was 1–2 µg L^{-1}.

A range of volatile phenol glycoconjugates (Table 2) was measured using high-performance liquid chromatography-tandem mass spectrometry (HPLC-MS/MS) according to stable isotope dilution analysis (SIDA) methods previously described [18,20]. The analysis was performed using an Agilent 1200 high-performance liquid chromatography (HPLC) equipped with a 1290 binary pump, coupled to an AB SCIEX Triple QuadTM 4500 tandem mass spectrometer, with a Turbo VTM ion source (Framingham, MA, USA). The preparation of the isotopically labeled internal standard d_3-syringol gentiobioside has been previously reported [18,20]. The limit of quantitation for volatile phenol glycosides was 1 µg kg^{-1}.

Table 2. List of glycoconjugates and volatile phenols, their abbreviation, and the sample in which they were measured.

Compound	Abbreviation/Label	Sample
Glycoconjugates		
Syringol gentiobiosides	SyGG	Berries/Wine
Syringol glucosides	SyMG	Berries/Wine
Syringol pentosylglucosides	SyPG	Berries/Wine
Cresol glucosylpentosides	CrPG	Berries/Wine
Cresol gentiobioside	CrGG	Berries
Cresol glucosides	CrMG	Berries
Cresol rutinosides	CrRG	Berries/Wine
Guaiacol pentosylglucosides	GuPG	Berries/Wine
Guaiacol gentiobiosides	GuGG	Berries/Wine
Guaiacol rutinosides	GuRG	Berries/Wine
Guaiacol glucosides	GuMG	Berries/Wine
Methylguaiacol pentosylglucosides	MGuPG	Berries/Wine
Methylguaiacol rutinosides	MGuRG	Berries/Wine
Methylguaiacol glucosides	MGuMG	Berries
Methylsyringol gentiobiosides	MSyGG	Berries/Wine
Methylsyringol pentosylglucosides	MSyPG	Berries/Wine
Phenol rutinosides	PhRG	Berries/Wine
Phenol gentiobiosides	PhGG	Berries/Wine
Phenol pentosylglucosides	PhPG	Berries/Wine
Phenol glucosides	PhMG	Berries/Wine

Table 2. Cont.

Compound	Abbreviation/Label	Sample
	Volatile Phenols	
Guaiacol	Guaiacol	Berries/Wine
4-Methylguaiacol	4-Methylguaiacol	Berries/Wine
Phenol	Phenol	Berries
o-Cresol	o-Cresol	Berries/Wine
Total m/p-cresols	Total m/p-cresol	Berries
m-Cresol	m-Cresol	Berries/Wine
p-Cresol	p-Cresol	Berries/Wine
Syringol	Syringol	Berries/Wine
4-Methylsyringol	4-Methylsyringol	Berries/Wine
Total cresols	Cresols	Berries

2.4. Sensory Evaluation-Consumer Test

A consumer test was conducted with participants (N = 31; age range: 21–59 years; 77% female and 23% male) constituted of staff and students from The University of Melbourne (UoM; Ethics ID: 1545786.2) that had been recruited via e-mail. According to the power analysis conducted using the SAS® Power and Sample Size v. 14.1 software (SAS Institute Inc., Cary, NC, USA), the number of participants was enough to find significant differences between samples (power: $1 - \beta > 0.99$). The session was carried out in the sensory laboratory of the Faculty of Veterinary and Agricultural Sciences (FVAS) in individual booths with uniform white light-emitting diode (LED) lights. Each booth was equipped with a tablet PC in which the Bio-Sensory Application (The University of Melbourne, Parkville, VIC, Australia) was set up with the questionnaire to gather consumer responses. The appearance, overall aroma, smoke aroma, bitterness, sweetness, acidity, astringency, a warming sensation, and overall liking were assessed on a likeness scale (i.e., dislike extremely—neither like nor dislike—like extremely). The levels of smoke aroma and perceived quality were rated on an intensity scale (i.e., absent-intense). Both liking and intensity measures were presented on a 15 cm non-structured continuous scale. In addition, emotional responses were recorded, using a 0–100 FaceScale, where 0 = sad ☹, 50 = neutral ☺, and 100 = happy ☺. Samples were randomly assigned a 3-digit code, and 10 mL samples were served at room temperature (20 °C) in International Standard Wine Tasting Glasses (Bormioli Luigi, Fidenza, Italy). Samples were served in random order to avoid bias. Plain water and water crackers were used as palate cleansers between samples.

2.5. Statistical Analysis and Machine Learning Modeling

Analysis of variance (ANOVA) was conducted on e-nose data using XLSTAT (ver. 19.3.2, Addinsoft Inc., New York, NY, USA), and Tukey's honest significant difference test (HSD; $\alpha = 0.05$) was used to assess significant differences between treatments.

Machine learning modeling was performed based on artificial neural networks (ANN) for both pattern recognition and regression models, using codes written in Matlab® R2019b (Mathworks, Inc., Natick, MA, USA) developed to test 17 different training algorithms. Five distinct models were developed using 20 data points from the peak of the e-nose outputs (nine sensors) as inputs. Model 1 (pattern recognition) used the scaled conjugate gradient training algorithm to classify the wine samples into the five different treatments: (i) LS, (ii) HS, (iii) HSM, (iv) C, and (v) CM. All four regression models were developed using the Levenberg Marquardt algorithm. Model 2 consisted of the use of the 20 glycoconjugates and 10 volatile phenols (Table 2) found in berries one hour after being exposed to smoke as targets. In comparison, Model 3 used the same 20 glycoconjugates and 10 volatile phenols in berries but measured at harvest. The targets used for Model 4 were 17 glycoconjugates and seven volatile phenols analyzed in the wine samples (Table 2). On the other hand, Model 5 was developed to predict 12 sensory responses, using the liking of (i) appearance, (ii) overall aroma, (iii) smoke aroma, (iv) bitterness, (v) sweetness, (vi) acidity, (vii) astringency, (viii) a warming sensation, (ix) overall liking, and (x) the intensity of (i) smoke aroma, (ii) perceived quality, and (iii) the FaceScale emotional response as targets.

All inputs and targets were normalized from −1 to 1. Data were divided randomly for all ANN models, with 60% of the data being used for the training stage, 20% for validation, and 20% for testing. Model 1 used a cross-entropy loss to test performance, while Models 2–5 were based on means squared error (MSE). Figure 1 shows the diagrams for Model A (Figure 1a), Models 2–4 (Figure 1b), and Model 5 (Figure 1c); all models consisted of a two-layer feedforward network with the hidden layer using a tan-sigmoid function and the output layer using softmax neurons (Model 1) and a linear transfer function (Models 2–5). A trimming test (data not shown) was performed to find the optimal number of neurons (3, 5, 7, 10) to get the best performance. Statistical data reported for regression models to assess under- or overfitting consist of the correlation coefficient (R), slope (b), MSE, and determination coefficient (R^2); the latter was calculated using the curve fitting tool found in Matlab®.

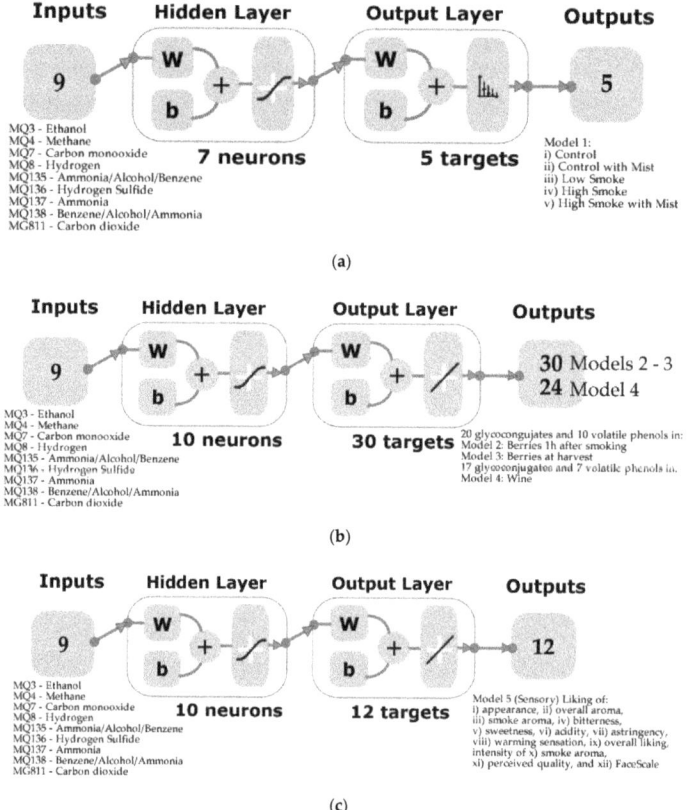

Figure 1. Model diagrams of the two-layer feedforward networks for (**a**) Model 1 for pattern recognition to classify samples into the five treatments using seven neurons, (**b**) Models 2–4 for regression to predict 20 glycoconjugates and 10 volatile phenols (Table 2) in Model 2: berries 1 h after smoke, Model 3: berries at harvest, and Model 4: wine, and (**c**) Model 5 for regression to predict 12 different sensory responses using 10 neurons. Abbreviations: W: weights, b: bias.

3. Results

3.1. Electronic Nose Results

Figure 2 shows the results from the ANOVA for the e-nose responses. It can be observed that there were significant differences ($p < 0.05$) between samples in the outputs from all nine sensors that integrated the e-nose. Ethanol gas (MQ3) presented the highest values for all wine samples with CM (mean = 4.07 V) being significantly different from HSM (mean = 3.85 V), HS (mean = 3.82 V), and C (mean = 3.92 V), and these from LS (mean = 3.66 V). Hydrogen sulfide (MQ136) was the lowest for all samples, and CM (mean = 0.34 V) was significantly different from all other samples (means = 0.23–0.27 V). The CO_2 sensor readings are inverse; therefore, higher Volts mean lower concentration; it can be observed that all the samples with smoke treatments (LS, HS, and HSM) had the lowest CO_2 and presented significant differences with control samples (CM and C).

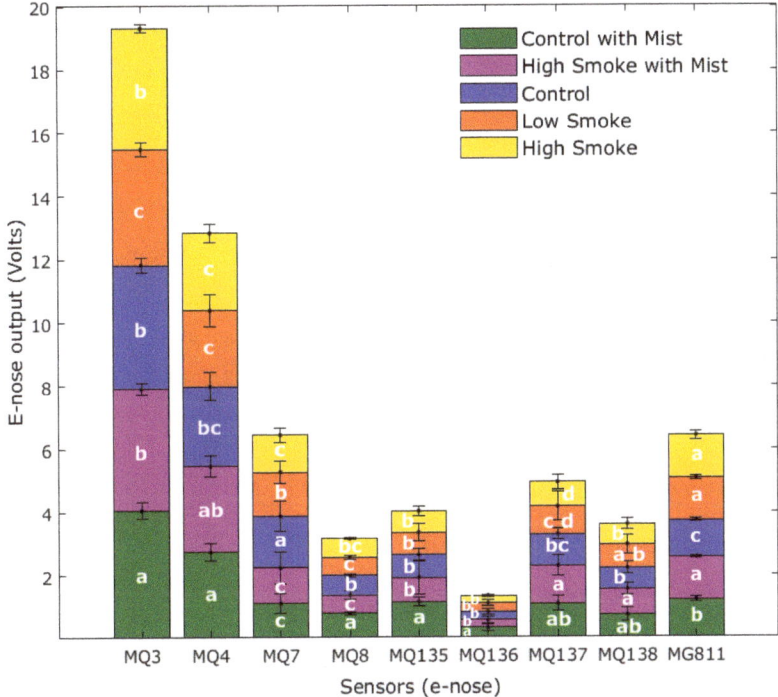

Figure 2. Mean values of the electronic nose outputs showing the letters of significance from the ANOVA and Tukey *post hoc* test ($\alpha = 0.05$). Sensors: MQ3 = ethanol, MQ4 = methane, MQ7 = carbon monoxide, MQ8 = hydrogen, MQ135 = ammonia/alcohol/benzene, MQ136 = hydrogen sulfide, MQ137 = ammonia, MQ138 = benzene/alcohol/ammonia, MG811 = carbon dioxide.

Table 3 shows the minimum, maximum, and average values of the glycoconjugates and volatile phenols detected in berries one hour after smoking, in berries at harvest, and wine. It can be observed that there was a wide range of values for all of the compounds, which indicated these were adequate samples to be used for machine learning modeling and to detect smoke contamination.

Table 3. Minimum (Min), maximum (Max), and mean values of the glycoconjugates (berries: µg kg^{-1}; wine: µg L^{-1}) and volatile phenols (µg L^{-1}) detected in berries and wine.

Compound	Berries 1 h After Smoking			Berries at Harvest			Wine		
	Min	Max	Mean	Min	Max	Mean	Min	Max	Mean
Syringol gentiobioside	2.37	56.93	15.42	6.30	772.81	186.55	10.43	582.11	152.58
Syringol monoglucoside	0.14	26.97	6.38	2.65	68.34	19.22	0.36	14.54	4.26
Syringol pentosylglucosides	0.76	4.52	1.79	6.41	369.14	88.76	1.70	103.37	27.73
Cresol glucosylpentoses	8.07	47.12	18.13	41.69	1395.52	382.63	0.40	17.67	5.28
Cresol gentiobioside	0.18	0.71	0.45	1.94	6.46	3.55	NA	NA	NA
Cresol monoglucoside	0.24	61.87	16.36	0	35.47	8.70	NA	NA	NA
Cresol rutinoside	1.62	13.34	4.90	3.11	122.07	38.35	2.91	133.85	40.55
Guaiacol pentosylglucosides	2.29	25.61	7.57	15.76	1233.46	268.39	5.30	330.36	80.47
Guaiacol gentiobioside	0.05	1.38	0.40	0.54	67.44	16.33	0.30	2.81	0.99
Guaiacol rutinoside	0	1.35	0.48	1.13	32.03	9.97	0	48.60	15.24
Guaiacol monoglucoside	0.03	30.04	7.07	1.22	30.25	7.15	0.12	12.60	3.46
Methylguaiacol pentosylglucosides	0.55	11.51	3.29	6.79	266.50	57.32	1.43	51.79	12.72
Methylguaiacol rutinoside	0.60	5.58	1.89	6.45	153.06	44.36	0.79	40.92	11.97
Methylguaiacol monoglucoside	0	0	0	0.94	11.52	3.89	NA	NA	NA
Methylsyringol gentiobioside	0.33	13.34	3.49	2.53	302.51	72.52	0.15	30.69	7.41
Methylsyringol pentosylglucosides	0.07	0.39	0.17	1.57	34.84	10.36	0.20	8.35	2.46
Phenol rutinoside	0.31	3.78	1.26	3.75	175.57	53.28	1.42	77.58	23.40
Phenol gentiobioside	0.01	0.61	0.15	0	28.54	6.57	0.08	6.22	1.70
Phenol pentosylglucosides	1.44	24.97	7.02	16.21	812.10	215.13	0.53	22.59	6.31
Phenol monoglucoside	0.04	2.55	0.63	0.99	21.52	5.65	0.74	43.48	11.86
Guaiacol	2.39	139.72	41.57	2.06	12.97	5.08	0	39.00	11.73
4-Methylguaiacol	3.54	27.72	9.50	3.52	4.45	3.80	0	5.00	1.40
Phenol	1.40	85.68	21.12	1.26	26.38	9.61	NA	NA	NA
o-Cresol	1.65	54.02	16.31	1.74	8.08	4.02	0	14.00	4.87
Total m/p-cresol	0.56	63.07	16.01	0.52	7.71	2.99	NA	NA	NA
m-Cresol	1.90	45.07	12.08	1.84	5.89	3.24	0	14.00	4.53
p-Cresol	0	18.00	4.38	0	2.04	0.44	0	9.00	2.60
Syringol	5.17	180.31	47.67	9.32	13.77	11.73	1.00	6.00	3.13
4-Methylsyringol	1.83	24.36	6.62	1.75	2.11	1.83	0	0	0
Total cresols	2.22	117.08	32.32	2.26	15.79	7.01	NA	NA	NA

Abbreviations: NA: Not applicable. Values <1 (µg L^{-1} and µg kg^{-1}) are considered as below the limit of detection. However, actual values were included in the modeling strategies.

Table 4 shows the minimum, maximum, and average values of the responses from the sensory session conducted with consumers when evaluating the wines. It can be observed that the results from all attributes were within the whole range of the scales used for liking and appearance (0–15) and FaceScale (0–100), which made the data suitable to be used for machine learning modeling.

Table 4. Minimum (Min), maximum (Max), and mean values of the sensory session responses for wine tasting.

Data/Sensory Attribute	Min	Max	Mean
Appearance liking	0.45	15.00	7.19
Overall aroma liking	0.30	14.85	6.21
Smoke aroma intensity	0	15.00	4.98
Smoke aroma liking	0	15.00	4.72
Bitter liking	0.30	15.00	5.98
Sweet liking	0	14.70	6.16
Acidity liking	0	14.70	6.23
Astringency liking	0.30	15.00	6.27
Warming liking	0.30	15.00	6.20
Overall liking	0.30	14.85	6.07
Perceived quality	0	14.85	5.66
FaceScale	0	99.00	42.15

3.2. Machine Learning Models

Table 5 shows the statistical results from Model 1 for the classification of the samples into the five different treatments. It can be observed that there was a high accuracy for all stages (>90%) and 97% for the overall model. According to the performance values, there were no signs of overfitting, as the training stage had a cross-entropy value lower than the validation and testing, and these two had similar performance. In Figure 3, the results from the receiver operating characteristic (ROC) curve are shown. This graph depicted the sensitivity (true positive rate) and specificity (false positive rate) of the overall model, with optimal operating points of 98%, 100%, 93%, 93%, and 98% for C, CM, LS, HS, and HSM, respectively.

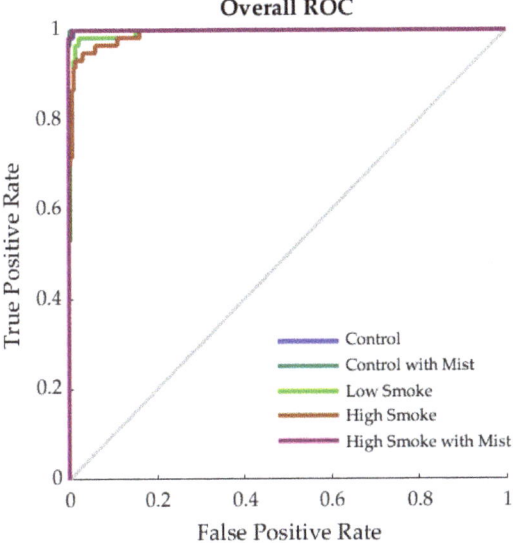

Figure 3. Receiver operating characteristic (ROC) curve for Model 1 to classify wine samples into the five different smoke treatments.

Table 5. Statistical results from the pattern recognition model (Model 1) to classify samples into five different treatments (control, control with mist, low smoke, high smoke, and high smoke with mist).

Stage Model 1	Samples	Accuracy	Error	Performance (Cross-Entropy)
Training	180	99%	1%	0.01
Validation	60	93%	7%	0.04
Testing	60	92%	8%	0.05
Overall	300	97%	3%	-

Table 6 depicts the statistical data for the four regression models. Model 2 had very high overall correlation and determination coefficients (R = 0.98; Figure 4a; R^2 = 0.95). The close value of the validation and training correlation coefficients (R = 0.96 and R = 0.98, respectively), along with the fact that the performance of the training stage (MSE = 0.01) was lower than that of the validation and testing (MSE = 0.03 and MSE = 0.02, respectively), showed that there were no signs of under- or overfitting. Models 3 and 4 had similar statistical values, both with high accuracy (Model 3: R = 0.99; Figure 4b; R^2 = 0.97; Model 4: R = 0.99; Figure 4c; R^2 = 0.98). These models also showed no signs of under- or overfitting. On the other hand, Model 5 also had a very high overall accuracy (R = 0.98; Figure 4d; R^2 = 0.96) with similar performance values for validation and testing (MSE = 0.04) and higher than that of the training stage (MSE = 0.02). All models presented a slope close to the unity (b ~ 1) for all stages (Figure 4).

Table 6. Statistical results from the four regression models (Models 2–4: glycoconjugates and volatile phenols; Model 5: sensory) showing the correlation coefficient (R), determination coefficient (R^2), slope (b), and performance based on means squared error (MSE) for each stage.

Stage/ Model 2 (Berries 1 h Smoke)	Samples	Observations	R	R^2	b	Performance (MSE)
Training	180	5400	0.98	0.96	0.96	0.01
Validation	60	1800	0.96	0.92	0.97	0.03
Testing	60	1800	0.97	0.95	0.97	0.02
Overall	300	9000	0.98	0.95	0.97	-
Stage/ Model 3 (Berries at Harvest)	Samples	Observations	R	R^2	b	Performance (MSE)
Training	180	5400	0.99	0.98	0.97	0.01
Validation	60	1800	0.98	0.95	0.96	0.02
Testing	60	1800	0.98	0.97	0.95	0.01
Overall	300	9000	0.99	0.97	0.96	-
Stage/ Model 4 (Wine)	Samples	Observations	R	R^2	b	Performance (MSE)
Training	180	4320	0.99	0.99	0.99	<0.01
Validation	60	1440	0.98	0.95	0.96	0.02
Testing	60	1440	0.98	0.96	0.95	0.01
Overall	300	7200	0.99	0.98	0.98	-

Table 6. Cont.

Stage/ Model 5 (Wine Sensory)	Samples	Observations	R	R^2	b	Performance (MSE)
Training	180	2160	0.98	0.97	0.97	0.02
Validation	60	720	0.97	0.94	0.97	0.04
Testing	60	720	0.97	0.94	0.97	0.04
Overall	300	3600	0.98	0.96	0.97	-

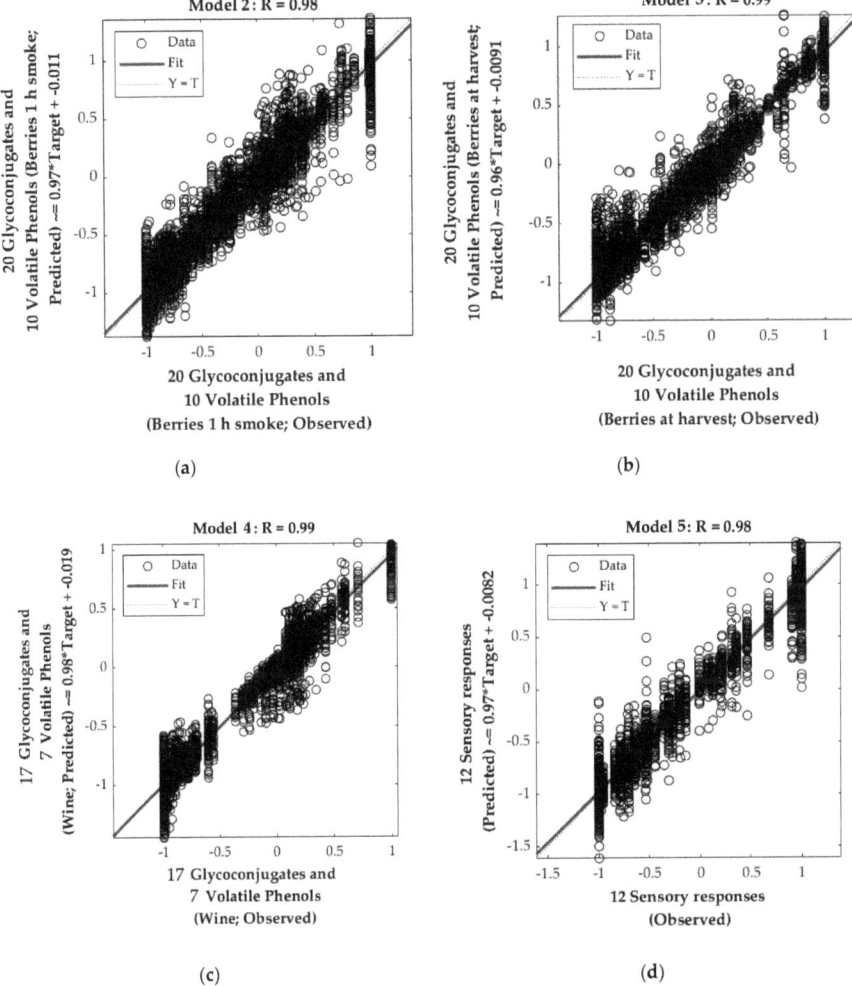

Figure 4. The overall correlation of the models to predict 20 glycoconjugates and 10 volatile phenols (Table 2) of (**a**) Model 2: berries after 1 h smoking, (**b**) Model 3: berries at harvest; (**c**) 17 glycoconjugates and seven volatile phenols of Model 4: wine. (**d**) Shows the Model 5 to predict 12 sensory descriptors obtained in a consumer test (Figure 1c).

4. Discussion

Nowadays, the only alternative for grape growers is to apply potential amelioration techniques before the bushfires and hope for the best since there are limited tools that can be applied in the field or at the winemaking stage, which can render results in near real-time for proper decision-making [35,36]. Recently, non-invasive devices have been proposed using infrared thermal imaging to assess contaminated grapevine canopies in the field and smoke taint in berries and wines using near-infrared spectroscopy [5,33]. The research presented in this paper has contributed to the potential implementation of new and emerging sensor technologies and modeling strategies using machine learning in the viticulture and winemaking industry. These low-cost e-noses could become a game-changer for the management of smoke contamination and taint in berries and wines due to bushfires.

In general, previous applications of e-noses in the wine industry have been implemented mainly for the analysis of grapes and crushing methods [37], improvement of maceration and fermentation processes [38], to monitor the aging of wine in barrels [39–41], geographical classification [42], wine spoilage [28,43,44], and to assess correlations with human perception through sensory evaluation [27,29,45]. However, most of these studies have been based on multivariate data analysis and correlation analysis.

Low-cost sensors presented in this research, developed by integrating an array of gas sensors [20], could be used in the winery to assess the level of grapevine smoke exposure. In the present study, models were developed to evaluate the effects of different amelioration techniques (Model 1) for berries immediately after the bushfire event (Model 2), at harvest time (Model 3), and in the actual wines (Model 4). Since smoke-derived glycoconjugates in berries are difficult to detect using e-noses due to the binding of these compounds with sugars in the berries, these assessments need to be performed after the winemaking process, in which the compounds are released through the maceration and fermentation processes.

A further model (Model 5) developed to assess sensory characteristics of wines rapidly and objectively, which can be implemented in parallel with successful amelioration techniques to reduce smoke taint, such as the addition of activated carbon to wines or fining agents [2,34]. For the latter case, Model 5 will offer a near real-time assessment of the techniques used.

The advantages of implementing these models coupled with low-cost sensor technology are that grape growers and winemakers will not depend on random sampling, which may not render representative results, or external laboratory services, which may not deliver results in a timely manner due to being overwhelmed by large sample volumes that are delivered when concurrent bushfires occur. Knowing the levels of smoke-derived compounds and the effects on consumer appreciation in the winemaking process offer the following advantages: (i) rapid and user-friendly smoke taint determination; (ii) potential implementation of techniques to reduce smoke taint using activated carbon or fining agents on samples and re-test using the e-nose and models developed; (iii) sensory panel not required for assessments/modifications, minimizing the time for the commercial release of wines and economic impacts of smoke taint.

Further applications of these low-cost e-noses can be implemented to assess the maturity of grapes in the field, specifically through the alcohol-based sensors. The latest research has shown that ethanol is released from grape berries when they become oxygen stressed [46]. So, being able to assess when cell death begins would be a useful tool in monitoring berry health and fruit ripening potential. These processes of berry cell death assessment can be done non-destructively by near-infrared spectroscopy and machine learning modeling [47] or by tracking ethanol release from grapevine bunches through the implementation of low-cost e-noses in the field using sensor networks or as a payload of low altitude unmanned aerial vehicle (UAV) surveys [48,49].

5. Conclusions

Low-cost e-nose sensor technology coupled with machine learning offers the advantage of easy implementation in field conditions using sensor networks or in the winery. Machine learning models obtained could make available valuable information to winemakers and winegrowers for the decision-making process to produce commercial wines by minimizing smoke taint. An artificial intelligence system can be implemented based on sensor technology and machine learning developed here to obtain the least tainted wine or to target specific sensory aroma profiles to take advantage of the decontamination process to maximize the likability of wines.

Author Contributions: Conceptualization, S.F. and C.G.V.; Data curation, S.F. and C.G.V.; Formal analysis, C.G.V.; Funding acquisition, R.R.U.; Investigation, S.F., V.S., C.G.V. and K.L.W.; Methodology, V.S., C.G.V., K.L.W., C.S. and R.R.U.; Project administration, S.F., C.G.V., and K.L.W.; Resources, S.F. and R.R.U.; Software, S.F., C.G.V. and R.R.U.; Supervision, S.F.; Validation, S.F., C.G.V. and N.L.; Visualization, S.F., C.G.V. and E.T.; Writing—original draft, S.F. and C.G.V.; Writing—review and editing, S.F., V.S., C.G.V., E.T., N.L., K.L.W., C.S. and R.R.U. All authors have read and agreed to the published version of the manuscript.

Funding: This research was supported by the Australian Research Councils Linkage Projects funding scheme (LP160101475).

Acknowledgments: The authors would like to acknowledge Bryce Widdicombe, Mimi Sun, and Jorge Gonzalez for their collaboration in the electronic nose development. C.S. was supported by the Australian Research Council Training Centre for Innovative Wine Production (www.arcwinecentre.org.au), which is part of the ARC's Industrial Transformation Research Program (Project No. ICI70100008), with support from Wine Australia and industry partners.

Conflicts of Interest: The authors declare no conflict of interest.

References

1. Kennison, K.; Wilkinson, K.L.; Pollnitz, A.; Williams, H.; Gibberd, M.R. Effect of smoke application to field–grown Merlot grapevines at key phenological growth stages on wine sensory and chemical properties. *Aust. J. Grape Wine Res.* **2011**, *17*, S5–S12. [CrossRef]
2. Ristic, R.; Fudge, A.L.; Pinchbeck, K.A.; De Bei, R.; Fuentes, S.; Hayasaka, Y.; Tyerman, S.D.; Wilkinson, K.L. Impact of grapevine exposure to smoke on vine physiology and the composition and sensory properties of wine. *Theor. Exp. Plant Physiol.* **2016**, *28*, 67–83. [CrossRef]
3. Szeto, C.; Ristic, R.; Capone, D.; Puglisi, C.; Pagay, V.; Culbert, J.; Jiang, W.; Herderich, M.; Tuke, J.; Wilkinson, K. Uptake and Glycosylation of Smoke-Derived Volatile Phenols by Cabernet Sauvignon Grapes and Their Subsequent Fate during Winemaking. *Molecules* **2020**, *25*, 3720. [CrossRef] [PubMed]
4. Bruyère, C.; Holland, G.; Prein, A.; Done, J.; Buckley, B.; Chan, P.; Leplastrier, M.; Dyer, A. *Severe Weather in a Changing Climate*; Insurance Australia Group and National Center for Atmospheric Research, November. Insurance Australia Group Limited, 2019. Available online: https://www.iag.com.au/sites/default/files/documents/Severe-weather-in-a-changing-climate-report-011119.pdf (accessed on 3 August 2020).
5. Fuentes, S.; Tongson, E.J.; De Bei, R.; Gonzalez Viejo, C.; Ristic, R.; Tyerman, S.; Wilkinson, K. Non-Invasive Tools to Detect Smoke Contamination in Grapevine Canopies, Berries and Wine: A Remote Sensing and Machine Learning Modeling Approach. *Sensors* **2019**, *19*, 3335. [CrossRef]
6. Dungey, K.A.; Hayasaka, Y.; Wilkinson, K.L. Quantitative analysis of glycoconjugate precursors of guaiacol in smoke-affected grapes using liquid chromatography–tandem mass spectrometry based stable isotope dilution analysis. *Food Chem.* **2011**, *126*, 801–806. [CrossRef]
7. Hayasaka, Y.; Parker, M.; Baldock, G.A.; Pardon, K.H.; Black, C.A.; Jeffery, D.W.; Herderich, M.J. Assessing the impact of smoke exposure in grapes: Development and validation of a HPLC-MS/MS method for the quantitative analysis of smoke-derived phenolic glycosides in grapes and wine. *J. Agric. Food Chem.* **2012**, *61*, 25–33. [CrossRef]
8. Cipriano, D.; Capelli, L. Evolution of Electronic Noses from Research Objects to Engineered Environmental Odour Monitoring Systems: A Review of Standardization Approaches. *Biosensors* **2019**, *9*, 75. [CrossRef]
9. Wilson, D.M.; DeWeerth, S.P. Odor discrimination using steady-state and transient characteristics of tin-oxide sensors. *Sens. Actuators B Chem.* **1995**, *28*, 123–128. [CrossRef]

10. Roussel, S.; Forsberg, G.; Steinmetz, V.; Grenier, P.; Bellon-Maurel, V. Optimisation of electronic nose measurements. Part I: Methodology of output feature selection. *J. Food Eng.* **1998**, *37*, 207–222. [CrossRef]
11. Carmel, L.; Levy, S.; Lancet, D.; Harel, D. A feature extraction method for chemical sensors in electronic noses. *Sens. Actuators B Chem.* **2003**, *93*, 67–76. [CrossRef]
12. Ayhan, B.; Kwan, C.; Zhou, J.; Kish, L.B.; Benkstein, K.D.; Rogers, P.H.; Semancik, S. Fluctuation enhanced sensing (FES) with a nanostructured, semiconducting metal oxide film for gas detection and classification. *Sens. Actuators B Chem.* **2013**, *188*, 651–660. [CrossRef]
13. Wojnowski, W.; Dymerski, T.; Gębicki, J.; Namieśnik, J. Electronic noses in medical diagnostics. *Curr. Med. Chem.* **2019**, *26*, 197–215. [CrossRef] [PubMed]
14. Young, R.C.; Buttner, W.J.; Linnell, B.R.; Ramesham, R. Electronic nose for space program applications. *Sens. Actuators B Chem.* **2003**, *93*, 7–16. [CrossRef]
15. Ryan, M.A.; Zhou, H.; Buehler, M.G.; Manatt, K.S.; Mowrey, V.S.; Jackson, S.P.; Kisor, A.K.; Shevade, A.V.; Homer, M.L. Monitoring space shuttle air quality using the jet propulsion laboratory electronic nose. *IEEE Sens. J.* **2004**, *4*, 337–347. [CrossRef]
16. Li, W.; Leung, H.; Kwan, C.; Linnell, B.R. E-nose vapor identification based on Dempster–Shafer fusion of multiple classifiers. *IEEE Trans. Instrum. Meas.* **2008**, *57*, 2273–2282. [CrossRef]
17. Peveler, W.J.; Parkin, I.P. *Electronic Noses: The Chemistry of Smell and Security*; Wortley, R., Sidebottom, A., Tilley, N., Laycock, Eds.; Routledge: Abingdon, Oxon, UK; New York, NY, USA, 2019; pp. 384–392. ISBN 9780415826266. [CrossRef]
18. Rusinek, R.; Siger, A.; Gawrysiak-Witulska, M.; Rokosik, E.; Malaga-Toboła, U.; Gancarz, M. Application of an electronic nose for determination of pre-pressing treatment of rapeseed based on the analysis of volatile compounds contained in pressed oil. *Int. J. Food Sci. Tech.* **2020**, *55*, 2161–2170. [CrossRef]
19. Liu, H.; Li, Q.; Yan, B.; Zhang, L.; Gu, Y. Bionic Electronic Nose Based on MOS Sensors Array and Machine Learning Algorithms Used for Wine Properties Detection. *Sensors* **2019**, *19*, 45. [CrossRef]
20. Gonzalez Viejo, C.; Fuentes, S.; Godbole, A.; Widdicombe, B.; Unnithan, R.R. Development of a low-cost e-nose to assess aroma profiles: An artificial intelligence application to assess beer quality. *Sens. Actuators B Chem.* **2020**, *308*, 127688. [CrossRef]
21. Gardner, J.W.; Bartlett, P.N. A brief history of electronic noses. *Sens. Actuators B Chem.* **1994**, *18*, 210–211. [CrossRef]
22. Turner, A.P.; Magan, N. Electronic noses and disease diagnostics. *Nat. Rev. Microbiol.* **2004**, *2*, 161–166. [CrossRef]
23. Schaller, E.; Bosset, J.O.; Escher, F. 'Electronic noses' and their application to food. *Lebensm-Wiss Technol* **1998**, *31*, 305–316. [CrossRef]
24. Wojnowski, W.; Majchrzak, T.; Dymerski, T.; Gębicki, J.; Namieśnik, J. Electronic noses: Powerful tools in meat quality assessment. *Meat. Sci.* **2017**, *131*, 119–131. [CrossRef] [PubMed]
25. Peris, M.; Escuder-Gilabert, L. Electronic noses and tongues to assess food authenticity and adulteration. *Trends Food Sci. Technol.* **2016**, *58*, 40–54. [CrossRef]
26. Rodríguez-Méndez, M.L.; De Saja, J.A.; González-Antón, R.; García-Hernández, C.; Medina-Plaza, C.; García-Cabezón, C.; Martín-Pedrosa, F. Electronic noses and tongues in wine industry. *Front. Bioeng. Biotechnol.* **2016**, *4*, 81. [CrossRef] [PubMed]
27. Lozano, J.; Santos, J.P.; Horrillo, M.C. *Electronic Noses and Tongues in Food Science*; Rodríguez Méndez, M.L., Ed.; Elsevier: Cambridge, MA, USA, 2016; pp. 137–148, 301–307. ISBN 9780128002438. [CrossRef]
28. Gamboa, J.C.R.; da Silva, A.J.; de Andrade Lima, L.L.; Ferreira, T.A. Wine quality rapid detection using a compact electronic nose system: Application focused on spoilage thresholds by acetic acid. *LWT* **2019**, *108*, 377–384. [CrossRef]
29. Gardner, D.M.; Duncan, S.E.; Zoecklein, B.W. Aroma characterization of Petit Manseng wines using sensory consensus training, SPME GC-MS, and electronic nose analysis. *Am. J. Enol. Vitic.* **2017**, *68*, 112–119. [CrossRef]
30. Han, F.; Zhang, D.; Aheto, J.H.; Feng, F.; Duan, T. Integration of a low-cost electronic nose and a voltammetric electronic tongue for red wines identification. *J. Food Sci.* **2020**, *8*, 4330–4339. [CrossRef]
31. Fan, H.; Hernandez Bennetts, V.; Schaffernicht, E.; Lilienthal, A.J. Towards gas discrimination and mapping in emergency response scenarios using a mobile robot with an electronic nose. *Sensors* **2019**, *19*, 685. [CrossRef]

32. Valente, J.; Almeida, R.; Kooistra, L. A Comprehensive Study of the Potential Application of Flying Ethylene-Sensitive Sensors for Ripeness Detection in Apple Orchards. *Sensors* **2019**, *19*, 372. [CrossRef]
33. Van der Hulst, L.; Munguia, P.; Culbert, J.A.; Ford, C.M.; Burton, R.A.; Wilkinson, K.L. Accumulation of volatile phenol glycoconjugates in grapes following grapevine exposure to smoke and potential mitigation of smoke taint by foliar application of kaolin. *Planta* **2019**, *249*, 941–952. [CrossRef]
34. Fudge, A.; Schiettecatte, M.; Ristic, R.; Hayasaka, Y.; Wilkinson, K.L. Amelioration of smoke taint in wine by treatment with commercial fining agents. *Aust. J. Grape Wine Res.* **2012**, *18*, 302–307. [CrossRef]
35. Fudge, A.; Ristic, R.; Wollan, D.; Wilkinson, K.L. Amelioration of smoke taint in wine by reverse osmosis and solid phase adsorption. *Aust. J. Grape Wine Res.* **2011**, *17*, S41–S48. [CrossRef]
36. Ristic, R.; Osidacz, P.; Pinchbeck, K.; Hayasaka, Y.; Fudge, A.; Wilkinson, K.L. The effect of winemaking techniques on the intensity of smoke taint in wine. *Aust. J. Grape Wine Res.* **2011**, *17*, S29–S40. [CrossRef]
37. Prieto, N.; Gay, M.; Vidal, S.; Aagaard, O.; De Saja, J.; Rodriguez-Mendez, M. Analysis of the influence of the type of closure in the organoleptic characteristics of a red wine by using an electronic panel. *Food Chem.* **2011**, *129*, 589–594. [CrossRef] [PubMed]
38. Pinheiro, C.; Rodrigues, C.M.; Schäfer, T.; Crespo, J.G. Monitoring the aroma production during wine–must fermentation with an electronic nose. *Biotechnol Bioeng* **2002**, *77*, 632–640. [CrossRef]
39. Wei, Y.J.; Yang, L.L.; Liang, Y.P.; Li, J.M. Application of electronic nose for detection of wine-aging methods. *Adv. Mater. Res.* **2014**, *875–877*, 2206–2213. [CrossRef]
40. Apetrei, I.; Rodríguez-Méndez, M.; Apetrei, C.; Nevares, I.; Del Alamo, M.; De Saja, J. Monitoring of evolution during red wine aging in oak barrels and alternative method by means of an electronic panel test. *Food Res. Int.* **2012**, *45*, 244–249. [CrossRef]
41. Lozano, J.; Arroyo, T.; Santos, J.; Cabellos, J.; Horrillo, M. Electronic nose for wine ageing detection. *Sens. Actuators B Chem.* **2008**, *133*, 180–186. [CrossRef]
42. Cynkar, W.; Dambergs, R.; Smith, P.; Cozzolino, D. Classification of Tempranillo wines according to geographic origin: Combination of mass spectrometry based electronic nose and chemometrics. *Anal. Chim. Acta* **2010**, *660*, 227–231. [CrossRef]
43. Macías, M.M.; Manso, A.G.; Orellana, C.J.G.; Velasco, H.M.G.; Caballero, R.G.; Chamizo, J.C.P. Acetic acid detection threshold in synthetic wine samples of a portable electronic nose. *Sensors* **2013**, *13*, 208–220. [CrossRef]
44. Wang, H.; Hu, Z.; Long, F.; Guo, C.; Yuan, Y.; Yue, T. Early detection of Zygosaccharomyces Rouxii—spawned spoilage in apple juice by electronic nose combined with chemometrics. *Int. J. Food Microbiol.* **2016**, *217*, 68–78. [CrossRef] [PubMed]
45. Aleixandre, M.; Cabellos, J.M.; Arroyo, T.; Horrillo, M. Quantification of Wine Mixtures with an electronic nose and a human Panel. *Front. Bioeng. Biotechnol.* **2018**, *6*, 14. [CrossRef] [PubMed]
46. Xiao, Z.; Rogiers, S.Y.; Sadras, V.O.; Tyerman, S.D. Hypoxia in grape berries: The role of seed respiration and lenticels on the berry pedicel and the possible link to cell death. *J. Exp. Bot.* **2018**, *69*, 2071–2083. [CrossRef] [PubMed]
47. Fuentes, S.; Tongson, E.; Chen, J.; Gonzalez Viejo, C. A Digital Approach to Evaluate the Effect of Berry Cell Death on Pinot Noir Wines' Quality Traits and Sensory Profiles Using Non-Destructive Near-Infrared Spectroscopy. *Beverages* **2020**, *6*, 39. [CrossRef]
48. Valente, J.; Munniks, S.; de Man, I.; Kooistra, L. Validation of a small flying e-nose system for air pollutants control: A plume detection case study from an agricultural machine. In Proceedings of the 2018 IEEE International Conference on Robotics and Biomimetics (ROBIO), Kuala Lumpur, Malaysia, 12–15 December 2018; pp. 1993–1998.
49. Muralidhara, B.; Geethanjali, B. Review on different technologies used in Agriculture. *Int. J. Pure Appl. Math.* **2018**, *119*, 4117–4134.

© 2020 by the authors. Licensee MDPI, Basel, Switzerland. This article is an open access article distributed under the terms and conditions of the Creative Commons Attribution (CC BY) license (http://creativecommons.org/licenses/by/4.0/).

Article

Classification of Smoke Contaminated Cabernet Sauvignon Berries and Leaves Based on Chemical Fingerprinting and Machine Learning Algorithms

Vasiliki Summerson [1], Claudia Gonzalez Viejo [1], Colleen Szeto [2,3], Kerry L. Wilkinson [2,3], Damir D. Torrico [4], Alexis Pang [1], Roberta De Bei [2] and Sigfredo Fuentes [1,*]

[1] Digital Agriculture, Food, and Wine Group, Faculty of Veterinary and Agricultural Sciences, The University of Melbourne, Parkville, VIC 3010, Australia; vsummerson@student.unimelb.edu.au (V.S.); cgonzalez2@unimelb.edu.au (C.G.V.); alexis.pang@unimelb.edu.au (A.P.)
[2] School of Agriculture, Food and Wine, The University of Adelaide, Waite Campus, PMB 1, Glen Osmond, SA 5064, Australia; colleen.szeto@adelaide.edu.au (C.S.); kerry.wilkinson@adelaide.edu.au (K.L.W.); roberta.debei@adelaide.edu.au (R.D.B.)
[3] The Australian Research Council Training Centre for Innovative Wine Production, PMB 1, Glen Osmond, SA 5064, Australia
[4] Department of Wine, Food and Molecular Biosciences, Faculty of Agriculture and Life Sciences, Lincoln University, Lincoln 7647, Canterbury, New Zealand; Damir.Torrico@lincoln.ac.nz
* Correspondence: sfuentes@unimelb.edu.au

Received: 22 August 2020; Accepted: 5 September 2020; Published: 7 September 2020

Abstract: Wildfires are an increasing problem worldwide, with their number and intensity predicted to rise due to climate change. When fires occur close to vineyards, this can result in grapevine smoke contamination and, subsequently, the development of smoke taint in wine. Currently, there are no in-field detection systems that growers can use to assess whether their grapevines have been contaminated by smoke. This study evaluated the use of near-infrared (NIR) spectroscopy as a chemical fingerprinting tool, coupled with machine learning, to create a rapid, non-destructive in-field detection system for assessing grapevine smoke contamination. Two artificial neural network models were developed using grapevine leaf spectra (Model 1) and grape spectra (Model 2) as inputs, and smoke treatments as targets. Both models displayed high overall accuracies in classifying the spectral readings according to the smoking treatments (Model 1: 98.00%; Model 2: 97.40%). Ultraviolet to visible spectroscopy was also used to assess the physiological performance and senescence of leaves, and the degree of ripening and anthocyanin content of grapes. The results showed that chemical fingerprinting and machine learning might offer a rapid, in-field detection system for grapevine smoke contamination that will enable growers to make timely decisions following a bushfire event, e.g., avoiding harvest of heavily contaminated grapes for winemaking or assisting with a sample collection of grapes for chemical analysis of smoke taint markers.

Keywords: smoke taint; remote sensing; climate change; near-infrared spectroscopy; volatile phenols

1. Introduction

The incidence and intensity of wildfires are increasing worldwide, mainly due to the effects of climate change [1–5]. Bushfires that occur near wine regions can result in grapevine smoke exposure, which can alter the chemical composition of grape berries. Wine produced from these smoke-affected grapes may exhibit unpalatable smoky aromas and flavors, such as "burnt wood", "ashy", and "burnt rubber" [6–9]. These undesirable characters have been attributed to smoke-derived volatile phenols (VPs), including guaiacol, 4-methylguaiacol, cresols, and syringol [7,10,11]. It is thought that these VPs accumulate primarily in the skin of grape berries following smoke exposure and, to a lesser extent, in the pulp and seeds [12–15]. Grapevine smoke exposure, and the resulting smoke taint in wine, have caused significant financial losses for grape growers and winemakers due to discarded grapes and unsaleable wine. For example, the 2009 Black Saturday bushfires in Victoria, Australia, were estimated to have caused AUD 300 million in lost revenue [16–19]. More recently, the Australian Grape and Wine Incorporated (AWGI) estimated an AUD 40 million loss from the 2019/2020 summer bushfires [20]. Vineyard smoke exposure, therefore, remains a significant issue for the wine industry, particularly given the increasing frequency and severity of bushfires [21].

Grapevine leaves have also been found to accumulate VPs, and a positive correlation has been demonstrated between the levels of smoke compounds detected in leaves and wine when they were included in the primary fermentation [13,22,23]. From a physiological point of view, smoke exposure has also been shown to decrease stomatal conductance in leaves, which may result from the reaction of carbon dioxide (CO_2) and carbon monoxide (CO) with water vapor in the substomatal cavity producing carbonic acid (H_2CO_3) [24,25]. Carbonic acid reduces the pH in the stomata, resulting in partial or complete stomatal closure [25,26]. Damage to leaf surfaces following smoke exposure has also been observed, with the development of necrotic lesions or, in extreme cases, total leaf necrosis [10,22,27]. This may be the result of ozone (O_3) present in smoke, which has been linked to chlorophyll destruction and accelerated leaf senescence [28,29].

Some chromatographic techniques such as gas chromatography-mass spectrometry (GC-MS) and high-performance liquid chromatography-tandem mass spectrometry (HPLC-MS/MS) have been developed to quantify levels of free and glycosidically bound VPs in grapes and wines [30–33]. While these techniques are currently used for qualitative and quantitative analysis and may assist growers in determining the level of smoke taint in the final wine, there are numerous shortcomings: sample preparation is time-consuming and destructive, and analyses require expensive reagents, standards, and equipment, as well as trained personnel. Furthermore, following a bushfire event, there may be long delays in the availability of results due to large numbers of samples being submitted to commercial laboratories for analysis [34,35]. Consequently, alternative methods of smoke taint analysis have recently been investigated and may offer non-destructive sample preparation, as well as accurate and rapid results.

The use of spectroscopic techniques has increased in recent years due to their ease of use, rapid results, minimal sample preparation, and non-destructive nature, all of which allow repeated measurements to be taken [34–39]. Furthermore, the development of smaller, handheld spectroscopic devices coupled with decreasing costs, has allowed these technologies to be more readily accessible and affordable to growers and farmers, while their portability allows for in-field use, reducing the risk of sample deterioration during transportation [39,40]. Ultraviolet (UV) to visible (Vis) spectroscopy involves the region between 200–780 nm, which can be used to analyze compounds containing organic acids, phenolic compounds, and pigments such as anthocyanins, carotenoids, and chlorophylls [41]. UV-Vis spectroscopy has been used to determine the contribution of chemical compounds towards the composition of extra virgin olive oils to determine the region in the Mediterranean it was produced, to optimize the aging process of Spanish wines, and to assess the impact of heating edible oils and to determine their acid level [42–44]. Near-infrared (NIR) spectroscopy between the light spectra regions of 780–2500 nm has been widely used in agricultural and food science applications, with NIR bands corresponding to overtones resulting from the vibrations of O-H, C-H, N-H, and S-H bonds [39,41]. Various spectroscopic techniques, most notably in the NIR region, have been used for numerous applications in viticulture, including the assessment of grape quality and ripeness as well as the authentication of geographical origin [38,45–50]. Research has also been conducted on the use of mid-infrared (MIR) spectroscopy (between 2500–25,000 nm) of the electromagnetic spectrum, as well as synchronous two-dimensional MIR correlation spectroscopy (2D-COS) for the classification of smoke tainted wines [34,35]. Both techniques showed potential for screening smoke tainted wine, with MIR spectroscopy achieving 61 and 70% classification rates for control and smoke affected wines, respectively. However, classification rates were affected by the degree of smoke taint, as well as compositional differences arising from the grape variety and oak maturation [34]. While this technology may help to assess wine samples for smoke taint, it does not provide an early, in-field detection system that could help growers identify which grapes may be contaminated before winemaking. At present, there is very little research investigating the in-field use of Vis-NIR spectroscopy for the classification of smoke-affected grapevine leaves and berries. Research by Fuentes and coworkers [19] developed a model using NIR spectroscopy between the region of 700–1100 nm to predict the levels of guaiacol glycoconjugates in berries and wine, and the levels of guaiacol in wine. These models may offer growers a non-destructive in-field detection system for grapevine smoke contamination. However, further research is required to determine the effectiveness of different NIR regions for monitoring smoke contamination.

Several chemometric techniques have been used to analyze spectral data, including partial least squares (PLS) regression, principal component analysis (PCA), and artificial neural networks (ANN), to name a few [41]. Of these techniques, ANNs have increased in popularity as classification, prediction, and clustering tools, particularly since they can better interpret the non-linear patterns of spectral data [51–54]. Machine learning (ML) modeling based on ANN can be trained from a set of given data known as 'inputs' or independent variables and form complex, non-linear relationships with these inputs and the 'targets' or dependent variables [54]. For example, preliminary ML models for the classification of smoke tainted grapevines have been developed using infra-red (IR) thermal imagery from canopies, which gave an indication of changes in stomatal conductance for classification of control and smoke-exposed grapevines [25]. In addition to this, another model has been proposed that aims to quantify levels of smoke derived compounds in grapes and wine using NIR spectroscopy measurements as inputs [25]. Furthermore, UV-Vis spectroscopy may offer insights into the degree of physiological performance of leaves as well as fruit ripening and quality through analyzing pigment content, such as chlorophylls, anthocyanins, and carotenoids [55–59].

The objective of this study was to investigate the use of NIR spectroscopy, coupled with ML modeling for the detection of grapevine smoke contamination. Grapevine leaves and berries were analyzed in the vineyard in a smoke trial using a NIR spectrometer, and the absorbance values were used as inputs to train different machine learning algorithms in order to create ANNs with the best classification performances. In addition to this, UV-Vis spectroscopy was used to assess the physiological performance and degree of senescence of leaves, as well as the degree of ripening and anthocyanin content of grapes. This may offer growers a rapid and non-destructive detection system that they can employ themselves to obtain real-time information regarding smoke exposure. This will facilitate timely decision-making around which fruit to sample for chemical analysis and/or to harvest to maintain wine quality.

2. Materials and Methods

2.1. Vineyard Site and Experimental Design for the Smoke Trial

The smoke trial was conducted in late January-early February during the 2018/2019 growing season, at the University of Adelaide's Waite Campus in Urrbrae, South Australia (34°58′ S, 138°38′ E). The trial, described previously by Szeto and colleagues [60], involved the application of smoke and/or in-canopy misting to Cabernet Sauvignon grapevines and comprised five different treatments: a control (C), i.e., neither misting nor smoke exposure; (ii) a control with misting (CM), i.e., in-canopy misting but no smoke exposure; (iii) a high-density smoke treatment (HS); (iv) a high-density smoke treatment with misting (HSM); and (v) a low-density smoke treatment without misting (LS). Treatments were applied to Cabernet Sauvignon grapevines planted in 1998 at 2.0 and 3.3 m vine and row spacings, and trained to a bilateral cordon, vertical shoot positioned trellis system (VSP), hand-pruned to a two-node spur system, with under vine drip irrigation (twice weekly, from fruit set to pre-harvest). Smoke treatments were applied (approximately seven days post-véraison, the period grapes are thought to be most susceptible to smoke contamination [10]) using a purpose-built smoke tent (Figure 1a,b) and experimental conditions reported previously [4,61]: low and high-density smoke treatments were achieved by burning different fuel loads (i.e., ~1.5 and 5 kg of barley straw, respectively). In-canopy misting was evaluated as a method for mitigating the uptake of smoke-derived volatile phenols by grapes and involved the continuous application of fine water droplets (65 µm) to the grapevine bunch zone using a purpose-built sprinkler system (delivering water at 11 L/h), as previously described [62]. Each treatment was applied to six vines from three adjacent panels, except the HS treatment, which comprised only five vines, with treatments separated by at least one buffer vine. LS, HS, and HSM treatments comprised duplicate applications of smoke to 1.5 panels/three vines at a time (except for one HS treatment). The in-canopy sprinkler system was turned on 5 min before the first HSM treatment was applied and off 15 min after the second HSM treatment was completed, such that CM and HSM grapevines were misted for approximately 2.5 h in total. The second and fifth vine from each treatment (the middle vines from smoke treatments) were then selected for physiological and NIR measurements.

Figure 1. Smoke treatments were applied to grapevines using a purpose-built smoke tent; grapevines were enclosed in the tent and exposed to smoke derived from the combustion of barley straw (a,b).

2.2. Physiological Measurements

The rate of photosynthesis (A), stomatal conductance (g_s), and transpiration (E) were determined using a portable infrared gas analyzer equipped with a broad leaf chamber (LCpro-SD, ADC Bioscientific Ltd., Hoddesdon, UK). Measurements were taken on three leaves of each side of the canopy per vine ($n = 12$ leaves per treatment) with a photosynthetic photon flux density of 1000 µmol m^{-2} s^{-1} supplied by a high efficiency, low heat output, mixed red-blue light-emitting diode (LED) array unit. Water vapor and CO_2 concentration in the chamber were set to ambient. Measurements were taken one day (24 h) after smoke treatments were applied, on clear, sunny days.

2.3. Determination of Volatile Phenols and Their Glycoconjugates in Grape Juice/Homogenate

The concentration of volatile phenols and their glycoconjugates were determined (in grape juice and homogenate, respectively) using analytical methods described previously [30,32,33,60]. Volatile phenols were measured by stable isotope dilution analysis (SIDA) [3,30,33], using an Agilent 6890 gas chromatograph coupled to a 5973-mass spectrometer (Agilent Technologies, Forest Hill, Vic., Australia). Isotopically labeled standards, i.e., d_4-guaiacol and d_3-syringol, were prepared in-house using methods outlined previously [3,30,33]. The limit of quantitation for volatile phenols was 1–2 µg/L. Volatile phenol glycoconjugates were also measured by SIDA [30,32], using an Agilent 1200 high-performance liquid chromatograph (HPLC) equipped with a 1290 binary pump, coupled to an AB SCIEX Triple Quad™ 4500 tandem mass spectrometer, with a Turbo V™ ion source (Framingham, MA, USA). The preparation of the isotopically labeled internal standard, i.e., d_3-syringol gentiobioside, has been reported previously [30,32]. The limit of quantitation for volatile phenol glycosides was 1 µg/kg.

2.4. Near-Infrared Data Collection

Grapevine leaf and berry spectra were collected one day after smoke exposure, using a microPHAZIR™ RX Analyzer (Thermo Fisher Scientific, Waltham, MA, USA), which had a spectral range of 1596 to 2396 nm at intervals of 7–9 nm. Prior to undertaking the measurements and after every 10–15 readings, the device was calibrated using a white background calibration standard (included with the device). The white background was placed on top of the leaf while measuring to avoid signal noise inclusion due to variation in light or environmental changes. Leaves and berries were also analyzed using the Lighting Passport Pro™ handheld spectrometer (Asensetek Incorporation, Xindian District, New Taipei City, Taiwan), which has a spectral range of 380–780 nm at intervals of 1 nm. Measurements were taken at approximately 3 cm from the leaves and berries. All measurements were conducted at ambient temperature between 9:00 a.m. and 6:00 p.m.

For the leaf spectral measurements, nine sunlit and nine shaded, mature, fully expanded leaves were selected (i.e., 18 leaves per vine, 36 leaves per treatment). Leaves were free of any visible signs of disease or blemishes. Each leaf was measured in three areas, in triplicate, using the microPHAZIR™ RX Analyzer, while three measurements per leaf were taken with the Lighting Passport Pro™ handheld spectrometer. For the berry spectra, two bunches were selected per vine, and nine berries (three from the top, middle, and bottom of each bunch) were measured, in triplicates using the microPHAZIR™ RX Analyzer (n = 540). On the other hand, twelve berries per treatment were analyzed using the Lighting Passport Pro™ (n = 180) while still attached to the bunch.

2.5. Calculating Spectral Indices

Spectral indices for the analysis of pigment content were calculated for both leaves and berries. Leaf spectra taken using the Lighting Passport Pro™ were used to calculate the normalized difference vegetation index (NDVI), normalized anthocyanin index (NAI), plant senescence reflectance index (PSRI), and carotenoid reflectance index (CRI) [56,57,59,63–65]. Berry spectra were used to calculate the NAI and PSRI. The calculations and wavelengths used for determining these indices are given in Table 1.

Table 1. Calculations for the spectral indices investigated in this study.

Index Name	Index Abbreviation	Equation	References
Normalized difference vegetation index	NDVI	$\frac{(I780-I660)}{(I780+I660)}$	[56,57]
Normalized anthocyanin index	NAI	$\frac{(I780-I570)}{(I780+I570)}$	[56,57]
Carotenoid reflectance index	CRI_{550}	$\frac{1}{I510} - \frac{1}{I550}$	[63,64]
Carotenoid reflectance index	CRI_{700}	$\frac{1}{I510} - \frac{1}{I700}$	[65]
Plant senescence reflectance index	PSRI	$\frac{I680-I500}{I750}$	[59]

2.6. Statistical Analysis

Physiological measurements, spectral indices, volatile phenols, and their glycoconjugates were analyzed by one-way analysis of variance (ANOVA) using Minitab®version 18.1 (Minitab Inc., State College, PA, USA). Mean comparisons were performed using the Fisher least significant difference (LSD) method as a *post-hoc* test at $\alpha = 0.05$. Near-infrared data were analyzed using The Unscrambler X version 10.3 software (CAMO Software, Oslo, Norway). Absorbance values for all wavelengths were plotted for both the microPHAZIR™ RX Analyzer and Lighting Passport Pro™ leaf and berry readings. Principal component analysis (PCA) was also performed using The Unscrambler X program. All microPHAZIR™ RX Analyzer measurements were pre-processed using the second derivative transformation, Savitzky–Golay derivation, and smoothing using The Unscrambler X version 10.3 software prior to the plotting of graphs and statistical analysis.

2.7. Artificial Neural Network Modeling

Three ANN models were developed for berry and leaf NIR readings, which were used as inputs to classify the different smoke treatments using customized code written in MATLAB®(version R2020a, MathWorks Inc., Natick, MA USA) (Figure 2). This code tested a total of 17 training algorithms in a loop to find the optimum in terms of accuracy and performance. Once the optimum training algorithm was identified, further training was performed to develop the most accurate ANN model. For both models, the Levenberg–Marquardt training algorithm was found to be the best algorithm, resulting in models with the highest accuracy and no signs of overfitting.

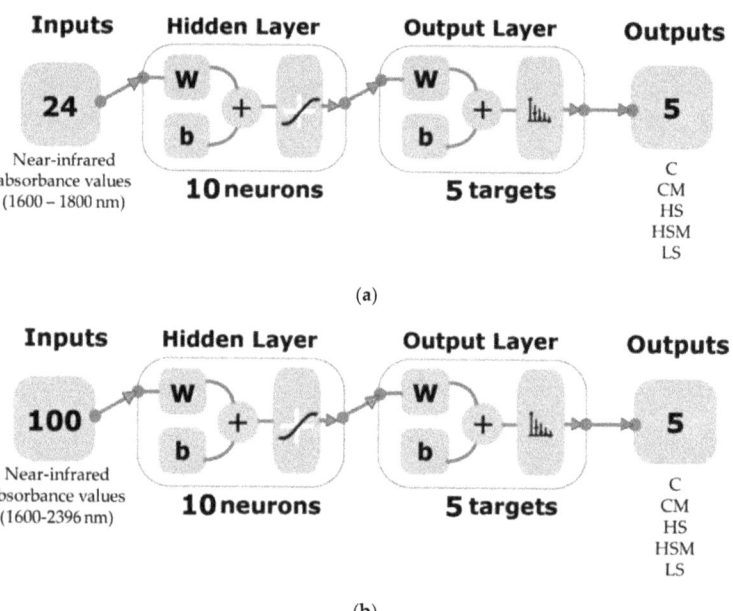

Figure 2. Two-layer feedforward network with ten hidden neurons and sigmoid function for the three classification models: (**a**). microPHAZIR™ leaf model (Model 1) and (**b**). microPHAZIR™ berry model (Model 2). Abbreviations: C = control without misting; CM = control with misting; HS = high density smoke without misting; HSM = high density smoke with misting; and LS = low density smoke.

Overtones within the 1596–1800 nm range were used as inputs for the microPHAZIR™ leaf model (Model 1). This region was selected to avoid water overtones and any classification resulting from the water status of the vines. The entire spectral range was used for the microPHAZIR™ berry model (Model 2) (1596–2396 nm). The two models were developed using a random data division with 70% ($n = 1134$ for Model 1 and 378 for Model 2) training, 15% ($n = 243$ for Model 1 and 81 for Model 2) for validation with a mean squared error (MSE) performance algorithm and 15% (n = 243 for Model 1 and 81 for Model 2) for testing with a default derivative function. Ten hidden neurons were selected for each of the two models after conducting a trimming exercise with three, five, and ten neurons.

3. Results

3.1. Physiological Measurements

Results of gas exchange parameters are shown in Table 2. The transpiration rate was lower for the HS treatment ($P < 0.005$) with a mean rate of 1.43 mmol m^{-2} s^{-1}, while no differences were observed in the other treatments. The CM and C treatments both had the highest g_s values with an average value of 0.15 mol m^{-2} s^{-1} for each, while HS and LS treatments had the lowest average g_s at 0.056 mol m^{-2} s^{-1} and 0.082 mol m^{-2} s^{-1} respectively. Mean rates of A were found to be highest in the C and CM treatments (10.77 µmol m^{-2} s^{-1} and 9.66 µmol m^{-2} s^{-1}, respectively), while the LS and HS treatments had the lowest (7.01 µmol m^{-2} s^{-1} and 5.59 µmol m^{-2} s^{-1}, respectively).

Table 2. Gas exchange parameters measured for the different smoke treatments.

Smoke Treatment	E (mmol m^{-2} s^{-1})		g_s (mol m^{-2} s^{-1})		A (µmol m^{-2} s^{-1})	
	Mean	SD	Mean	SD	Mean	SD
C	2.48 a	0.70	0.15 a	0.05	10.77 a	3.46
CM	2.31 a	0.54	0.15 a	0.05	9.66 ab	2.31
HS	1.43 b	0.62	0.06 c	0.03	5.59 d	2.8
HSM	2.06 a	0.44	0.10 b	0.03	8.15 bc	1.97
LS	2.18 a	0.78	0.08 bc	0.03	7.01 cd	2.42

Abbreviations: C = control without misting; CM = control with misting; HS = high density smoke without misting; HSM = high density smoke with misting; and LS = low density smoke; SD = standard deviation. Means followed by different letters are significantly different based on Fisher least significant difference (LSD) post hoc test ($\alpha = 0.05$).

3.2. Levels of Smoke Taint Marker Compounds in Grape Juice/Homogenate

Differences in volatile phenol concentrations between HS and HSM treatments were found for guaiacol, 4-methylsyringol, and syringol ($P < 0.05$; Table S1). In particular, 4-methylsyringol and syringol had the largest differences in concentrations amongst the smoke treatments, with the HS treatment exhibiting the highest mean values (17 and 126 µg/L, respectively) followed by the HSM treatment (9 and 59 µg/L, respectively) while the CM treatments exhibited the lowest mean values (2 and 8 µg/L), which displayed the lowest mean value. There were no differences between the HS and HSM treatments, nor between the C, CM, and LS treatments for 4-methylguaiacol, phenol, and total cresols; however, HS and HSM grapes had significantly higher volatile phenol concentrations than C, CM, and LS grapes.

Some differences in volatile phenol glycoconjugate levels could be seen amongst the five smoke treatments. Some glycoconjugates displayed differences between the HS and HSM treatments. There was no difference in GuRG levels between the LS, HS, and HSM treatments, with no levels detected in the C and CM treatments. The HS smoke treatment had the highest levels of PhRG, PhGG, CrPG, SyGG, and SyPG, followed by the HSM and LS treatments and then the C and CM treatments. Interestingly the C and HS treatments had the highest level of CrGG followed by the CM and HSM treatment, while the LS treatment had the lowest concentration.

3.3. NIR Absorbance Patterns for Leaves and Berries

Absorbance spectra for the averages of replicates for both raw and transformed leaf absorbance spectra are depicted in Figures 3 and 4. For the microPHAZIR[TM] RX Analyzer leaf absorbances, clear differences in spectral readings were observed for each smoking treatment. A peak was observed at approximately 1784–1793 nm (Figure 3a), while for the transformed data (Figure 3b), large peaks are present between 1596–1647 nm.

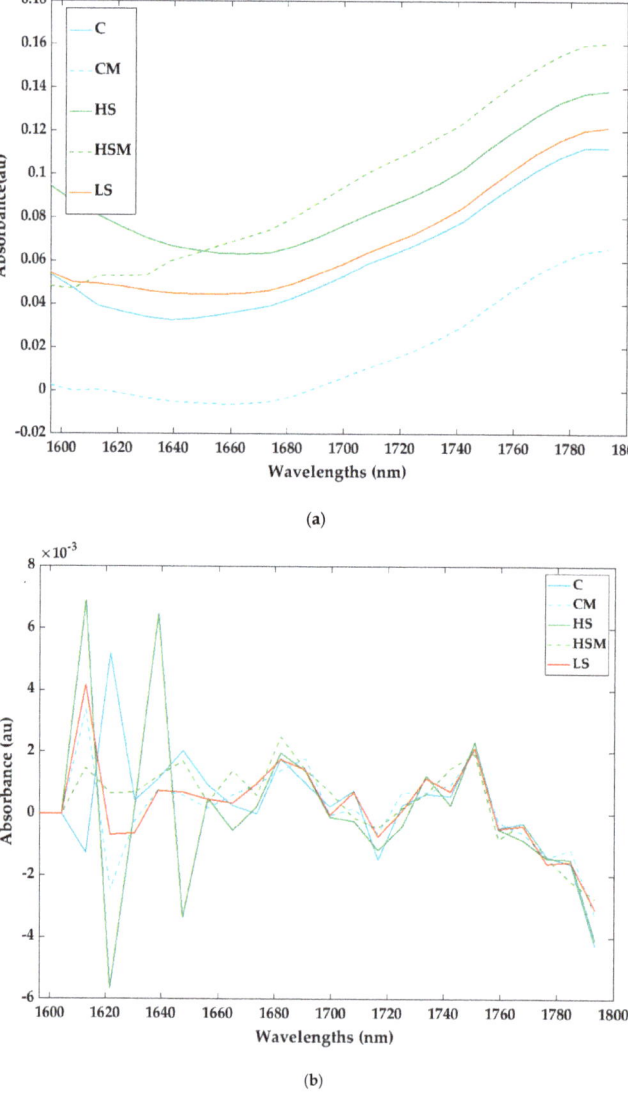

Figure 3. Raw leaf absorbance (**a**) and second derivative spectra (**b**) measured with the microPHAZIR[TM] near-infrared (NIR) analyzer for the different smoke and misting treatments. Abbreviations: C = control without misting; CM = control with misting; HS = high density smoke without misting; HSM = high density smoke with misting; and LS = low density smoke.

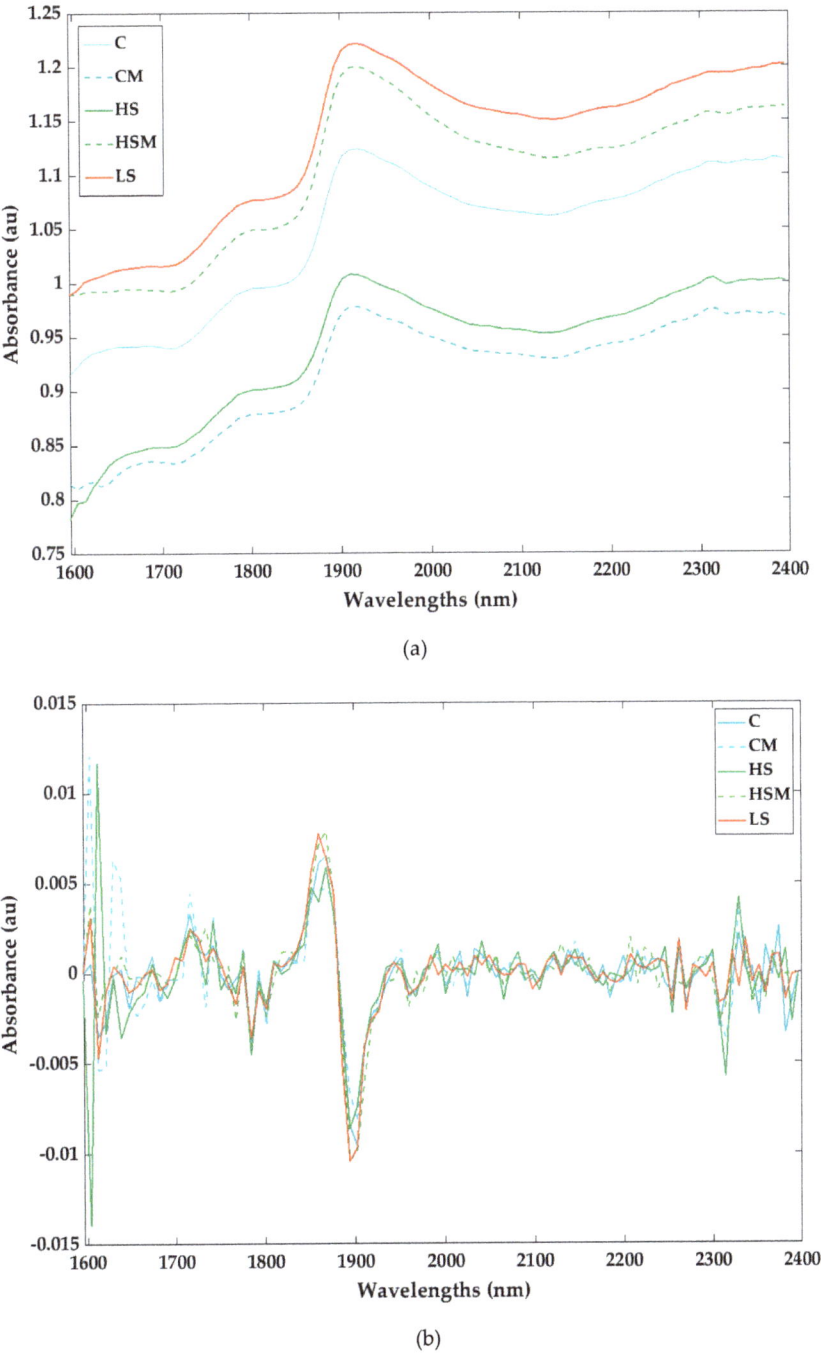

Figure 4. Raw berry absorbance (**a**) and second derivative spectra (**b**) measured with the microPHAZIR[TM] NIR analyzer for the different smoke and misting treatments. Abbreviations: C = control without misting; CM = control with misting; HS = high density smoke without misting; HSM = high density smoke with misting; and LS = low density smoke.

Differences in absorption readings were also found for the microPHAZIRTM RX Analyzer berry absorbance spectra (Figure 4a). Peaks were originally observed at approximately 1785 and 1902 nm, but in the transformed data (Figure 4b), large peaks were observed between approximately 1596–1640 nm and 1820–1940 nm.

3.4. Principal Component Analysis

Figure 5a shows the principal component analysis (PCA) for the microPHAZIR™ RX Analyzer leaf spectra with absorbance values between 1600–1800 nm. The first principal component (PC1) accounted for 62% of the data variability, while principal component two (PC2) accounted for 24%. Hence, 86% of the total variability was explained by these PCs. There was no clear separation of the different smoke treatments when modeled with the microPHAZIR™ leaf spectra. PC1 was represented by wavelengths between 1604–1621 nm and between 1621–1647 nm (loadings shown in Figure 5b). PC2 was represented by wavelengths between 1613–1647 nm, as well as 1604 nm.

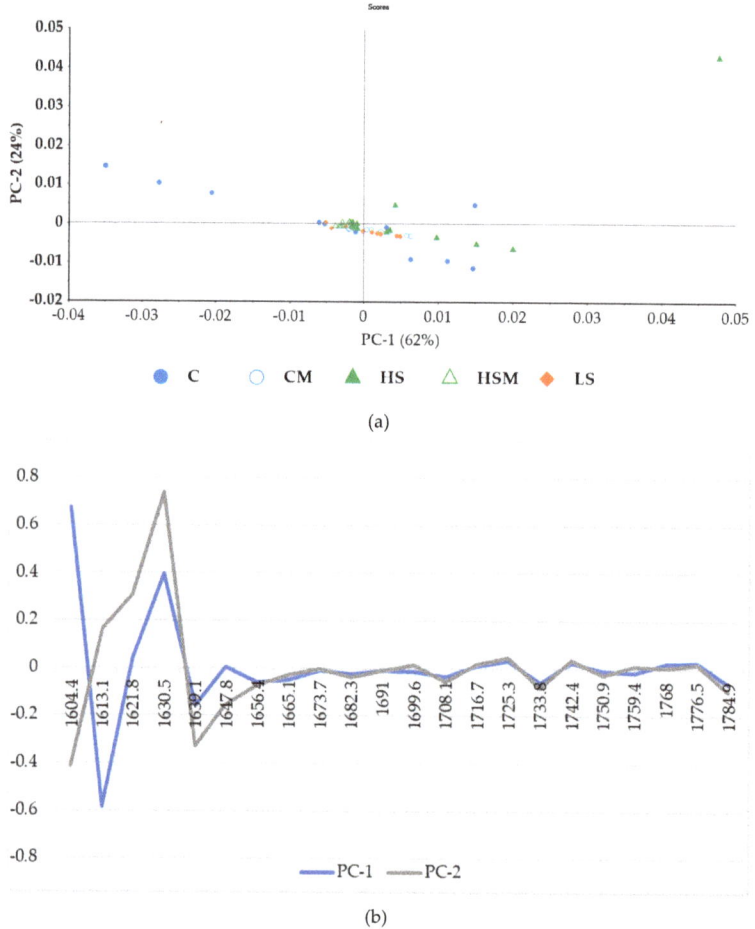

Figure 5. Principal component analysis (PCA) for the microPHAZIR™ leaf absorbance values between 1600–1800 nm (**a**) and loadings (**b**). Abbreviations: C = control without misting; CM = control with misting; HS = high density smoke without misting; HSM = high density smoke with misting; and LS = low density smoke.

Figure 6a shows the PCA for the microPHAZIR™ RX Analyzer berry spectra, where 59% of the data variability was described by PC1, while PC2 accounted for 10% of the data variability; thus, a total of 69% of the total data variability was explained by the first two components of the PCA. As with the microPHAZIR™ RX Analyzer leaf spectra, most of the smoke treatments overlapped quadrants. The CM treatment was grouped primarily in the upper right quadrant, while C and LS treatments were grouped primarily in the lower right. The HS treatment was located primarily in the upper right and left quadrants, while the HSM treatment was grouped in the left upper and lower quadrants. PC1 one was represented by the wavelength region 1604–1622. PC2 was represented by the wavelengths between 1630–1647 nm and 2374–2389 nm (loadings shown in Figure 6b).

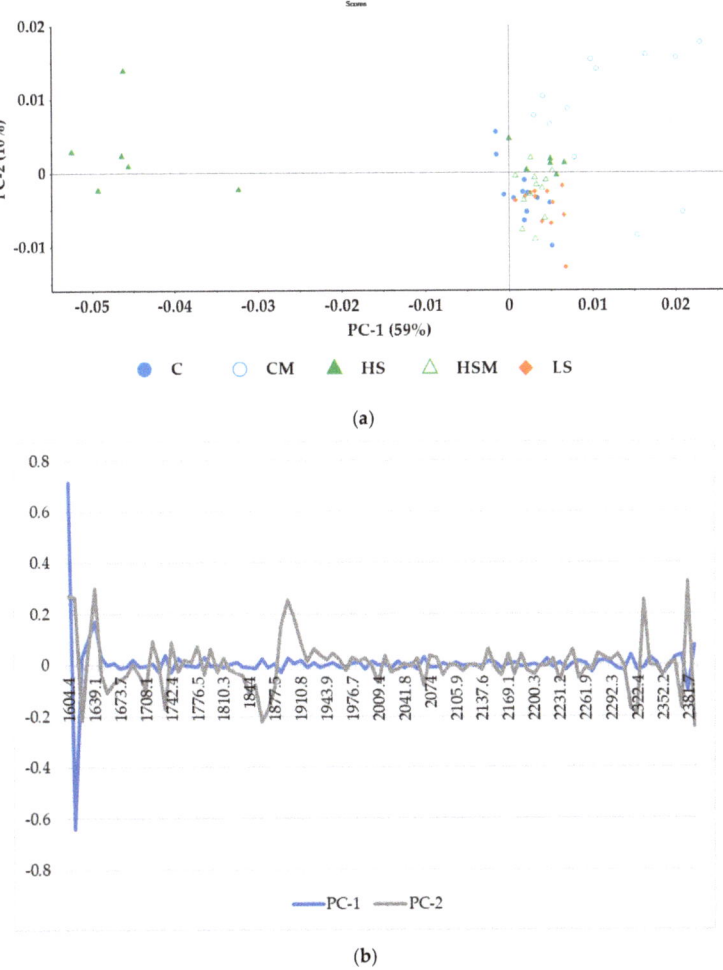

Figure 6. Principal component analysis (PCA) for the microPHAZIR™ berry absorbance values between 1600–2396 nm (**a**) and loadings (**b**). Abbreviations: C = control without misting; CM = control with misting; HS = high density smoke without misting; HSM = high density smoke with misting; and LS = low density smoke.

3.5. Spectral Indices

Results for the spectral indices are shown in Table 3. In the case of the leaf NDVI and NAI, the HS and C treatments had the lowest mean values (0.72 and 0.64 for the HS treatment and 0.84 and 0.74 for the C) ($P < 0.05$). There were no differences for the remaining treatments. For the leaf PSRI, the HS treatment had the highest mean value at 0.065, with no differences for the remaining treatments. For the leaf CRI_{500}, the LS and HS treatments had the highest values at 1.45 and 1.20, respectively, and for the CRI_{700}, the LS treatments had the highest mean values at 1.76, with no differences for the remaining treatments.

In the case of the berry NAI, the HS and LS treatments had the highest mean values with 0.88 and 0.87, with both the C and LS treatments having the lowest mean values of 0.80 and 0.75. For the PSRI, both the LS and C treatments had the highest mean values of 0.02, while the HSM had the lowest value at −0.02.

Table 3. Means and standard deviation (SD) of spectral indices calculated for leaves and berries.

Treatment	Leaf										Berry			
	NDVI		NAI		PSRI		CRI500		CRI700		NAI		PSRI	
	Mean	SD	Mean	SD	Mean	SD	Mean	SD	Mean	SD	Mean	SD	Mean	SD
CM	0.85 [a]	0.10	0.77 [a]	0.11	0.00 [b]	0.01	0.70 [b]	0.64	0.82 [b]	0.79	-	-	-	-
C	0.84 [ab]	0.08²	0.74 [ab]	0.11	0.01 [b]	0.02	0.67 [b]	0.78	0.77 [b]	0.87	0.80 [b]	0.07	0.02 [a]	0.02
HS	0.72 [b]	0.50	0.64 [b]	0.49	0.07 [a]	0.19	1.20 [a]	0.24	0.82 [b]	0.62	0.88 [a]	0.04	0.00 [b]	0.00
HSM	0.87 [a]	0.11	0.79 [a]	0.11	0.00 [b]	0.02	0.48 [b]	0.06	0.58 [b]	0.45	0.75 [b]	0.10	−0.02 [c]	0.00
LS	0.92 [a]	0.04	0.84 [a]	0.08	0.00 [b]	0.01	1.45 [a]	1.08	1.76 [a]	1.40	0.87 [a]	0.05	0.02 [a]	0.01

Abbreviations: C = control without misting; CM = control with misting; HS = high density smoke without misting; HSM = high density smoke with misting; and LS = low density smoke. Means followed by different letters are statistically significant based on Fisher's least significant difference (LSD) post hoc test ($\alpha = 0.05$).

3.6. Artificial Neural Network Models

Table 4 shows the confusion matrices for the two models developed using the spectral readings as inputs and the experimental treatments as targets. Both models displayed high accuracy in classifying the spectral readings according to the treatments, with an overall accuracy of 98% for the microPHAZIR™ leaf model (Model 1) and 97.4% for the microPHAZIR™ berry model (Model 2). Models 1 and 2 presented validation accuracies (94% and 93%, respectively) close to those of the training stage (100% both models). Furthermore, performance values for training (Models 1 and 2: MSE < 0.01) were lower than the other stages and validation (Model 1: MSE = 0.02; Model 2: MSE = 0.03) and testing (Model 1: MSE = 0.02; Model 2: MSE = 0.04) were similar; this indicates that there were no signs of overfitting for both Model 1 and Model 2.

Figure 7 depicts the receiver operating characteristic (ROC) curves for the two ANN models developed. All models showed high true-positive rates (sensitivity) and low false-positive rates (specificity) for classifying the spectral readings according to the experimental treatment, which can also be observed in the last column of each confusion matrix. For Model 2, the HS treatment had the highest sensitivity (100%), followed by the CM and HSM treatments (99.1% each) and LS treatment (96.3%). The C treatment had the lowest sensitivity of 92.6% for this model. For Model 1, the C treatment had the highest sensitivity (99.1%), followed by the LS treatment (98.8%), HS treatment (97.8%), and CM treatment (97.5%), while the HSM had the lowest sensitivity of 96.9%.

Table 4. Statistical results for the artificial neural networks pattern recognition models. Model 1: microPHAZIRTM for leaves, and Model 2: microPHAZIRTM for berries. Performance is based on means squared error (MSE).

Stage	Samples (n)	Accuracy %	Error %	Performance (MSE)
		Model 1		
Training	1131	100	0	0.00
Validation	243	94.2	5.8	0.02
Testing	243	92.6	7.4	0.02
Overall	1617	98.0	2	-
		Model 2		
Training	378	100	0	0.00
Validation	81	92.6	7.4	0.03
Testing	81	90.1	9.9	0.04
Overall	540	97.4	2.6	-

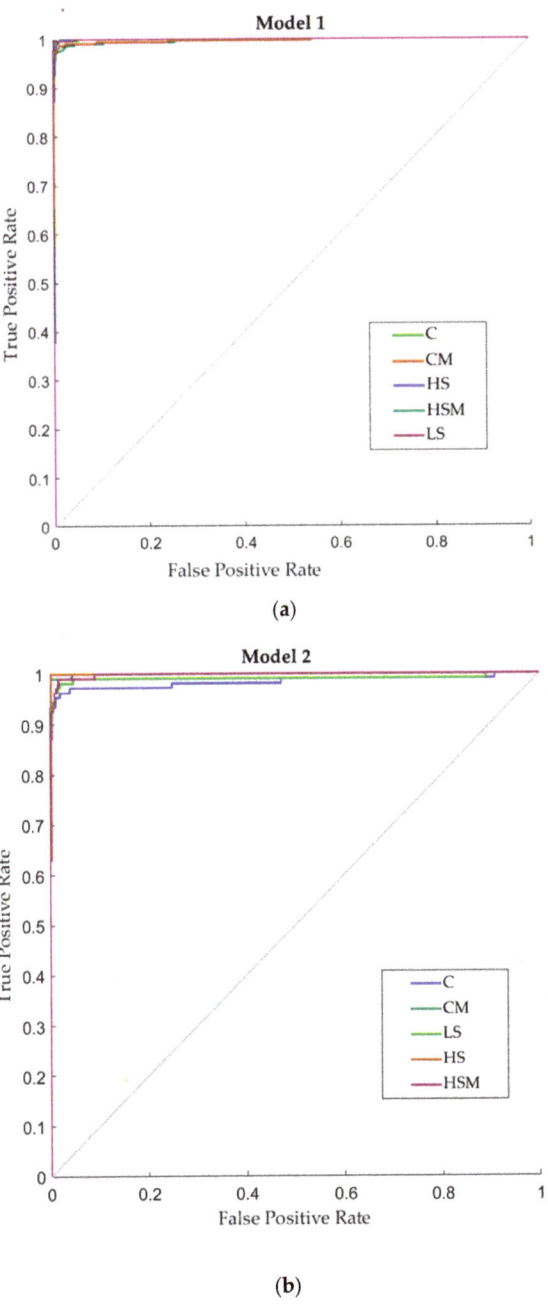

Figure 7. Receiver operating characteristic (ROC) curves for the two models developed (a) the microPHAZIR™ leaf model, (b) the microPHAZIR™ berry model. Colored lines represent the different smoking treatments. Abbreviations: C = control without misting; CM = control with misting; HS = high density smoke without misting; HSM = high density smoke with misting; and LS = low density smoke.

4. Discussion

4.1. Physiological Measurements

Leaf gas exchange parameters were measured the day after smoking. The three smoke treatments showed significant reductions in g_s, in particular, the high-density smoke without misting (HS) treatment, which showed the lowest average reading for g_s (Table 2). Stomatal closure is one of the first responses to smoke exposure undertaken by plants [6,26], and a study by Ristic and colleagues [26] found that the time required for g_s to recover following one hour of smoke exposure for Cabernet Sauvignon grapevines was approximately 6–10 days. A previous study by Bell et al. [6] found that g_s of potted Cabernet Sauvignon grapevines had returned to 60% of pre-smoke exposure rate following fifteen min exposure to smoke using Tasmanian blue gum (*Eucalyptus globulus* L.) leaves as fuel, while rates had returned to 80% of pre-smoke values following exposure to smoke derived from Coast Live Oak (*Quercus agrifolia* Née) leaves. This indicates that in addition to the type of fuel used, the intensity of smoke exposure may also affect the extent of stomatal closure and, hence, reduction in g_s. It is, therefore, not surprising that the HS treatment had the lowest g_s. However, it is interesting that the low smoke treatment (LS) had lower g_s than the high smoke with misting treatment (HSM), which indicates that misting may have reduced the effect of smoke exposure on g_s. During a bushfire, the type of fuel burnt will vary depending on the region and the type of plant species native to the area, as well as the amount of smoke exposure due to land topography and wind vectors; therefore, the effect on g_s may vary [17,18,23,66]. While misting only partially prevented the uptake of volatile phenols and glycoconjugates in grapes [60], it did appear to have a physiological effect. It is evident that misting reduced the effect of smoke exposure on g_s. Smoke contains a complex mixture of gases such as sulfur dioxide (SO_2), O_3, and nitrogen dioxide (NO_2), as well as dust particles that have been shown to inhibit photosynthesis and affect stomatal opening [6,26,29]. Stomata are the primary point of entry for these gases and dust particles [6]; therefore, misting may help prevent the uptake of dust and other particles by trapping them in water that has condensed on the leaf surface, preventing their entrance into the stomata. The present water may also act as a solvent for gases such as SO_2 and NO_2, thereby incorporating them into a solution that then may drip off the leaf surface. In addition to this, smoke exposure may trigger stomatal closure by producing high vapor pressure deficits [26,29]. The presence of misting may help reduce the leaf-to-air vapor pressure difference produced by smoke exposure, thereby reducing the impact on g_s. Misting also appeared to reduce the effect of smoke exposure on transpiration rate (E) as there were no differences between the two control treatments and the LS and HSM treatments. Only the HS treatment had significantly reduced E. Mean rates of photosynthesis (A) followed similar patterns to g_s, with the HS treatment having the lowest value, followed by the LS treatment and then the HSM treatment, while the control without misting (C) had the highest rate of A. This indicates that while misting may have reduced A in the control treatments, it may also help reduce the effects of smoke exposure on A.

4.2. Near-Infrared Spectroscopy Patterns and Principal Component Analysis

From the PCA biplots (Figures 5 and 6) and spectra (Figures 3 and 4) generated in the current study, it is evident that smoke exposure alters the NIR spectral signals of grapevine leaves and berries, and this may prove useful for the detection of grapevine smoke contamination. For the microPHAZIR™ RX Analyzer leaf spectra, high loadings (Figure 5b.) were observed for the wavelength regions between 1604–1621, 1621–1647, and 1613–1647 nm, all of which correspond to C-H stretching of sugars and aromatic compounds [67–70]. For the microPHAZIR™ RX Analyzer berry spectra, high loadings (Figure 6b.) were observed for the wavelength regions between 1604–1622, 1630–1647, and 2374–2389 nm, which correspond to C-H stretching of sugars, such as glucose, as well as aromatic hydrocarbons, which may be due to the presence of smoke-derived volatile phenols, such as guaiacols, cresols, and syringols, and their glycoconjugates [67,68,71,72].

4.3. Spectral Indices

4.3.1. Leaf

The normalized difference vegetation index (NDVI) gives an indication of plant vigor and fruit ripening resulting from relative changes in chlorophyll content. It is based on the variation between the maximum absorption of red by chlorophyll pigments and the maximum reflectance in the infrared caused by leaf cellular structure [56,57,73–75]. Similarly, relative changes in anthocyanin content are expressed as the normalized anthocyanin index (NAI). Both the NDVI and NAI are expressed as a normalized value between −1 (lack of green or redness) to +1 (green or red) [56,57]. Not surprisingly, HS leaves had the lowest NDVI and NAI values. Previous studies investigating the effects of pollution on leaf pigments found a decrease in photosynthetic pigments following exposure to pollutants, including sulfur dioxide (SO_2), carbon dioxide (CO_2), nitrogen dioxide (NO_2), and ozone (O_3) [59,76,77]. These studies are often used as comparisons for investigating the effects of smoke exposure on leaves as compounds in air pollution can also found in smoke [6,22]. There were no differences in NDVI and NAI values between the LS, HSM, and control treatments (C and CM), indicating that misting may reduce the effects of smoke exposure on leaf pigments, and low levels of smoke exposure for one hour may also have no effect. Longer periods of smoke exposure (days or weeks, as is often the case with wildfires) may be required to cause a noticeable change in leaf pigments.

The plant senescence reflectance index (PSRI) gives an indication of the stage of leaf senescence and fruit ripening through assessing changes in carotenoid accumulation and their proportion to chlorophyll. Values range from −1 to +1, with higher values indicating increased stress and carotenoid accumulation [55,63–65,78]. The PSRI was highest for the HS treatment, indicating heightened stress and leaf senescence. This also corresponds with the high CRI_{500} value for this smoke treatment, indicating increased carotenoid accumulation.

4.3.2. Berries

Research by Noestheden et al. [5] found that smoke exposure induced changes in phenylpropanoid metabolites in Pinot Noir berries and wine, some of which are associated with the color and mouthfeel of the wine. Berries exposed to HS and LS treatments had the highest mean NAI values, indicating that smoke exposure may increase anthocyanin content, possibly due to an increase in phenolic accumulation as a stress response induced by exposure to ozone present in smoke [5,79,80]. The HSM treatment had a low NAI value, indicating that misting may reduce anthocyanin concentrations through increased irrigation. Castellarin et al. [81] found that early (before véraison) and late (after the onset of ripening) season, water deficits increased anthocyanin accumulation during ripening. The application of in-canopy misting may reduce water stress and, therefore, reduce anthocyanin accumulation.

Interestingly the HSM followed by the HS treatments had the lowest PSRI values. As carotenoid concentrations in grapes generally decrease during véraison, this may have resulted in lower PSRI values. Therefore, the PSRI may not be suitable for assessing the degree of ripening in grape berries.

4.4. ANN Modeling

Both ANN models classified leaf and berry readings as a function of smoke exposure with high accuracy. The microPHAZIR™ leaf model (model 1) had the highest positive classification, with 98% accuracy (Table 4). The NIR region selected for use in Model 1 was between 1600–1800 nm in order to minimize any possible interference due to the absorption spectra of water in the region of approximately 1930 nm [69]. Furthermore, the region between 1680–1690 nm is associated with aromatic C-H stretching [67]; as such, any patterns observed by the ANN would most likely be due to the presence of smoke-derived volatile phenols. Research by Kennison [22] found a positive correlation between levels of smoke-derived compounds found in leaves and levels in wine; this ANN model developed may, therefore, offer a rapid, in-field method for assessing grapevine smoke contamination. It also demonstrates great promise for further research into the use of NIR spectroscopy coupled with unmanned aerial vehicles (UAVs) with Global Positioning System (GPS) trackers, which could fly over vineyards to scan grapevine canopies and provide maps of smoke contaminated regions.

The microPHAZIR™ berry model (model 2) also had a high overall accuracy in classifying grape berries according to smoke treatment (97.4%). For Model 2, the entire wavelength range between 1600–2396 nm was used. This includes the C-H stretching of aromatic compounds at 1680 nm, O-H stretching at 1930 nm associated with glucose, cellulose, and water, and C=O second overtone associated with carboxylic acids and water between 1900–1910 nm [67,69]. As NIR measurements were conducted in-field on whole berries, this offers a non-destructive tool for assessing grapevine smoke contamination. Whole grapes may be used for assessment as smoke compounds have been found to occur primarily in grape skins [3,25]. Furthermore, the Lighting Passport™ smart handheld spectrometer may be of interest to growers due to its affordability compared to other spectrometers. It is also very small and lightweight, making it easy to undertake measurements in-field, and it can be connected to smartphones via Bluetooth, where data can be stored and retrieved for later analysis [82].

The two ANN models more accurately differentiated the spectral readings relative to PCA. This may be because ANNs are better suited to handle complex, non-linear data, and more readily find patterns or relationships between data than other forms of analysis [53,83–85]. Research by Janik et al. [53] found that the combination of ANNs with partial least squares (PLS) or PCA overcomes issues of non-linearity as well as increasing the accuracy of regression models in predicting total anthocyanin concentrations in red grape homogenates. This may also explain why Model 2 was able to accurately differentiate the berry spectral readings from C, CM, and LS treatments, despite analysis of variance indicating there were no statistically significant differences.

As smoke exposure altered the chemical fingerprinting of grapevine leaves and berries, the ANN models were able to detect changes in the spectral patterns and then classify the readings as a function of experimental treatments. This may offer grape growers a rapid method of assessing the level of smoke contamination in grape berries and leaves, with a high level of accuracy and precision. This may assist growers in deciding which berry samples to send for further chemical analysis to quantify the levels of smoke compounds in grapes and predict the level of smoke taint in the final wine, or they may decide to avoid harvesting heavily contaminated grapes for winemaking. Furthermore, as this method is non-destructive, repeated measurements are possible. By knowing the level of smoke contamination, growers can make informed decisions.

While the ANN models developed were able to classify Cabernet Sauvignon leaf and berry spectra accurately, further research is required to assess whether these models can be used for other grape varieties, as differences in berry composition and leaf physiology may affect the accuracy of classification [6,34]. Previous research evaluated MIR spectroscopy for the classification of smoke tainted wines found compositional differences due to grape variety prevailed over differences resulting from low levels of smoke exposure [34]. Furthermore, the physiological responses of different grape varieties to smoke were found to vary, both in magnitude and in recovery time [6,26]. Thus, further testing of these models using berry and leaf spectra from different grapevine varieties is required.

5. Conclusions

Results from this study indicate that smoke exposure alters the NIR spectra of Cabernet Sauvignon grapevine leaves and berries. As a result, accurate classification models can be developed using ANN modeling. Artificial neural networks are better at classifying non-linear or complex data than traditional techniques, such as principal component analysis. Furthermore, the use of UV-Vis spectroscopy may offer insights into the physiological performance of leaves and the quality and degree of ripening of grapes. These techniques may assist grape growers in identifying grapevines that have been contaminated by smoke, thereby informing decision-making to avoid harvesting and processing heavily contaminated grapes and/or the need for mitigation techniques to manage the risk of smoke taint in resulting wine. Further testing of the ANN models developed in the current study is required to assess their accuracy in classifying grapevine leaf and berry spectra from other grape varieties.

Supplementary Materials: The following are available online at http://www.mdpi.com/1424-8220/20/18/5099/s1, Table S1: Concentrations of volatile phenols in grape juice (µg/L) and their glycoconjugates in grape homogenate (µg/kg) one hour after smoke treatments.

Author Contributions: Conceptualization, V.S., and S.F.; data curation, V.S., C.G.V., and S.F.; formal analysis, V.S.; funding acquisition, S.F.; investigation, V.S., and C.G.V.; methodology, V.S., C.G.V., C.S., K.L.W., and S.F.; project administration, K.L.W., and S.F.; resources, K.L.W., R.D.B., and S.F.; software, C.G.V., and S.F.; supervision, D.D.T., A.P., and S.F.; validation, C.G.V., K.L.W., and S.F.; visualization, V.S., C.G.V., D.D.T., and S.F.; writing—original draft, V.S.; writing—review and editing, V.S., C.G.V., C.S., K.L.W., D.D.T., A.P., R.D.B., and S.F. All authors have read and agreed to the published version of the manuscript.

Funding: This research received no external funding.

Acknowledgments: This research was supported through the Australian Government Research Training Program Scholarship, as well as the Digital Viticulture program funded by the University of Melbourne's Networked Society Institute, Australia. C.S. was supported by the Australian Research Council Training Centre for Innovative Wine Production (www.arcwinecentre.org.au), which is funded as part of the ARC's Industrial Transformation Research Program (Project No. ICI70100008), with support from Wine Australia and industry partners. The authors greatly acknowledge the Digital Agriculture, Food, and Wine Group.

Conflicts of Interest: The authors declare no conflict of interest. The funders had no role in the design of the study; in the collection, analyses, or interpretation of data; in the writing of the manuscript, or in the decision to publish the results.

References

1. Cain, N.; Hancock, F.; Rogers, P.; Downey, M. The effect of grape variety and smoking duration on the accumulation of smoke taint compounds in wine. *Wine Vitic. J.* **2013**, *28*, 48–49.
2. CSIRO. Australian Government Bureau of Meteorology. *State Clim.* **2018**, *2018*, 5.
3. Dungey, K.A.; Hayasaka, Y.; Wilkinson, K.L. Quantitative analysis of glycoconjugate precursors of guaiacol in smoke-affected grapes using liquid chromatography–tandem mass spectrometry based stable isotope dilution analysis. *Food Chem.* **2011**, *126*, 801–806. [CrossRef]
4. Kennison, K.; Gibberd, M.; Pollnitz, A.; Wilkinson, K. Smoke-derived taint in wine: The release of smoke-derived volatile phenols during fermentation of merlot juice following grapevine exposure to smoke. *J. Agric. Food Chem.* **2008**, *56*, 7379–7383. [CrossRef] [PubMed]
5. Noestheden, M.; Noyovitz, B.; Riordan-Short, S.; Dennis, E.G.; Zandberg, W.F. Smoke from simulated forest fire alters secondary metabolites in *Vitis vinifera* L. Berries and wine. *Planta* **2018**, *248*, 1537–1550. [CrossRef]
6. Bell, T.; Stephens, S.; Moritz, M. Short-term physiological effects of smoke on grapevine leaves. *Int. J. Wildland Fire* **2013**, *22*, 933–946. [CrossRef]
7. De Vries, C.; Mokwena, L.; Buica, A.; McKay, M. Determination of volatile phenol in *Cabernet sauvignon* wines, made from smoke-affected grapes, by using hs-spme GC-MS. *S. Afr. J. Enol. Vitic.* **2016**, *37*, 15–21. [CrossRef]
8. Noestheden, M.; Dennis, E.G.; Romero-Montalvo, E.; DiLabio, G.A.; Zandberg, W.F. Detailed characterization of glycosylated sensory-active volatile phenols in smoke-exposed grapes and wine. *Food Chem.* **2018**, *259*, 147–156. [CrossRef]
9. Parker, M.; Osidacz, P.; Baldock, G.A.; Hayasaka, Y.; Black, C.A.; Pardon, K.H.; Jeffery, D.W.; Geue, J.P.; Herderich, M.J.; Francis, I.L. Contribution of several volatile phenols and their glycoconjugates to smoke-related sensory properties of red wine. *J. Agric. Food Chem.* **2012**, *60*, 2629–2637. [CrossRef]
10. Kennison, K.; Wilkinson, K.; Pollnitz, A.; Williams, H.; Gibberd, M. Effect of smoke application to field-grown Merlot grapevines at key phenological growth stages on wine sensory and chemical properties. *Aust. J. Grape Wine Res.* **2011**, *17*, 5–12. [CrossRef]
11. Ristic, R.; van der Hulst, L.; Capone, D.; Wilkinson, K. Impact of bottle aging on smoke-tainted wines from different grape cultivars. *J. Agric. Food Chem.* **2017**, *65*, 4146–4152. [CrossRef] [PubMed]
12. Härtl, K.; Schwab, W. Smoke taint in wine-how smoke-derived volatiles accumulate in grapevines. *Wines Vines* **2018**, *99*, 62–64.
13. Hayasaka, Y.; Baldock, G.; Pardon, K.; Jeffery, D.; Herderich, M. Investigation into the formation of guaiacol conjugates in berries and leaves of grapevine *Vitis vinifera* L. Cv. *Cabernet sauvignon* using stable isotope tracers combined with hplc-ms and ms/ms analysis. *J. Agric. Food Chem.* **2010**, *58*, 2076–2081. [CrossRef] [PubMed]
14. Hoj, P.; Pretorius, I.; Blair, R. Investigations conducted into the nature and amelioration of taints in grapes and wine, caused by smoke resulting from bushfires. *Aust. Wine Res. Inst. Annu. Rep.* **2003**, 37–39.

15. Kelly, D.; Zerihun, A.; Singh, D.; Vitzthum von Eckstaedt, C.; Gibberd, M.; Grice, K.; Downey, M. Exposure of grapes to smoke of vegetation with varying lignin composition and accretion of lignin derived putative smoke taint compounds in wine. *Food Chem.* **2012**, *135*, 787–798. [CrossRef]
16. Singh, D.; Chong, H.; Pitt, K.; Cleary, M.; Dokoozlian, N.; Downey, M. Guaiacol and 4-methylguaiacol accumulate in wines made from smoke-affected fruit because of hydrolysis of their conjugates. *Aust. J. Grape Wine Res.* **2011**, *17*, S13–S21. [CrossRef]
17. Noestheden, M.; Thiessen, K.; Dennis, E.G.; Tiet, B.; Zandberg, W.F. Quantitating organoleptic volatile phenols in smoke-exposed *Vitis vinifera* berries. *J. Agric. Food Chem.* **2017**, *65*, 8418–8425. [CrossRef]
18. Krstic, M.; Johnson, D.; Herderich, M. Review of smoke taint in wine: Smoke-derived volatile phenols and their glycosidic metabolites in grapes and vines as biomarkers for smoke exposure and their role in the sensory perception of smoke taint. *Aust. J. Grape Wine Res.* **2015**, *21*, 537–553. [CrossRef]
19. Kennison, K.; Wilkinson, K.; Williams, H.; Smith, J.; Gibberd, M. Smoke-derived taint in wine: Effect of postharvest smoke exposure of grapes on the chemical composition and sensory characteristics of wine. *J. Agric. Food Chem.* **2007**, *55*, 10897–10901. [CrossRef]
20. Claughton, C.; Jeffery, C.; Pritchard, M.; Hough, C.; Wheaton, C. Wine Industry's 'Black Summer' as Cost of Smoke Taint, Burnt Vineyards, and Lost Sales Add up. *ABC News*, 28 February 2020.
21. Dutta, R.; Das, A.; Aryal, J. Big data integration shows Australian bushfire frequency is increasing significantly. *R. Soc. Open Sci.* **2016**, *3*, 150241. [CrossRef]
22. Kennison, K. *Bushfire Generated Smoke Taint in Grapes and Wine. Final Report to Grape and Wine Research and Development Corporation*; RD 05/02–3; Department of Agriculture and Food Western Australia: Kalgoorlie, Australia, 2009.
23. Simos, C. The implications of smoke taint and management practices. *Aust. Vitic. Jan./Feb.* **2008**, *12*, 77–80.
24. Fuentes, S.; Tongson, E. Advances in smoke contamination detection systems for grapevine canopies and berries. *Wine Vitic. J.* **2017**, *32*, 36.
25. Fuentes, S.; Tongson, E.J.; De Bei, R.; Gonzalez Viejo, C.; Ristic, R.; Tyerman, S.; Wilkinson, K. Non-invasive tools to detect smoke contamination in grapevine canopies, berries and wine: A remote sensing and machine learning modeling approach. *Sensors* **2019**, *19*, 3335. [CrossRef] [PubMed]
26. Ristic, R.; Fudge, A.; Pinchbeck, K.; De Bei, R.; Fuentes, S.; Hayasaka, Y.; Tyerman, S.; Wilkinson, K. Impact of grapevine exposure to smoke on vine physiology and the composition and sensory properties of wine. *Theor. Exp. Plant Physiol.* **2016**, *28*, 67–83. [CrossRef]
27. Collins, C.; Gao, H.; Wilkinson, K. An observational study into the recovery of grapevines (*Vitis vinifera* L.) following a bushfire. *Am. J. Enol. Vitic.* **2014**, *65*, 285–292. [CrossRef]
28. Mikkelsen, T.N.; Heide-Jørgensen, H.S. Acceleration of leaf senescence in fagus sylvatica l. By low levels of tropospheric ozone demonstrated by leaf colour, chlorophyll fluorescence and chloroplast ultrastructure. *Trees* **1996**, *10*, 145–156. [CrossRef]
29. Calder, J.; Lifferth, G.; Moritz, M.; St Clair, S. Physiological effects of smoke exposure on deciduous and conifer tree species. *Int. J. For. Res.* **2010**, *2010*, 438930. [CrossRef]
30. Hayasaka, Y.; Baldock, G.; Parker, M.; Pardon, K.; Black, C.; Herderich, M.; Jeffery, D. Glycosylation of smoke-derived volatile phenols in grapes as a consequence of grapevine exposure to bushfire smoke. *J. Agric. Food Chem.* **2010**, *58*, 10989–10998. [CrossRef]
31. Hayasaka, Y.; Dungey, K.; Baldock, G.; Kennison, K.; Wilkinson, K. Identification of a β-d-glucopyranoside precursor to guaiacol in grape juice following grapevine exposure to smoke. *Anal. Chim. Acta* **2010**, *660*, 143–148. [CrossRef]
32. Hayasaka, Y.; Parker, M.; Baldock, G.A.; Pardon, K.H.; Black, C.A.; Jeffery, D.W.; Herderich, M.J. Assessing the impact of smoke exposure in grapes: Development and validation of a HPLC-MS/MS method for the quantitative analysis of smoke-derived phenolic glycosides in grapes and wine. *J. Agric. Food Chem.* **2013**, *61*, 25–33. [CrossRef]
33. Pollnitz, A.P.; Pardon, K.H.; Sykes, M.; Sefton, M.A. The effects of sample preparation and gas chromatograph injection techniques on the accuracy of measuring guaiacol, 4-methylguaiacol and other volatile oak compounds in oak extracts by stable isotope dilution analyses. *J. Agric. Food Chem.* **2004**, *52*, 3244–3252. [CrossRef] [PubMed]
34. Fudge, A.; Wilkinson, K.; Ristic, R.; Cozzolino, D. Classification of smoke tainted wines using mid-infrared spectroscopy and chemometrics. *J. Agric. Food Chem.* **2012**, *60*, 52–59. [CrossRef] [PubMed]

35. Fudge, A.; Wilkinson, K.; Ristic, R.; Cozzolino, D. Synchronous two-dimensional MIR correlation spectroscopy (2d-cos) as a novel method for screening smoke tainted wine. *Food Chem.* **2013**, *139*, 115–119. [CrossRef] [PubMed]
36. van der Hulst, L.; Munguia, P.; Culbert, J.A.; Ford, C.M.; Burton, R.A.; Wilkinson, K.L. Accumulation of volatile phenol glycoconjugates in grapes following grapevine exposure to smoke and potential mitigation of smoke taint by foliar application of kaolin. *Planta* **2019**, *249*, 941–952. [CrossRef] [PubMed]
37. Barbin, D.F.; Felicio, A.L.D.S.M.; Sun, D.-W.; Nixdorf, S.L.; Hirooka, E.Y. Application of infrared spectral techniques on quality and compositional attributes of coffee: An overview. *Food Res. Int.* **2014**, *61*, 23–32. [CrossRef]
38. Urraca, R.; Sanz-Garcia, A.; Tardaguila, J.; Diago, M.P. Estimation of total soluble solids in grape berries using a handheld NIR spectrometer under field conditions. *J. Sci. Food Agric.* **2016**, *96*, 3007–3016. [CrossRef] [PubMed]
39. Dos Santos, C.A.T.; Lopo, M.; Ricardo, N.; Lopes, J. A review on the applications of portable near-infrared spectrometers in the agro-food industry. *Appl. Spectrosc.* **2013**, *67*, 1215–1233. [CrossRef]
40. Hall, A. Remote sensing applications for viticultural terroir analysis. *Elements* **2018**, *14*, 185–190. [CrossRef]
41. Yu, J.; Wang, H.; Zhan, J.; Huang, W. Review of recent UV–Vis and infrared spectroscopy researches on wine detection and discrimination. *Appl. Spectrosc. Rev.* **2018**, *53*, 65–86. [CrossRef]
42. Zhang, W.; Li, N.; Feng, Y.; Su, S.; Li, T.; Liang, B. A unique quantitative method of acid value of edible oils and studying the impact of heating on edible oils by UV–Vis spectrometry. *Food Chem.* **2015**, *185*, 326–332. [CrossRef]
43. Ferreiro-González, M.; Ruiz-Rodríguez, A.; Barbero, G.F.; Ayuso, J.; Álvarez, J.A.; Palma, M.; Barroso, C.G. FT-IR, Vis spectroscopy, color and multivariate analysis for the control of ageing processes in distinctive spanish wines. *Food Chem.* **2019**, *277*, 6–11. [CrossRef]
44. Alves, F.C.G.B.S.; Coqueiro, A.; Março, P.H.; Valderrama, P. Evaluation of olive oils from the mediterranean region by UV–Vis spectroscopy and independent component analysis. *Food Chem.* **2019**, *273*, 124–129. [CrossRef] [PubMed]
45. Mandrile, L.; Zeppa, G.; Giovannozzi, A.M.; Rossi, A.M. Controlling protected designation of origin of wine by raman spectroscopy. *Food Chem.* **2016**, *211*, 260–267. [CrossRef]
46. Larraín, M.; Guesalaga, A.R.; Agosín, E. A multipurpose portable instrument for determining ripeness in wine grapes using NIR spectroscopy. *IEEE Trans. Instrum. Meas.* **2008**, *57*, 294–302. [CrossRef]
47. González-Caballero, V.; Pérez-Marín, D.; López, M.-I.; Sánchez, M.-T. Optimization of NIR spectral data management for quality control of grape bunches during on-vine ripening. *Sensors* **2011**, *11*, 6109–6124. [CrossRef] [PubMed]
48. Barnaba, F.E.; Bellincontro, A.; Mencarelli, F. Portable NIR-AOTF spectroscopy combined with winery FTIR spectroscopy for an easy, rapid, in-field monitoring of sangiovese grape quality. *J. Sci. Food Agric.* **2014**, *94*, 1071–1077. [CrossRef]
49. Fernández-Novales, J.; López, M.-I.; Sánchez, M.-T.; Morales, J.; González-Caballero, V. Shortwave-near infrared spectroscopy for determination of reducing sugar content during grape ripening, winemaking, and aging of white and red wines. *Food Res. Int.* **2009**, *42*, 285–291. [CrossRef]
50. Cao, F.; Wu, D.; He, Y. Soluble solids content and ph prediction and varieties discrimination of grapes based on visible–near infrared spectroscopy. *Comput. Electron. Agric.* **2010**, *71*, S15–S18. [CrossRef]
51. Dolatabadi, Z.; Elhami Rad, A.H.; Farzaneh, V.; Akhlaghi Feizabad, S.H.; Estiri, S.H.; Bakhshabadi, H. Modeling of the lycopene extraction from tomato pulps. *Food Chem.* **2016**, *190*, 968–973. [CrossRef]
52. Gumus, Z.P.; Ertas, H.; Yasar, E.; Gumus, O. Classification of olive oils using chromatography, principal component analysis and artificial neural network modelling. *J. Food Meas. Charact.* **2018**, *12*, 1325–1333. [CrossRef]
53. Janik, L.J.; Cozzolino, D.; Dambergs, R.; Cynkar, W.; Gishen, M. The prediction of total anthocyanin concentration in red-grape homogenates using visible-near-infrared spectroscopy and artificial neural networks. *Anal. Chim. Acta* **2007**, *594*, 107–118. [CrossRef] [PubMed]
54. Pero, M.; Askari, G.; Skåra, T.; Skipnes, D.; Kiani, H. Change in the color of heat-treated, vacuum-packed broccoli stems and florets during storage: Effects of process conditions and modeling by an artificial neural network. *J. Sci. Food Agric.* **2018**, *98*, 4151–4159. [CrossRef] [PubMed]

55. Merzlyak, M.N.; Gitelson, A.A.; Chivkunova, O.B.; Rakitin, V.Y. Non-destructive optical detection of pigment changes during leaf senescence and fruit ripening. *Physiol. Plant.* **1999**, *106*, 135–141. [CrossRef]
56. Overbeck, V.; Schmitz, M.; Blanke, M. Non-destructive sensor-based prediction of maturity and optimum harvest date of sweet cherry fruit. *Sensors* **2017**, *17*, 277. [CrossRef] [PubMed]
57. Solomakhin, A.A.; Blanke, M.M. Overcoming adverse effects of hailnets on fruit quality and microclimate in an apple orchard. *J. Sci. Food Agric.* **2007**, *87*, 2625–2637. [CrossRef]
58. Merzlyak, M.N.; Solovchenko, A.E.; Gitelson, A.A. Reflectance spectral features and non-destructive estimation of chlorophyll, carotenoid and anthocyanin content in apple fruit. *Postharvest Biol. Technol.* **2003**, *27*, 197–211. [CrossRef]
59. Sims, D.A.; Gamon, J.A. Relationships between leaf pigment content and spectral reflectance across a wide range of species, leaf structures and developmental stages. *Remote Sens. Environ.* **2002**, *81*, 337–354. [CrossRef]
60. Szeto, C.; Ristic, R.; Capone, D.; Puglisi, C.; Pagay, V.; Culbert, J.; Jiang, W.; Herderich, M.; Tuke, J.; Wilkinson, K. Uptake and glycosylation of smoke-derived volatile phenols by *Cabernet sauvignon* grapes and their subsequent fate during winemaking. *Molecules* **2020**, *25*, 3720. [CrossRef]
61. Ristic, R.; Osidacz, P.; Pinchbeck, K.; Hayasaka, Y.; Fudge, A.; Wilkinson, K. The effect of winemaking techniques on the intensity of smoke taint in wine. *Aust. J. Grape Wine Res.* **2011**, *17*, S29–S40. [CrossRef]
62. Caravia, L.; Pagay, V.; Collins, C.; Tyerman, S.D. Application of sprinkler cooling within the bunch zone during ripening of *Cabernet sauvignon* berries to reduce the impact of high temperature. *Aust. J. Grape Wine Res.* **2017**, *23*, 48–57. [CrossRef]
63. Mate, A.R.; Deshmukh, R.R. Analysis of effects of air pollution on chlorophyll, water, carotenoid and anthocyanin content of tree leaves using spectral indices. *Int. J. Eng. Sci* **2016**, *6*, 5465–5474.
64. Gitelson, A.A.; Zur, Y.; Chivkunova, O.B.; Merzlyak, M.N. Assessing carotenoid content in plant leaves with reflectance spectroscopy. *Photochem. Photobiol.* **2002**, *75*, 272–281. [CrossRef]
65. García-Estévez, I.; Quijada-Morín, N.; Rivas-Gonzalo, J.C.; Martínez-Fernández, J.; Sánchez, N.; Herrero-Jiménez, C.M.; Escribano-Bailón, M.T. Relationship between hyperspectral indices, agronomic parameters and phenolic composition of *Vitis vinifera* cv *Tempranillo* grapes. *J. Sci. Food Agric.* **2017**, *97*, 4066–4074. [CrossRef] [PubMed]
66. De Vries, C.; Buica, A.; McKay, J.B.M. The impact of smoke from vegetation fires on sensory characteristics of *Cabernet sauvignon* wines made from affected grapes. *S. Afr. J. Enol. Vitic.* **2016**, *37*, 22–30. [CrossRef]
67. Boido, E.; Fariña, L.; Carrau, F.; Dellacassa, E.; Cozzolino, D. Characterization of glycosylated aroma compounds in tannat grapes and feasibility of the near infrared spectroscopy application for their prediction. *Food Anal. Methods* **2013**, *6*, 100–111. [CrossRef]
68. Cynkar, W.; Cozzolino, D.; Dambergs, R.; Janik, L.; Gishen, M. Effect of variety, vintage and winery on the prediction by visible and near infrared spectroscopy of the concentration of glycosylated compounds (g-g) in white grape juice. *Aust. J. Grape Wine Res.* **2007**, *13*, 101–105. [CrossRef]
69. Burns, D.; Ciurczak, E. *Handbook of Near-Infrared Analysis*; CRC Press: Boca Raton, FL, USA, 2007.
70. Osborne, B.G.; Fearn, T.; Hindle, P.H. *Practical NIR Spectroscopy with Applications in Food and Beverage Analysis*; Longman Scientific & Technical: Harlow, UK, 1993; Volume 2.
71. Cozzolino, D.; Cynkar, W.; Dambergs, R.; Janik, L.; Gishen, M. Effect of both homogenisation and storage on the spectra of red grapes and on the measurement of total anthocyanins, total soluble solids and pH by visual near infrared spectroscopy. *J. Near Infrared Spectrosc.* **2005**, *13*, 213–223. [CrossRef]
72. Martelo-Vidal, M.J.; Vazquez, M. Evaluation of ultraviolet, visible, and near infrared spectroscopy for the analysis of wine compounds. *Czech J. Food Sci.* **2014**, *32*, 37–47. [CrossRef]
73. Haboudane, D.; Miller, J.R.; Pattey, E.; Zarco-Tejada, P.J.; Strachan, I.B. Hyperspectral vegetation indices and novel algorithms for predicting green lai of crop canopies: Modeling and validation in the context of precision agriculture. *Remote Sens. Environ.* **2004**, *90*, 337–352. [CrossRef]
74. Hall, A.; Lamb, D.; Holzapfel, B.; Louis, J. Optical remote sensing applications in viticulture—A review. *Aust. J. Grape Wine Res.* **2002**, *8*, 36–47. [CrossRef]
75. Lamb, D.W.; Weedon, M.M.; Bramley, R.G.V. Using remote sensing to predict grape phenolics and colour at harvest in a *Cabernet sauvignon* vineyard: Timing observations against vine phenology and optimising image resolution. *Aust. J. Grape Wine Res.* **2004**, *10*, 46–54. [CrossRef]

76. Iqbal, M.; Jura-Morawiec, J.; Włoch, W.; Mahmooduzzafar. Foliar characteristics, cambial activity and wood formation in *Azadirachta indica* A. Juss. As affected by coal–smoke pollution. *Flora Morphol. Distrib. Funct. Ecol. Plants* **2010**, *205*, 61–71. [CrossRef]
77. Nighat, F.; Iqbal, M. Stomatal conductance, photosynthetic rate, and pigment content in *Ruellia tuberosa* leaves as affected by coal-smoke pollution. *Biol. Plant.* **2000**, *43*, 263–267. [CrossRef]
78. Ren, S.; Chen, X.; An, S. Assessing plant senescence reflectance index-retrieved vegetation phenology and its spatiotemporal response to climate change in the inner mongolian grassland. *Int. J. Biometeorol.* **2017**, *61*, 601–612. [CrossRef] [PubMed]
79. Sandermann, H.; Ernst, D.; Heller, W.; Langebartels, C. Ozone: An abiotic elicitor of plant defence reactions. *Trends Plant Sci.* **1998**, *3*, 47–50. [CrossRef]
80. Bellincontro, A.; Catelli, C.; Cotarella, R.; Mencarelli, F. Postharvest ozone fumigation of Petit Verdot grapes to prevent the use of sulfites and to increase anthocyanin in wine. *Aust. J. Grape Wine Res.* **2017**, *23*, 200–206. [CrossRef]
81. Castellarin, S.D.; Matthews, M.A.; Di Gaspero, G.; Gambetta, G.A. Water deficits accelerate ripening and induce changes in gene expression regulating flavonoid biosynthesis in grape berries. *Planta* **2007**, *227*, 101–112. [CrossRef]
82. Allied Scientific Pro. Lighting Passport the World's First Smart Handheld Spectrometer. 2017. Available online: https://lightingpassport.alliedscientificpro.com (accessed on 10 August 2020).
83. Chandraratne, M.; Kulasiri, D.; Samarasinghe, S. Classification of lamb carcass using machine vision: Comparison of statistical and neural network analyses. *J. Food Eng.* **2007**, *82*, 26–34. [CrossRef]
84. Al-Alawi, S.M.; Abdul-Wahab, S.A.; Bakheit, C.S. Combining principal component regression and artificial neural networks for more accurate predictions of ground-level ozone. *Environ. Model. Softw.* **2008**, *23*, 396–403. [CrossRef]
85. Diamantopoulou, M.J.; Milios, E. Modelling total volume of dominant pine trees in reforestations via multivariate analysis and artificial neural network models. *Biosyst. Eng.* **2010**, *105*, 306–315. [CrossRef]

© 2020 by the authors. Licensee MDPI, Basel, Switzerland. This article is an open access article distributed under the terms and conditions of the Creative Commons Attribution (CC BY) license (http://creativecommons.org/licenses/by/4.0/).

Article

Early Detection of Aphid Infestation and Insect-Plant Interaction Assessment in Wheat Using a Low-Cost Electronic Nose (E-Nose), Near-Infrared Spectroscopy and Machine Learning Modeling

Sigfredo Fuentes [1,*], Eden Tongson [1], Ranjith R. Unnithan [2] and Claudia Gonzalez Viejo [1]

1. Digital Agriculture Food and Wine Group, School of Agriculture and Food, Faculty of Veterinary and Agricultural Sciences, University of Melbourne, Melbourne, VIC 3010, Australia; eden.tongson@unimelb.edu.au (E.T.); cgonzalez2@unimelb.edu.au (C.G.V.)
2. Department of Electrical and Electronic Engineering, School of Engineering, University of Melbourne, Melbourne, VIC 3010, Australia; r.ranjith@unimelb.edu.au
* Correspondence: sfuentes@unimelb.edu.au

Citation: Fuentes, S.; Tongson, E.; Unnithan, R.R.; Gonzalez Viejo, C. Early Detection of Aphid Infestation and Insect-Plant Interaction Assessment in Wheat Using a Low-Cost Electronic Nose (E-Nose), Near-Infrared Spectroscopy and Machine Learning Modeling. *Sensors* **2021**, *21*, 5948. https://doi.org/10.3390/s21175948

Academic Editor: Asim Biswas

Received: 28 July 2021
Accepted: 1 September 2021
Published: 4 September 2021

Publisher's Note: MDPI stays neutral with regard to jurisdictional claims in published maps and institutional affiliations.

Copyright: © 2021 by the authors. Licensee MDPI, Basel, Switzerland. This article is an open access article distributed under the terms and conditions of the Creative Commons Attribution (CC BY) license (https://creativecommons.org/licenses/by/4.0/).

Abstract: Advances in early insect detection have been reported using digital technologies through camera systems, sensor networks, and remote sensing coupled with machine learning (ML) modeling. However, up to date, there is no cost-effective system to monitor insect presence accurately and insect-plant interactions. This paper presents results on the implementation of near-infrared spectroscopy (NIR) and a low-cost electronic nose (e-nose) coupled with machine learning. Several artificial neural network (ANN) models were developed based on classification to detect the level of infestation and regression to predict insect numbers for both e-nose and NIR inputs, and plant physiological response based on e-nose to predict photosynthesis rate (A), transpiration (E) and stomatal conductance (gs). Results showed high accuracy for classification models ranging within 96.5–99.3% for NIR and between 94.2–99.2% using e-nose data as inputs. For regression models, high correlation coefficients were obtained for physiological parameters (gs, E and A) using e-nose data from all samples as inputs (R = 0.86) and R = 0.94 considering only control plants (no insect presence). Finally, R = 0.97 for NIR and R = 0.99 for e-nose data as inputs were obtained to predict number of insects. Performances for all models developed showed no signs of overfitting. In this paper, a field-based system using unmanned aerial vehicles with the e-nose as payload was proposed and described for deployment of ML models to aid growers in pest management practices.

Keywords: remote sensing; volatile compounds; artificial neural networks; photosynthesis modeling; plant water status modeling

1. Introduction

Early detection of insect infestation in crops is critical for decision-making related to pest management and alerting potential infestation to neighboring susceptible crops. One of the most common agronomical assessments for detrimental insect infestation in crops is visual at determined and critical periods of the crop development in synchronicity with the insect's population dynamics [1] and migrations [2]. The next step for more practical monitoring is using pheromone traps [3], which can be used for more ecological pest management [4]. Some of these pheromone traps have been integrated with digital technologies, such as video cameras [5] to assess effectiveness [6] and implementing computer vision for pest identification and automatic counting using machine learning [7–12]. Some of these systems are web-based and used to support agronomical decision-making in developing countries [8].

Even though these applications are certainly an advancement in automated pest monitoring and management, they still rely on sentinel locations within the crop field.

The latter could translate into an economic limitation for extensive crops, which require a significant number of monitoring nodes and increasing complexity of the sensor network. Furthermore, these monitoring and counting systems do not give much information on the insect-plant interaction, insect natural predator's interaction, or detrimental effects or symptomatology from the plant's perspective.

Other remote sensing techniques have been implemented for pest detection in crops [13] based on sensor networks [14], IoT for moths [15], hyperspectral imaging based on airborne [16], satellite [17], and unmanned aerial vehicles (UAV) [18], among others. These systems offer the advantage of increased spatial resolution and potential temporal resolution in airborne and UAV platforms. However, there are some disadvantages related to the plant-based nature of remote sensing monitoring. The first disadvantage is related to monitoring and modeling based on plant symptomatology in response to insect attacks, often assessed late, with detrimental implications in yield and quality of produce. Another disadvantage is that there is no assurance that symptomatology targeted using remote sensing to detect insects are entirely related to the specific biotic stress of interest. Some plants may have other biotic and abiotic symptomatology, such as water, salinity, and other insect interaction stresses. These issues could create biases in models developed and hinder capabilities of deployment of models to other locations.

Hence, there is a need for a digital system that considers the early detection of the pest of interest and early interaction with the host plant. To understand the specific insect-plant interactions for machine learning modeling purposes, controlled experiments must be considered before deployment in field conditions. Furthermore, a digital system based on volatile compounds could offer advantages compared to other systems. The implementation of electronic noses (e-noses) for insect detection have been proposed for disease detection and diagnosis [19] and pest detection [20], specifically for cotton [21], as a portable e-nose development, and specifically for aphid detection on tomato plants [22] using four low-cost gas sensors and comparing with gas chromatography results. In wheat, some authors have also used commercial e-noses to detect mite infestation [23] to predict the age and insect damage during storage using linear discriminant analysis [24], and to detect rusty grain beetle, *Cryptolestes ferrugineus*, and red flour beetle in wheat [25]. Some studies have also combined computer vision systems and e-noses for pests in agriculture [26]. There is an increasing interest in developing compact, portable, and low-cost e-noses for these purposes [27]. However, most of these new researches are focused only on the detection of the variation in volatile compounds related to the insect presence and the interaction between insects and plants [25,28–30], and in some researches, combining e-nose and computer vision [26] for insect detection and identification, but so far, no attempt has been made to separate them through comprehensive modeling on these separate processes.

This paper proposed the implementation of a newly developed e-nose comprised of nine gas sensors described by Gonzalez Viejo et al. (2020) [31] and near-infrared spectroscopy (NIR) for the early detection of aphids (*Rhophalosiphum padi*) on wheat plants in controlled conditions. Raw data from the e-nose and NIR were used as inputs for machine learning algorithms to develop different classification models to detect insect's presence at different phenological stages and regression models to predict the number of insects and physiological responses of plants based on gas-exchange measurements. Furthermore, a deployment system was proposed to validate these models in the field using the e-nose as a UAV payload to test different flying altitudes for detection sensitivity purposes. The latter system could have several advantages compared to research done so far by addressing the gaps discussed above.

The implementation of the proposed system can be highly beneficial to growers being able to provide high temporal and spatial resolution for more precise and targeted decision-making. Furthermore, the deployment of this system could support not only pest detection and management but also other agronomical activities, such as plant water status and irrigation scheduling and the detection of other biotic and abiotic stresses.

2. Materials and Methods

2.1. Plant and Insect Material, and Experimental Design Description

Wheat seeds of Kittyhawk variety (Pacific Seeds, Toowoomba, QLD, Australia) were surface sterilized with 0.8% sodium hypochlorite and were pre-germinated in the dark at 4 °C for 48 h, followed by lit conditions (17–25 °C) for 72 h. The germinated seeds were transferred individually to Jiffy-7® pellets (Jiffy Products S.L. (Private) Ltd., Mirigama, Sri Lanka). The seedlings were further grown to a two-leaf stage (GS12) prior to transplanting in pots.

The plants were grown in a non-circulating passive hydroponic method based on Kratky [32]. The wheat seedlings were transplanted into 3 Li (190 mm × 170 mm) hydroponic pots (Anti-Spiral Pot, Garden City Plastics, Dandenong South, VIC, Australia) filled with expanded clay pebbles (CANNA Aqua Clay Pebbles, Subiaco, WA, Australia) as substrate, with three seedlings placed equidistant in each pot. Duplicate pots were placed in a black plastic tub filled with modified Hoagland nutrient solution [33] up to root submergence level. The nutrient solutions were replaced every two weeks throughout the experiment. Each tub of hydroponic set-up is placed inside an insect rearing tent (Bug-Dorm, Australian Entomological Supplies Pty., Ltd., South Murwillumbah, NSW, Australia) constructed with nylon mesh with 160 μm aperture. The plants were maintained inside a growth room (Biosciences Glasshouse Complex, The University of Melbourne, Parkville, VIC, Australia) with 16 h daylight/8 h night and 20–25 °C controlled automatically.

Oat aphids (*Rhophalosiphum padi*) were obtained from laboratory cultures of Pest & Environmental Adaptation Research Group, School of Biosciences, The University of Melbourne, Australia. The starting colony was allowed to reproduce for population increase in a rearing tent supplied with wheat plants (in a similar hydroponic set-up described above). Adult *R. padi* were randomly selected from the colony plants and introduced into the experimental plants, approximately at stem elongation stage with third leaf emerged (GS32). Three treatments were determined based on the economic threshold for winter cereals which is an average of 15 aphids per tiller on 50% of tillers [34]: high load (15 aphids per tiller in 50% of tillers), medium load (10 aphids per tiller in 50% of tillers), and low aphid load (5 aphids per tiller in 50% of tillers). The aphids were carefully transferred into the wheat plants with a fine natural bristle brush. For simplicity, days referred in models developed correspond to days after infestation at the wheat phenological stage GS32.

A total of eight experimental set-ups were made with duplicate set-ups for each treatment (low, medium, and high aphid load) and two aphid-free set-ups as controls. Each experimental set-up was composed of one insect rearing tent, containing two pots with each pot planted with three wheat plants, maintained in hydroponics as described above and shown in Figure 1. These were randomly arranged inside the growth room.

Insect population models (adults) were developed using initial insect infestations and exponential growth models applicable to sigmoidal population insect growth [35,36]. Curves were adjusted by image analysis and manual insect counting per leaf, and extrapolation per plant in the middle and end of the experiment to account for insect mortality (data not shown).

2.2. Physiological Measurements

Plant physiological parameters such as stomatal conductance (gs; mol H_2O m^{-2} s^{-1}), transpiration (E; mmol H_2O m^{-2} s^{-1}), and photosynthesis (A; μmol CO_2 m^{-2} s^{-1}) were measured using a Li-6400 XT open gas exchange system (Li-Cor Inc., Environmental Sciences, Lincoln, NE, USA). Measurements were made on the youngest fully expanded leaves, repeated three times in different tillers of each plant (n = 18 per tent; n = 36 per treatment).

Figure 1. Wheat plants were grown in non-circulating passive hydroponic system (**left**) and contained in insect rearing tents (**right**).

2.3. Near-Infrared Spectroscopy Measurements

A single leaf of each wheat plant (three per pot and six leaves per tent) was measured on six different spots ($n = 36$ per tent; $n = 72$ per treatment) using a handheld near-infrared (NIR) spectroscopy device (MicroPHAZIR™ RX; Thermo Fisher Scientific, Waltham, MA, USA). This device measures the absorbance values within the 1596–2396 nm wavelength range. A blank reference was used as background to calibrate the device every 10 measurements and was placed on the top of the leaf while measuring to avoid recording noise from the environment (Figure 2). The raw absorbance values were used for all analyses presented in this study.

Figure 2. Photosynthetic gas exchange (**left**) and near-infrared (**right**) devices while taking measurements.

2.4. Electronic Nose Measurements

A portable and low-cost electronic nose (e-nose) developed by the Digital Agriculture Food and Wine Group and the Department of Electrical and Electronic Engineering from The University of Melbourne was used to assess volatile compounds produced by the control plants and treatments with aphids. This e-nose consists of an array of nine sensors sensitive to different gases: (i) MQ3 (alcohol), (ii) MQ4 (methane: CH_4), (iii) MQ7 (carbon monoxide: CO), (iv) MQ8 (hydrogen: H_2), (v) MQ135 (ammonia/alcohol/benzene), (vi) MQ136 (hydrogen sulfide: H_2S), (vii) MQ137 (ammonia), (viii) MQ138 (benzene/alcohol/ammonia), and (ix) MG811 (carbon dioxide: CO_2), as well as a humidity and temperature sensor to measure the environment conditions (Figure 3; Henan Hanwei Electronics Co., Ltd., Henan, China). The e-nose was calibrated for ~30 s prior to recording each measurement to ensure all sensors reached the baseline and then placed inside the tent on top of the plants to record data for 1.5 min; each tent was measured in triplicates. The output data (Volts) were then analyzed using a code written in MATLAB® R2020a (Mathworks Inc., Natick, MA, USA) to extract the mean values of ten segments from the highest peak of the curves as described by Gonzalez Viejo et al. [37].

(a)   (b)

(c)

Figure 3. Electronic nose (e-nose) showing (**a**) the front part with gas sensors and their model ID (Henan Hanwei Electronics Co., Ltd., Henan, China) and (**b**) the back part which holds the humidity/temperature sensor; (**c**) Shows the e-nose positioning while taking measurements.

2.5. Statistical Analysis and Machine Learning Modeling

Physiological and e-nose data were analyzed using ANOVA to assess significant differences ($p < 0.05$) between samples; additionally, a Tukey honestly significant difference (HSD) *post hoc* test ($\alpha = 0.05$) was conducted using XLSTAT v.2020.3.1 (Addinsoft, New York, NY, USA). These data were then analyzed for significant correlations ($p < 0.05$) based on covariance using MATLAB® R2020a and represented with a matrix.

Several machine learning models based on artificial neural networks (ANN) were developed with three different purposes to (i) predict physiological data using e-nose outputs and the infestation level (control: 0, low: 0.25, medium: 0.75, and high: 1) as inputs using data from all treatments (Model 1), and only the baseline and control treatments (Model 2), (ii) classify samples into the different infestation treatments (control, low, medium, and high) using the NIR absorbance values (Models 3–7), and e-nose outputs (Models 8–12) as inputs, and (iii) predict the number of aphids using the NIR absorbance values (Model 13) and e-nose outputs (Model 14) as inputs. All models were constructed using a customized code written in MATLAB® R2020a to test 17 different training algorithms in a loop and find the best models based on accuracy and performance [38,39]. Furthermore, a neuron trimming test (3, 5, 7, and 10 neurons) was performed to assess the most optimal number of neurons to avoid under- or over-fitting of the models (data not shown). The regression models (i, iii) consisted of a feedforward network with a hidden (tan-sigmoid function) and an output (linear transfer function) layer. On the other hand, the classification models (ii) consisted of a feedforward network with a hidden (tan-sigmoid function) and an output (Softmax neurons) layer.

The best models to predict the physiological data (photosynthesis, stomatal conductance, transpiration) were developed using the Bayesian Regularization training algorithm for regression modeling. For this, two models were developed: Model 1 using as inputs the e-nose outputs and infestation level (control: 0, low: 0.25, medium: 0.75, and high: 1) from all measurements and treatments (general model), and Model 2 using the e-nose outputs from samples with no insects such as the baseline and control (Figure 4a). Data were divided randomly as 70% for training and 30% for testing using a performance algorithm based on means squared error (MSE).

Models to classify the samples into the different treatments (Figure 4b) using the NIR absorbance values as inputs were developed using the Levenberg–Marquardt training algorithm. One model was developed per day of measurement as Model 3 (baseline + Day 3), Model 4 (Day 7), Model 5 (Day 10), Model 6 (Day 14), and Model 7 (Day 17) to assess the level of infestation at different stages. Data were divided randomly as 70% for training, 15% for validation using a performance algorithm based on MSE, and 15% for testing. On the other hand, models to classify the samples into the different treatments using the e-nose outputs as inputs were constructed using the Bayesian Regularization training algorithm. Same as the previous, one model was developed per day of measurements as Model 8 (baseline + Day 3), Model 9 (Day 7), Model 10 (Day 10), Model 11 (Day 14), and Model 12 (Day 17). Data were also divided randomly as 70% for training and 30% for the testing stage using the MSE performance algorithm.

The Bayesian Regularization training algorithm produced the best models to predict the number of aphids using the NIR absorbance values (Model 13) and e-nose outputs (Model 14) from days 7 to 17 as inputs (Figure 4c). A random data division was used as 70% for training and 30% for testing with an MSE performance algorithm.

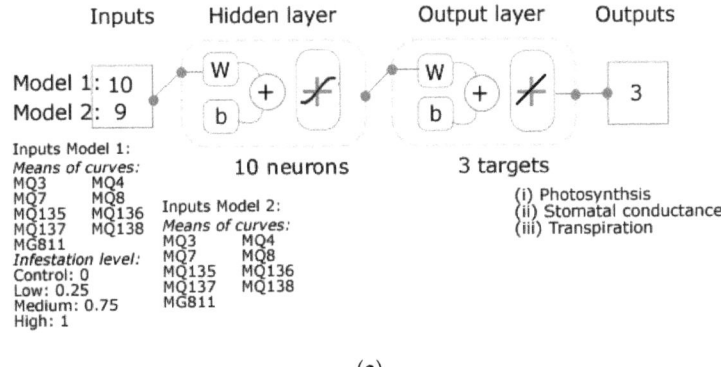

(a)

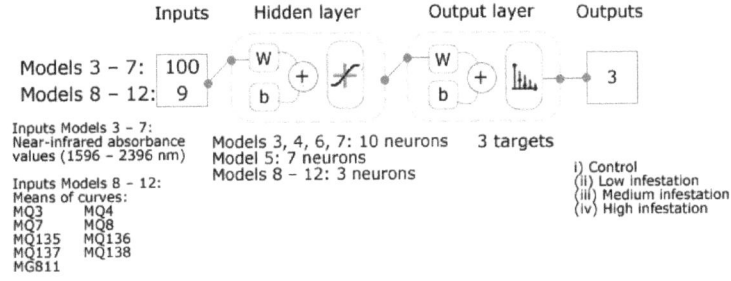

(b)

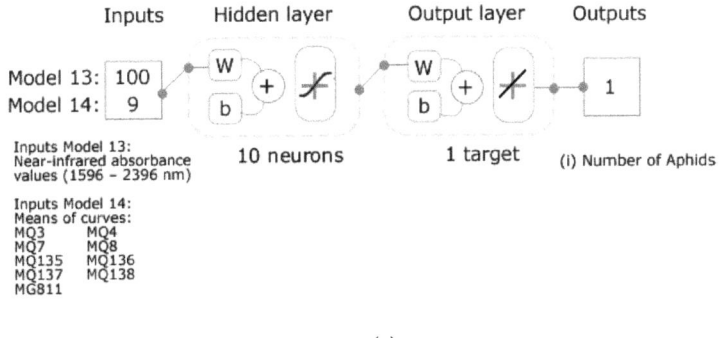

(c)

Figure 4. Diagrams of machine learning models based on artificial neural networks showing (a) the structure of regression Models 1 and 2; (b) Pattern recognition Models 3 to 12, and (c) Regression Models 13 and 14. Abbreviations: W: weights; b: bias; electronic nose sensors MQ3: alcohol; MQ4: methane; MQ7: carbon monoxide; MQ8: hydrogen; MQ135: ammonia/alcohol/benzene; MQ136: hydrogen sulfide; MQ137: ammonia; MQ138: benzene/alcohol/ammonia; MG811: carbon dioxide.

For Models 3–14, six support vector machine (SVM) algorithms (i) linear, (ii) quadratic, (iii) cubic, (iv) fine Gaussian, (v) medium Gaussian, and (vi) coarse Gaussian were also tested to compare results with ANN and find the best models. These algorithms were run using the Classification and Regression Learner applications in MATLAB® Statistics and Machine Learning Toolbox 12.1. Accuracy percentage for classification and correlation coefficient (R) and MSE for regression models were considered to compare the different ML methods/algorithms. However, only accuracy percentage and R values are reported in results due to their lower accuracy compared to ANN. These algorithms were not tested for Models 1 and 2 because SVM algorithms are unable to construct multi-target models, which makes them inefficient for further deployment.

3. Results

Table 1 shows non-significant differences ($p > 0.05$) between treatments for baseline measurements of any physiological parameters. For photosynthesis, at days 10 and 17, the control was significantly higher ($p < 0.05$; 12.47 and 12.57 µmol m^{-2} s^{-1}, respectively) than the infested treatments. Similarly, stomatal conductance was significantly higher for control at days 7, 10, and 17 (0.51, 0.55, and 0.62 mol m^{-2} s^{-1}, respectively). On the other hand, transpiration was significantly higher for control in all measurement days (day 3–17) with values within the 3.60–6.00 mmol m^{-2} s^{-1} range.

Figure 5a shows that the non-infested plants presented higher absorbance values at Days 10–17, especially within the 1900–2000 nm range, with Day 7 being the lowest. For the infested treatments (Figure 5b), the major overtones were also within the 1900–2000 nm range. The lowest absorbance values were at Day 7 for all low, medium, and high treatments, while the highest values were found at Day 17 for the medium and high treatments.

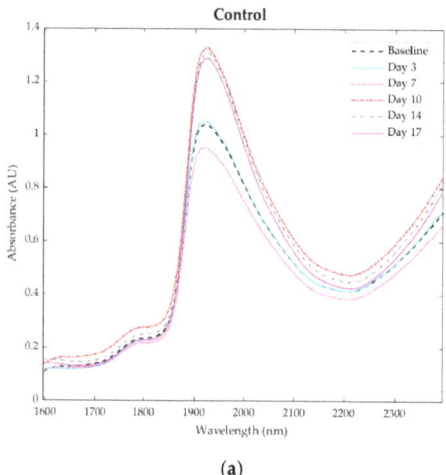
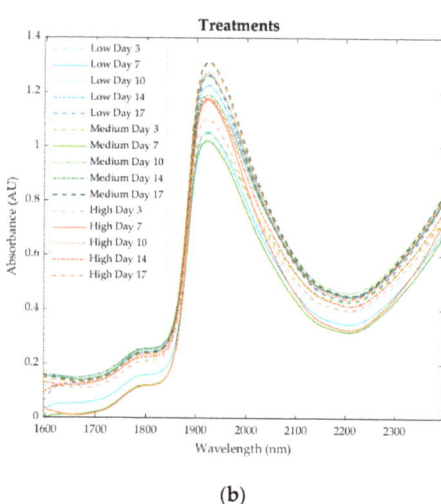

Figure 5. Near-infrared curves showing the absorbance values within the 1596–2396 nm wavelength range for (**a**) the control measurements and (**b**) the treatments (low, medium, high infestation) measured at different dates.

Figure 6 shows there were significant differences ($p < 0.05$) between treatments in all measurement days for all sensors, except for MQ4 (CH$_4$) on Day 3, MQ136 (H$_2$S) on Days 3, 14, and 17, and MQ8 (H$_2$) at Days 14 and 17. It can be observed that the highest values were found in sensors MG811 (CO$_2$), MQ4 (CH$_4$), MQ3 (alcohol), and MQ7 (CO).

Table 1. Results from the physiological data of the four treatments. Numbers on the top represent the mean values, while numbers at the bottom are the standard error.

Sample/Parameter		Photosynthesis (μmol CO_2 m^{-2} s^{-1})						Stomatal Conductance (mol H_2O m^{-2} s^{-1})						Transpiration (mmol H_2O m^{-2} s^{-1})					
Measurement	BL	D3	D7	D10	D14	D17	BL	D3	D7	D10	D14	D17	BL	D3	D7	D10	D14	D17	
Control	6.78	9.18 [a]	13.78 [a]	12.47 [a]	13.22 [a]	12.75 [a]	0.16	0.32 [a]	0.51 [a]	0.55 [a]	0.50 [a]	0.62 [a]	2.35	3.60 [a]	4.84 [a]	4.16 [a]	4.00 [a]	6.00 [a]	
	±0.32	±0.01	±0.13	±0.30	±0.02	±0.12	±0.16	±0.01	±0.05	±0.25	±0.02	±0.09	±0.24	±0.02	±0.06	±0.28	±0.02	±0.06	
Low	4.50	9.65 [ab]	11.34 [b]	10.49 [b]	11.42 [b]	10.27 [b]	0.07	0.28 [a]	0.35 [c]	0.36 [b]	0.37 [bc]	0.40 [c]	1.29	3.09 [b]	3.81 [c]	3.53 [b]	2.95 [c]	4.96 [b]	
	±0.23	±0.01	±0.09	±0.40	±0.02	±0.13	±0.41	±0.03	±0.16	±0.27	±0.02	±0.12	±0.27	±0.02	±0.10	±0.23	±0.02	±0.11	
Medium	7.03	7.52 [c]	12.19 [b]	10.37 [b]	10.70 [b]	10.84 [b]	0.15	0.16 [b]	0.44 [ab]	0.36 [b]	0.33 [c]	0.51 [b]	2.19	2.11 [c]	4.34 [b]	3.12 [b]	2.79 [c]	5.36 [b]	
	±0.34	±0.02	±0.16	±0.65	±0.02	±0.20	±0.19	±0.02	±0.13	±0.33	±0.03	±0.17	±0.34	±0.03	±0.19	±0.24	±0.03	±0.15	
High	7.03	10.93 [a]	13.49 [a]	10.67 [b]	13.25 [a]	11.07 [b]	0.18	0.27 [a]	0.44 [b]	0.35 [b]	0.44 [ab]	0.50 [b]	2.47	2.73 [b]	4.43 [b]	3.21 [b]	3.53 [b]	5.34 [b]	
	±0.37	±0.02	±0.15	±0.34	±0.01	±0.09	±0.22	±0.02	±0.07	±0.21	±0.02	±0.10	±0.25	±0.02	±0.09	±0.30	±0.02	±0.11	

Abbreviations: BL: baseline; D: Day. Different letters denote significant differences between treatments according to ANOVA ($p < 0.05$) and Tukey honestly significant difference *post hoc* test ($\alpha = 0.05$).

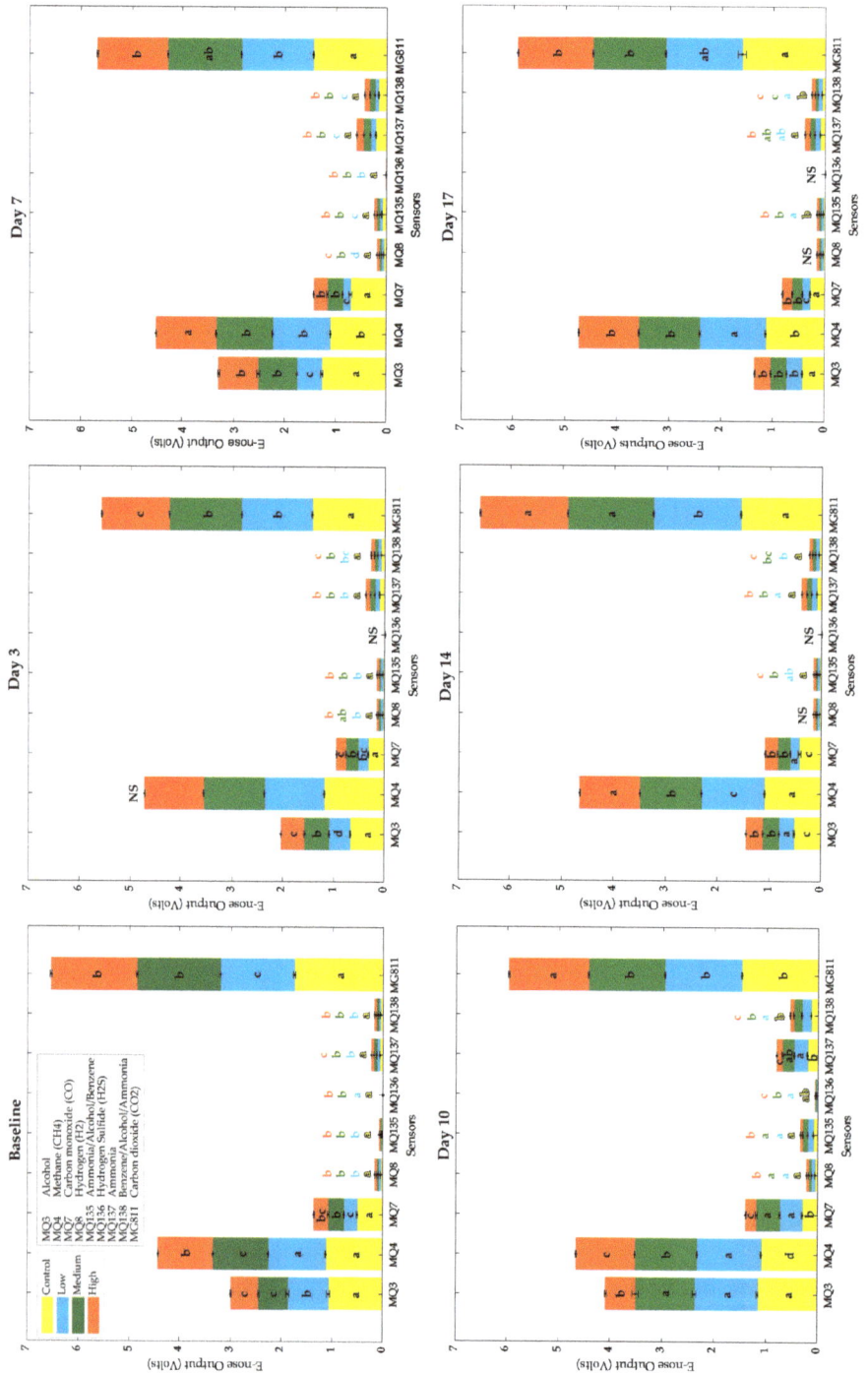

Figure 6. Electronic nose (e-nose) outputs (means) for each integrated sensor at each day of measurements. Error bars are based on standard error, and different letters denote significant differences between treatments according to ANOVA ($p < 0.05$) and Tukey honestly significant difference post hoc test ($\alpha = 0.05$).

Figure 7 shows that both photosynthesis and transpiration had a positive and significant correlation ($p < 0.05$) with MQ3 (alcohol; $r = 0.45$ and $r = 0.65$, respectively), and MQ7 (CO; $r = 0.55$ and $r = 0.71$, respectively), and a negative correlation with number of aphids ($r = -0.44$ and $r = -0.59$, respectively) and MQ4 (CH_4; $r = -0.51$ and $r = -0.45$, respectively). Similarly, stomatal conductance had a positive correlation with MQ3 ($r = 0.65$) and MQ7 ($r = 0.69$), and a negative correlation with number of aphids ($r = -0.56$). Transpiration and stomatal conductance were also positively correlated with MQ8 (H_2; $r = 0.43$). On the other hand, number of aphids had a positive correlation with MQ4 ($r = 0.52$) and a negative correlation with MQ3, MQ7, MQ8, MQ135, MQ136, MQ137, and MQ138 with correlations within the $r = -0.56$–-0.81 range.

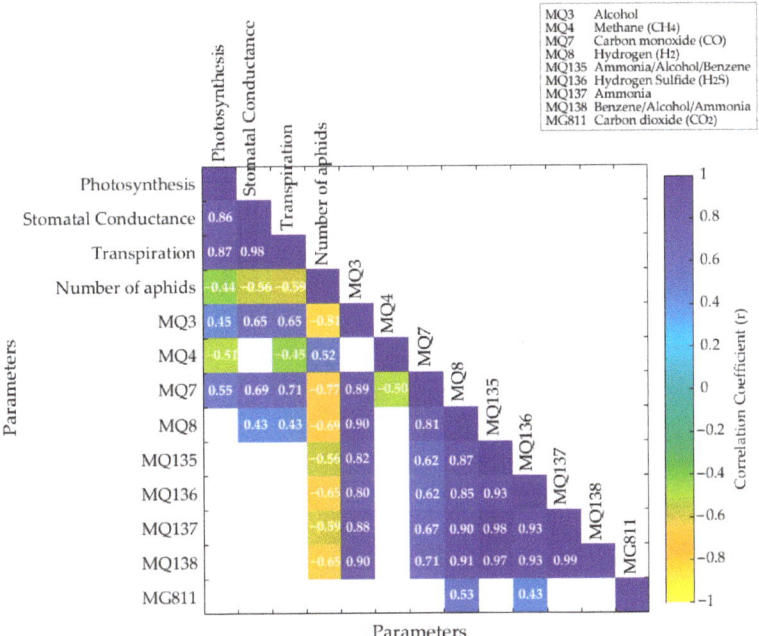

Figure 7. Matrix showing the significant correlations ($p < 0.05$) between the physiological data, number of aphids, and the electronic nose sensors. Color bar represents the negative (yellow) to positive (blue) correlations. Numbers within the boxes denote the correlation coefficients (r).

Table 2 shows the results from the machine learning regression models to predict physiological data (photosynthesis, stomatal conductance, and transpiration) using the e-nose outputs and infestation level as inputs. Model 1 was constructed as a general model using data from all treatments, and measurement days had an overall correlation coefficient $R = 0.86$. It had no signs of under- or overfitting as the MSE value of the training stage (MSE = 0.05) was lower than the testing (MSE = 0.06); however, the slope values were medium ($b = 0.76$). On the other hand, Model 2, which was developed using only the data from non-infested plants (baseline and controls), had high overall accuracy ($R = 0.94$) with high slope values ($b = 0.90$) and no signs of under- or overfitting with training MSE = 0.02 lower than testing MSE = 0.04. The overall models are shown in Figure 8, where data points from Model 1 (Figure 8a) are more dispersed and had 5% of outliers (216 out of 4320) based on the 95% prediction bounds. Model 2 (Figure 8b) also presented 5% of outliers (81 out of 1620), but the slope was closer to the unity ($b = 0.90$). It can also be observed that for Model 2, most of the outliers were from stomatal conductance, while in Model 1, they were more similar for the three targets.

Table 2. Machine learning regression models based on artificial neural networks (Bayesian Regularization) to predict physiological data using the electronic nose outputs as inputs. Abbreviations: R: correlation coefficient; b: slope; MSE: means squared error.

Stage	Samples	Observations	R	b	Performance (MSE)
Model 1—General (all treatments and measurement days)—10 neurons					
Training	1008	3024	0.87	0.75	0.05
Testing	432	1296	0.83	0.75	0.06
Overall	1440	4320	0.86	0.75	-
Model 2—Baseline and control—10 neurons					
Training	378	1134	0.95	0.90	0.02
Testing	162	486	0.93	0.90	0.04
Overall	540	1620	0.94	0.90	-

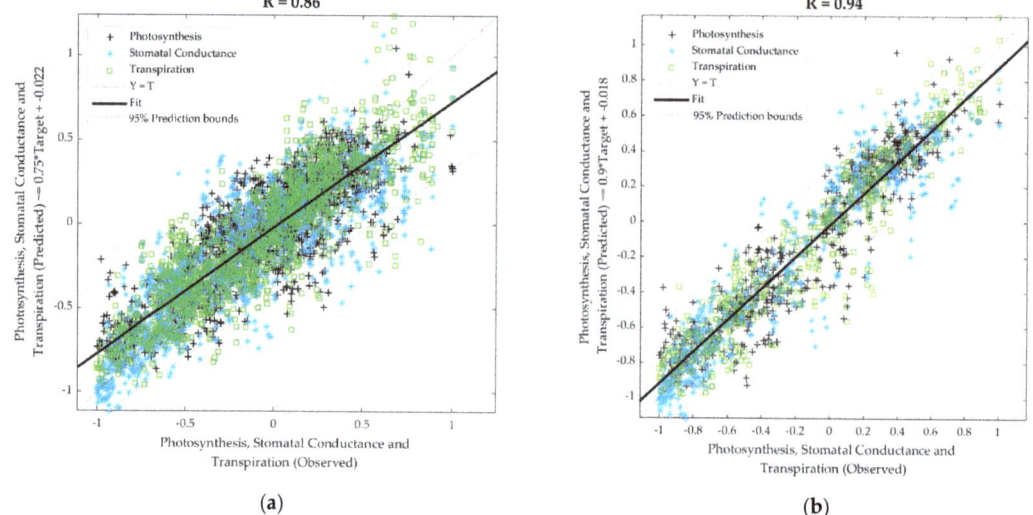

Figure 8. Overall regression models to predict physiological data using (**a**) the electronic nose outputs and infestation level as inputs for general data using all treatments at all measurement days and (**b**) using the electronic nose outputs as inputs with the baseline and control data (non-infested). Abbreviations: R: correlation coefficient; T: targets.

Table 3 shows the results from the pattern recognition models to classify samples into the different treatments (control, low, medium, and high) using the NIR absorbance values as inputs. It can be observed that Model 3 was constructed using data from the baseline and Day 3, and Model 6 was developed with data from Day 14; both had a very high overall accuracy of 97%, being the lowest in accuracy compared to the other days of measurement. Model 4 was developed using data from Day 7 presented a higher overall accuracy of 98%. On the other hand, Models 5 and 7 had the highest overall accuracy (99%), with Model 5 being the best as it was constructed using a lower number of neurons (Model 5: 7 neurons; Model 7: 10 neurons). None of the five models presented any signs of under- or overfitting, and the MSE values of training (MSE < 0.01 for all) were lower than the validation and testing, and the latter stages had similar MSE values.

Table 3. Machine learning pattern recognition models based on artificial neural networks (Levenberg–Marquardt) to classify samples into infestation treatment levels using the near-infrared absorbance values as inputs. Abbreviations: MSE: means squared error.

Stage	Samples	Accuracy	Error	Performance (MSE)
Model 3—Baseline + Day 3—10 neurons				
Training	404	100%	0.0%	<0.01
Validation	86	88.4%	11.6%	0.05
Testing	86	88.4%	11.6%	0.05
Overall	576	96.5%	3.5%	-
Model 4—Day 7—10 neurons				
Training	202	100%	0.0%	<0.01
Validation	43	95.3%	4.7%	0.02
Testing	43	93.0%	7.0%	0.02
Overall	288	98.3%	1.7%	-
Model 5—Day 10—7 neurons				
Training	202	100%	0.0%	<0.01
Validation	43	97.7%	2.3%	0.01
Testing	43	95.3%	4.7%	0.02
Overall	288	99.0%	1.0%	-
Model 6—Day 14—10 neurons				
Training	202	100%	0.0%	<0.01
Validation	43	90.7%	9.3%	0.05
Testing	43	86.0%	14.0%	0.04
Overall	288	96.5%	3.5%	-
Model 7—Day 17—10 neurons				
Training	202	100%	0.0%	<0.01
Validation	43	97.7%	2.3%	0.01
Testing	43	97.7%	2.3%	0.01
Overall	288	99.3%	0.7%	-

Accuracy results for Models 3–7 using SVM were lower than those from ANN Levenberg–Marquardt algorithm (Table 3). Results were within the following ranges: (i) Linear SVM (Models 3–7: 56–74%), (ii) Quadratic SVM (Models 3–7: 80–95%), (iii) Cubic SVM (Models 3–7: 90–92%, 88–98%), (iv) Fine Gaussian SVM (Models 3–7: 82–83%), (v) Medium Gaussian SVM (Models 3–7: 58–65%), and (vi) Coarse Gaussian SVM (Models 3–7: 41–45%). As can be seen, the model with the highest accuracy was with quadratic SVM (98%). However, this is lower than the ANN models, which presented the highest accuracy of 99.3%.

Table 4 shows the results from the pattern recognition models to classify samples into the different treatments (control, low, medium, and high) using the e-nose outputs as inputs. It can be observed that Models 8, 9, and 11 were developed using data from days 3, 7, and 14, respectively, and had very high overall accuracy (98%). Whilst Model 10 constructed with data from Day 10 presented the highest overall accuracy (99%). On the other hand, Model 12, developed using data from the last day of measurements (Day 17), presented high overall accuracy of 94%; however, it was the lowest compared to models from previous days. All of the models were constructed using three neurons, and none of them presented signs of under- or overfitting as the MSE values of the training stage (MSE < 0.01) were lower than the testing.

Table 4. Machine learning pattern recognition models based on artificial neural networks (Bayesian Regularization) to classify samples into infestation treatment levels using the electronic nose outputs as inputs. Abbreviations: MSE: means squared error.

Stage	Samples	Accuracy	Error	Performance (MSE)
Model 8—Baseline + Day 3—3 neurons				
Training	336	99.7%	0.3%	<0.01
Testing	144	95.1%	4.9%	0.02
Overall	480	98.3%	1.7%	-
Model 9—Day 7—3 neurons				
Training	168	100%	0.0%	<0.01
Testing	72	94.4%	5.6%	0.03
Overall	240	98.3%	1.7%	-
Model 10—Day 10—3 neurons				
Training	168	100%	0.0%	<0.01
Testing	72	97.2%	2.8%	0.01
Overall	240	99.2%	0.8%	-
Model 11—Day 14—3 neurons				
Training	168	98.8%	1.2%	<0.01
Testing	72	97.2%	2.8%	0.02
Overall	240	98.3%	1.7%	-
Model 12—Day 17—3 neurons				
Training	168	97.6%	2.4%	<0.01
Testing	72	86.1%	13.9%	0.06
Overall	240	94.2%	5.8%	-

Accuracy results for Models 8–12 using SVM were lower than those from ANN Bayesian Regularization algorithm (Table 4). Results were within the following ranges: (i) Linear SVM (Models 8–12: 75–85%), (ii) Quadratic SVM (Models 8–12: 84–96%), (iii) Cubic SVM (Models 8–12: 88–98%), (iv) Fine Gaussian SVM (Models 8–12: 89–93%), (v) Medium Gaussian SVM (Models 8–12: 85–94%), and (vi) Coarse Gaussian SVM (Models 8–12: 72–85%). As can be observed, the model with the highest accuracy was cubic SVM (98%). However, this was lower than the ANN models, which presented the highest accuracy of 99.2%.

Table 5 shows the results from regression models to predict the number of aphids using data from Days 7 to 17. It can be observed that Model 13, developed using NIR absorbance values as inputs, had a very high overall correlation coefficient (R = 0.97). However, Model 14, constructed with the e-nose outputs as inputs, presented higher accuracy (R = 0.99). Both models had very high overall slope values (b = 0.97), and none showed any signs of under- or overfitting based on the performance values. From the overall models, Model 13 (Figure 9a) had 4.98% of outliers (43 out of 864) based on the 95% prediction bounds with the highest number of outliers due to the low infestation treatment. Similarly, Model 14 (Figure 9b) presented 5% of outliers (36 out of 720); however, for this model, the highest number of outliers was due to the medium infestation treatment. The difference in the number of aphids (target) between both models relies on the samples/measurements as NIR was measured on each plant, while e-nose was measured per tent.

Table 5. Machine learning regression models based on artificial neural networks (Bayesian Regularization) to predict the number of aphids' data using the near-infrared absorbance values (Model 13) and electronic nose outputs (Model 14) from Days 7 to 17 as inputs. Abbreviations: R: correlation coefficient; b: slope; MSE: means squared error.

Stage	Samples	Observations	R	Slope	Performance (MSE)
Model 13—NIR Day 7–Day 17—10 neurons					
Training	605	605	0.99	0.97	555
Testing	259	259	0.94	0.98	3078
Overall	864	864	0.97	0.97	-
Model 14—E-Nose Day 7–Day 17—10 neurons					
Training	504	504	0.99	0.98	20,014
Testing	216	216	0.98	0.94	40,125
Overall	720	720	0.99	0.97	-

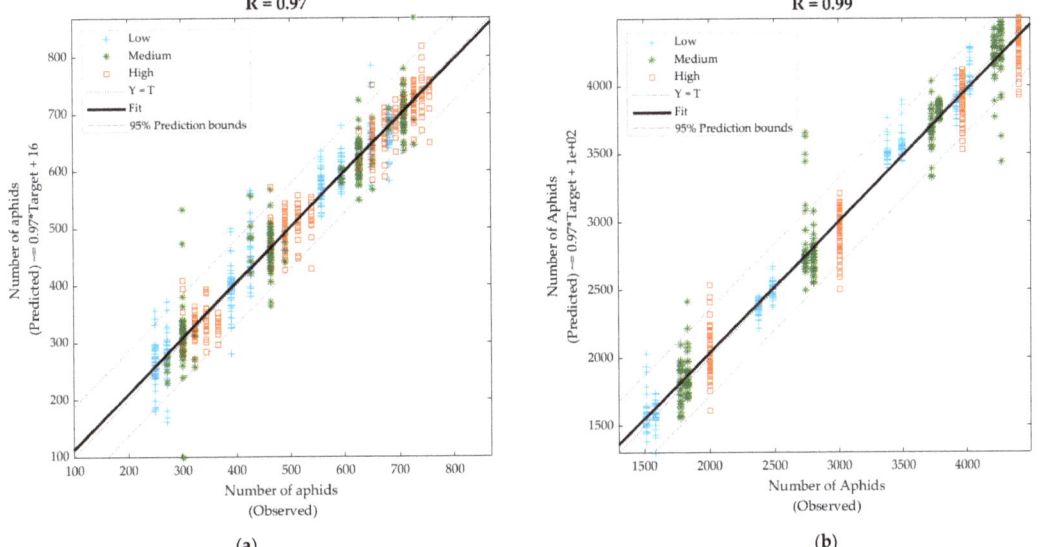

Figure 9. Overall regression models to predict the number of aphids using (**a**) the near-infrared absorbance values and (**b**) the electronic nose outputs as inputs with data from Days 7–17. Abbreviations: R: correlation coefficient; T: targets.

Correlation coefficients for Models 13 and 14 using SVM were lower than those from ANN Bayesian Regularization algorithm (Table 5). Results from regression SVM were the following: (i) Linear SVM (Model 13: R = 0.68; Model 14: R = 0.80), (ii) Quadratic SVM (Model 13: R = 0.85; Model 14: R = 0.91), (iii) Cubic SVM (Model 13: R = 0.95; Model 14: R = 0.89), (iv) Fine Gaussian SVM (Model 13: R = 0.80; Model 14: R = 0.97), (v) Medium Gaussian SVM (Model 13: R = 0.73; Model 14: R = 0.92), and (vi) Coarse Gaussian SVM (Model 13: R = 0.60; Model 14: R = 0.70). It can be observed that for the model developed using NIR inputs, i.e., Model 13, the highest accuracy was with cubic SVM (R = 0.95), while for the model developed using e-nose inputs, i.e., Model 14, the highest accuracy was obtained with medium Gaussian SVM (R = 0.97). However, these were presented with lower accuracies than the ANN models which presented R values of 0.97 (Model 13) and 0.99 (Model 14).

4. Discussion

4.1. Physiological Response of Plants to Insect Infestation

Having no statistical differences in physiological data for the baseline with no insects for all plants (Table 1) helped ascertain that those initial conditions were similar for all the plants considered in the study, and no other stresses were present. Differences in physiological data after the introduction of insects in different treatments followed a variable pattern with not much difference for the photosynthetic rate (A), which is expected since plants compensate by either maintaining or increasing in some conditions A due to abiotic [40,41] or biotic stresses such as aphid attack [42].

In the case of stomatal conductance (gs) and transpiration, there were decreasing values according to the level of insect infestation, which is in accordance with previous studies, which have shown that gs is the most sensitive parameter to other stresses, such as water stress [43,44], pathogen-based stress [45], and water stress–aphid interactions in wheat [46].

4.2. Chemical Fingerprinting and Volatile Compounds' Response to Insect Infestation

The NIR measurements offer a chemical fingerprint of the different leaf samples monitored, including the baseline measurements (Figure 5a) and treatments (Figure 5b) for the different days of the experimental trial. The main variations observable are in the overtones corresponding to hydrogen peroxide (H_2O_2) in the range of 1596 and 1650 nm [47,48] with similar absorbance levels for all treatments, which may explain the lower effect on photosynthesis reductions. The overtones for water content (status) are shown in the major peak within 1900–2000 nm (1940 nm) [49]. Furthermore, overtones of aromatic compounds can be found in the range of the NIR instrument sensitivity, at 1660 nm, 1672 nm, and 1685 nm [50,51]. Compounds with amide functional groups are at 1920 nm, 1960–1980 nm, 2000–2050 nm, and 2110–2160 nm [51,52]. Overtones of urea, which is an important amide compound, are found at 1990 nm, 2030 nm, and 2070 nm [51]; this was expected to be found in the samples as it is a nitrogen component contained in fertilizers added to the hydroponic solution, which is translocated through the plants.

In the case of e-nose (Figure 6), the baseline data were similar for plants and all tents measured. However, some differences between tents were statistically significant, contrary to the physiological data measured by gas exchange (Table 1). This can be explained by the sensitivity and responsiveness of e-nose sensors (every 0.5 s), which depend on small eddies formed in the growth chamber. Some sensors were more stable than others, such as MQ4 for Day 3, corresponding to methane sensitivity. The differences in sensor readings for subsequent days are expected, and it is assumed that their patterns are related to the interaction between aphids and plants and the increased number of insects in time and plant growth/decline, even small changes in the MG811 (CO_2), which corresponds to photosynthetic activity.

When analyzing the correlations between physiological parameters and the sensors that compose the e-nose (Figure 7), it can be seen that, as expected, there is a positive and direct correlation between photosynthesis, stomatal conductance, and transpiration. On the contrary, there is an inverse correlation between physiological parameters and the number of insects, which corresponds to the decline of the plants or response to insect activity. Alcohol has been documented to be produced in plants as an allelopathic response to insect infestation. However, salivary proteins from aphids are able to stop this process when feeding on the plant sap [53,54]. The latter effect may explain the inverse correlation between the number of insects and the MQ3 sensor response signal. Furthermore, methane (MQ4) signal response increase may be due to activity of insects and anaerobic digestion [55], which explains the inverse correlations with physiological parameters and positive correlations with the number of insects. The carbon monoxide sensor (MQ7) had a similar response as the alcohol sensor (MQ3), shown by the high correlation coefficient (r = 0.89) and MQ8 (hydrogen). Most of the other sensors (MQ135, MQ136, MQ137, and MQ138) had an inverse correlation with the number of insects. The inverse correlation between the ammonia sensors (MQ135, MQ137, and MQ138) may be due to the capacity of aphids to assimilate ammonia into amino acids [56]. Finally, the levels of CO_2 (MG811) were not significantly affected by the interaction between insects and plants. The correlations among the different sensors from the e-nose have been reported in previous research [31], which explains in detail the e-nose used in this study.

4.3. Machine Learning Models Developed

The plant physiological machine learning model developed from e-nose data as inputs and LiCOR data as targets for all plants and treatments (Figure 8a), and only control plants (Figure 8b) showed high correlation coefficients and no signs of overfitting. As far as authors' knowledge, this is the first time these models are presented, which use a low-cost e-nose compared to an established gas exchange method for plant physiological measurements. The LiCOR instrument has been used as a validation method for several remote sensing techniques for other crops [57–60]. The accuracy of the models obtained may not be surprising since both systems, LiCOR and the e-nose, measure gas exchange in different ways. Furthermore, these models are supported by the correlations between different sensors and physiological parameters (Figure 7). The lower correlation found in the ML model, including all plants (R = 0.86), may be explained by the higher variability of the data due to the interaction between plant and insect. Both models may be used to assess the level of the effect of plant-insect interaction on physiological parameters and for further applications to assess plant water stress [61], irrigation scheduling, and the physiological effect of other biotic or abiotic stresses, such as salinity, other insects, plant diseases, and environmental stress such as heatwaves, cold temperatures, and smoke contamination due to bushfires [62].

The accuracy of classification ML models based on NIR and e-nose data as inputs and level of insect infestation as targets was high and similar with over 94% accuracy for all models and dates, with slightly higher accuracies for ML models based on e-nose (Tables 3 and 4, respectively). Within the most important are Models 3 and 8, respectively, since they can be considered for early detection only after three days of insect introduction to the plants' environments and the corresponding treatments in a critical and vulnerable wheat phenological stage. In these models specifically, the baseline data from all plants were used as control, which explained the higher number of samples (576 and 480, respectively). Even though there was unbalanced data for the treatments as classifiers, the models were able to recognize non-infected plants with 96.5% and 98.3%, respectively. All further models can be used either to monitor insect activity or to verify the effectiveness of control methods using either chemical pesticides [63], organic pesticides [64,65], and natural predators through integrated pest management (IPM) [66,67].

4.4. Deployment Method for ML Models Developed Proposed Using UAV

One of the main advantages of creating AI models for the early detection of pests using growth chambers is that data can be obtained in control conditions. Hence the ML models developed do not include stresses related to other biotic or abiotic factors. The similarity of models developed using NIR and e-nose validate the effectiveness of the low-cost instrumentation proposed by comparing them with more established instruments, such as NIR spectroscopy; other studies have used, as a validation point, gas chromatography [20,22].

One advantage of the NIR models, especially for insect number detection, is that they are based on the different patterns of chemical fingerprinting resulting from the plant-insect interactions. Hence, this instrument can be used as a validation method to deploy the ML models developed in this study to field conditions. NIR measurements in plant leaves take just seconds and can be made on a grid of 10 × 10 m in a wheat field instead of visual insect counting, which is extremely difficult and time-consuming [68,69]. The latter can also be assessed using mathematical modeling strategies based on population models [36,70] or through smartphone devices and machine vision [71], image analysis and machine learning [72], and deep learning [73]. However, the e-nose model with R = 0.99 was more adequate, accurate, practical, and is a low cost method. Even though ANN models were selected as best compared to SVM, the authors also have the latter models available for deployment depending on future usage needs.

The deployment method for the e-nose proposed is as a payload for a UAV (Figure 10); the advantage of the e-nose is that it weighs only 200 g, and power can be accessed via the UAV. To assess the sensitivity and efficacy of the models, it is proposed to start flights at 5, 15, 20, and 50 m from the crop's surface to test the ML models.

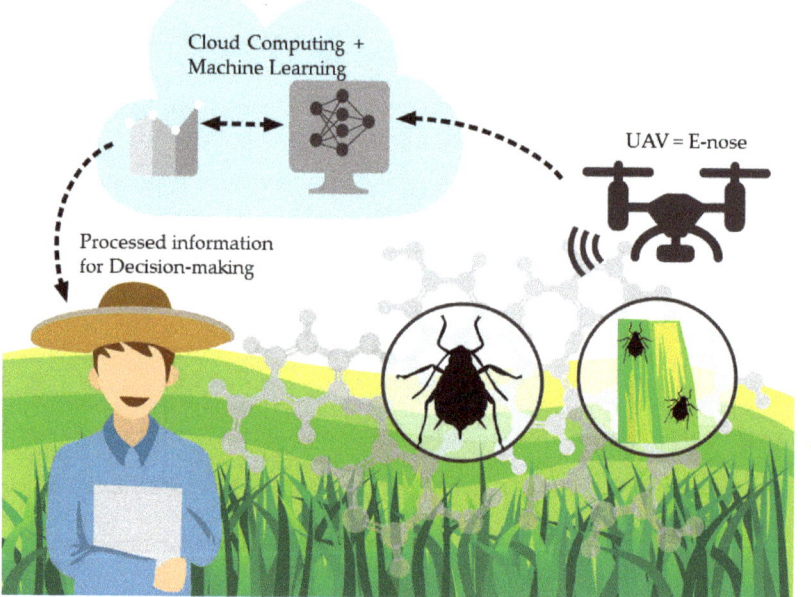

Figure 10. Diagram showing the proposed validation and deployment of machine learning models developed for early detection of aphids in wheat fields using an unmanned aerial vehicle and the e-nose as payload.

5. Conclusions

The low cost and accuracy of the models presented in this study could make the early detection of insect infestation in crop fields feasible using the UAV system proposed. The data and models used in this study can be used as a base for deployments in wheat fields and validation points considering other insects of interest. Further models developed following the phenological stages of plants can be used as testing systems for agronomical management practices for insect control, such as chemical and organic product applications, the introduction of natural predators, and integrated pest management tools. Furthermore, plant physiology models based on the low-cost e-nose opens the use of models to assess other biotic and abiotic stress effects on plants for further management practices such as fertilization, irrigation scheduling, and the general effect of climate change and climatic anomalies, such as heatwaves, frosts, and smoke contamination due to bushfires.

Author Contributions: Conceptualization, S.F., E.T. and C.G.V.; Data curation, C.G.V.; Formal analysis, C.G.V., S.F.; Funding acquisition, S.F.; Investigation, S.F., E.T. and C.G.V.; Methodology, S.F., E.T. and C.G.V.; Project administration, S.F. and C.G.V.; Resources, S.F. and R.R.U.; Software, S.F., R.R.U. and C.G.V.; Validation, S.F. and C.G.V.; Visualization, S.F., E.T. and C.G.V.; Writing—original draft, S.F., E.T. and C.G.V.; Writing—review and editing, S.F., E.T., R.R.U. and C.G.V. All authors have read and agreed to the published version of the manuscript.

Funding: This research was supported by seed funding from (i) the Australian Grain Pest Innovation Program, an investment from the Grains Research and Development Corporation (GRDC), Australia (ID: 2062311) and (ii) Bayer Grants4Ag sustainability-focused program (ID: 106027).

Institutional Review Board Statement: Not applicable.

Informed Consent Statement: Not applicable.

Data Availability Statement: Data and intellectual property belong to The University of Melbourne; any sharing needs to be evaluated and approved by the University.

Acknowledgments: This research was supported by the Australian Research Council's Linkage Projects funding scheme (LP160101475). The authors would like to acknowledge Bryce Widdicombe, Mimi Sun, and Jorge Gonzalez from the School of Engineering, Department of Electrical and Electronic Engineering of The University of Melbourne for their collaboration in electronic nose development.

Conflicts of Interest: The authors declare no conflict of interest.

References

1. Wollmann, J.; Schlesener, D.C.H.; Ferreira, M.S.; Krüger, A.; Bernardi, D.; Garcia, J.A.B.; Nunes, A.M.; Garcia, M.S.; Garcia, F. Population Dynamics of Drosophila suzukii (Diptera: Drosophilidae) in Berry Crops in Southern Brazil. *Neotrop. Èntomol.* **2019**, *48*, 699–705. [CrossRef] [PubMed]
2. Satterfield, D.A.; Sillett, T.S.; Chapman, J.W.; Altizer, S.; Marra, P.P. Seasonal insect migrations: Massive, influential, and overlooked. *Front. Ecol. Environ.* **2020**, *18*, 335–344. [CrossRef]
3. Saha, T.; Chandran, N. Chemical ecology and pest management: A review. *Int. J. Cardiovasc. Sci.* **2017**, *5*, 618–621.
4. Kim, K.; Huang, Q.; Lei, C. Advances in insect phototaxis and application to pest management: A review. *Pest Manag. Sci.* **2019**, *75*, 3135–3143. [CrossRef] [PubMed]
5. Preti, M.; Verheggen, F.; Angeli, S. Insect pest monitoring with camera-equipped traps: Strengths and limitations. *J. Pest Sci.* **2020**, *94*, 203–217. [CrossRef]
6. Milosavljević, I.; Hoddle, C.D.; Mafra-Neto, A.; Gómez-Marco, F.; Hoddle, M.S. Use of Digital Video Cameras to Determine the Efficacy of Two Trap Types for Capturing Rhynchophorus palmarum (Coleoptera: Curculionidae). *J. Econ. Èntomol.* **2020**, *113*, 3028–3031. [CrossRef] [PubMed]
7. Remboski, T.B.; de Souza, W.D.; de Aguiar, M.S.; Ferreira, P.R., Jr. Identification of Fruit Fly in Intelligent Traps Using Techniques of Digital Image Processing and Machine Learning. In Proceedings of the 33rd Annual ACM Symposium on Applied Computing, Pau, France, 9–13 April 2018; pp. 260–267.
8. Chulu, F.; Phiri, J.; Nyirenda, M.; Kabemba, M.M.; Nkunika, P.; Chiwamba, S. Developing an automatic identification and early warning and monitoring web based system of fall army worm based on machine learning in developing countries. *Zamb. ICT J.* **2019**, *3*, 13–20. [CrossRef]
9. Barbedo, J.G.A. Detecting and Classifying Pests in Crops Using Proximal Images and Machine Learning: A Review. *AI* **2020**, *1*, 312–328. [CrossRef]
10. Marković, D.; Vujičić, D.; Tanasković, S.; Đorđević, B.; Ranđić, S.; Stamenković, Z. Prediction of Pest Insect Appearance Using Sensors and Machine Learning. *Sensors* **2021**, *21*, 4846. [CrossRef]
11. Kasinathan, T.; Uyyala, S.R. Machine learning ensemble with image processing for pest identification and classification in field crops. *Neural Comput. Appl.* **2021**, 1–14. [CrossRef]
12. Lima, M.C.F.; Leandro, M.E.D.D.A.; Valero, C.; Coronel, L.C.P.; Bazzo, C.O.G. Automatic Detection and Monitoring of Insect Pests—A Review. *Agriculture* **2020**, *10*, 161. [CrossRef]
13. Zhang, J.; Huang, Y.; Pu, R.; González-Moreno, P.; Yuan, L.; Wu, K.; Huang, W. Monitoring plant diseases and pests through remote sensing technology: A review. *Comput. Electron. Agric.* **2019**, *165*, 104943. [CrossRef]
14. Velásquez, D.; Sánchez, A.; Sarmiento, S.; Toro, M.; Maiza, M.; Sierra, B. A Method for Detecting Coffee Leaf Rust through Wireless Sensor Networks, Remote Sensing, and Deep Learning: Case Study of the Caturra Variety in Colombia. *Appl. Sci.* **2020**, *10*, 697. [CrossRef]
15. Brunelli, D.; Albanese, A.; D'Acunto, D.; Nardello, M. Energy Neutral Machine Learning Based IoT Device for Pest Detection in Precision Agriculture. *IEEE Internet Things Mag.* **2019**, *2*, 10–13. [CrossRef]
16. Poblete, T.; Camino, C.; Beck, P.; Hornero, A.; Kattenborn, T.; Saponari, M.; Boscia, D.; Navas-Cortes, J.; Zarco-Tejada, P. Detection of Xylella fastidiosa infection symptoms with airborne multispectral and thermal imagery: Assessing bandset reduction performance from hyperspectral analysis. *ISPRS J. Photogramm. Remote. Sens.* **2020**, *162*, 27–40. [CrossRef]
17. Hornero, A.; Hernández-Clemente, R.; North, P.; Beck, P.; Boscia, D.; Navas-Cortes, J.; Zarco-Tejada, P. Monitoring the incidence of Xylella fastidiosa infection in olive orchards using ground-based evaluations, airborne imaging spectroscopy and Sentinel-2 time series through 3-D radiative transfer modelling. *Remote. Sens. Environ.* **2019**, *236*, 111480. [CrossRef]

18. Aasen, H.; Honkavaara, E.; Lucieer, A.; Zarco-Tejada, P.J. Quantitative Remote Sensing at Ultra-High Resolution with UAV Spectroscopy: A Review of Sensor Technology, Measurement Procedures, and Data Correction Workflows. *Remote. Sens.* **2018**, *10*, 1091. [CrossRef]
19. Cellini, A.; Blasioli, S.; Biondi, E.; Bertaccini, A.; Braschi, I.; Spinelli, F. Potential Applications and Limitations of Electronic Nose Devices for Plant Disease Diagnosis. *Sensors* **2017**, *17*, 2596. [CrossRef]
20. Cui, S.; Ling, P.; Zhu, H.; Keener, H.M. Plant Pest Detection Using an Artificial Nose System: A Review. *Sensors* **2018**, *18*, 378. [CrossRef]
21. Lampson, B.D.; Khalilian, A.; Greene, J.K.; Han, Y.J.; Degenhardt, D.C. Development of a Portable Electronic Nose for Detection of Cotton Damaged by Nezara viridula (Hemiptera: Pentatomidae). *J. Insects* **2014**, *2014*, 1–8. [CrossRef] [PubMed]
22. Cui, S.; Inocente, E.A.A.; Acosta, N.; Keener, H.M.; Zhu, H.; Ling, P.P. Development of Fast E-nose System for Early-Stage Diagnosis of Aphid-Stressed Tomato Plants. *Sensors* **2019**, *19*, 3480. [CrossRef]
23. Ridgway, C.; Chambers, J.; Portero-Larragueta, E.; Prosser, O. Detection of mite infestation in wheat by electronic nose with transient flow sampling. *J. Sci. Food Agric.* **1999**, *79*, 2067–2074. [CrossRef]
24. Zhang, H.; Wang, J. Detection of age and insect damage incurred by wheat, with an electronic nose. *J. Stored Prod. Res.* **2007**, *43*, 489–495. [CrossRef]
25. Wu, J.; Jayas, D.; Zhang, Q.; White, N.; York, R. Feasibility of the application of electronic nose technology to detect insect infestation in wheat. *Can. Biosyst. Eng.* **2013**, *55*, 3.1–3.9. [CrossRef]
26. Ahouandjinou, S.A.R.M.; Kiki, M.P.A.F.; Badoussi, P.E.N.A.; Assogba, K.M. A Multi-level Smart Monitoring System by Combining an E-Nose and Image Processing for Early Detection of FAW Pest in Agriculture. In Proceedings of the Innovations and Interdisciplinary Solutions for Underserved Areas: 4th EAI International Conference, InterSol 2020, Nairobi, Kenya, 8–9 March 2020; pp. 20–32. [CrossRef]
27. Cheng, L.; Meng, Q.-H.; Lilienthal, A.J.; Qi, P.-F. Development of compact electronic noses: A review. *Meas. Sci. Technol.* **2021**, *32*, 062002. [CrossRef]
28. Poland, T.M.; Rassati, D. Improved biosecurity surveillance of non-native forest insects: A review of current methods. *J. Pest Sci.* **2018**, *92*, 37–49. [CrossRef]
29. Lan, Y.-B.; Zheng, X.-Z.; Westbrook, J.K.; López, J.; Lacey, R.; Hoffmann, W.C. Identification of Stink Bugs Using an Electronic Nose. *J. Bionic Eng.* **2008**, *5*, 172–180. [CrossRef]
30. Zhou, B.; Wang, J. Use of electronic nose technology for identifying rice infestation by Nilaparvata lugens. *Sens. Actuators B Chem.* **2011**, *160*, 15–21. [CrossRef]
31. Viejo, C.G.; Fuentes, S.; Godbole, A.; Widdicombe, B.; Unnithan, R.R. Development of a low-cost e-nose to assess aroma profiles: An artificial intelligence application to assess beer quality. *Sens. Actuators B Chem.* **2020**, *308*, 127688. [CrossRef]
32. Kratky, B. A suspended pot, non-circulating hydroponic method. *Acta Hortic.* **2004**, *648*, 83–89. [CrossRef]
33. Shavrukov, Y.; Genc, Y.; Hayes, J. *The Use of Hydroponics in Abiotic Stress Tolerance Research*; InTech Rijeka: Rijeka, Croatia, 2012.
34. McDonald, G.; Umina, P.; Hangartner, S. Oat Aphid Rhophalosiphum padi. Available online: https://cesaraustralia.com/pestnotes/aphids/oat-aphid/ (accessed on 25 November 2020).
35. Jarošík, V.; Honěk, A.; Tichopad, A. Comparison of Field Population Growths of Three Cereal Aphid Species on Winter Wheat. *Plant Prot. Sci.* **2011**, *39*, 61–64. [CrossRef]
36. Costamagna, A.C.; Van Der Werf, W.; Bianchi, F.J.J.A.; Landis, D.A. An exponential growth model with decreasing r captures bottom-up effects on the population growth of Aphis glycines Matsumura (Hemiptera: Aphididae). *Agric. For. Èntomol.* **2007**, *9*, 297–305. [CrossRef]
37. Viejo, C.G.; Tongson, E.; Fuentes, S. Integrating a Low-Cost Electronic Nose and Machine Learning Modelling to Assess Coffee Aroma Profile and Intensity. *Sensors* **2021**, *21*, 2016. [CrossRef] [PubMed]
38. Viejo, C.G.; Torrico, D.D.; Dunshea, F.R.; Fuentes, S. Emerging Technologies Based on Artificial Intelligence to Assess the Quality and Consumer Preference of Beverages. *Beverages* **2019**, *5*, 62. [CrossRef]
39. Viejo, C.G.; Torrico, D.D.; Dunshea, F.R.; Fuentes, S. Development of Artificial Neural Network Models to Assess Beer Acceptability Based on Sensory Properties Using a Robotic Pourer: A Comparative Model Approach to Achieve an Artificial Intelligence System. *Beverages* **2019**, *5*, 33. [CrossRef]
40. Sehar, Z.; Jahan, B.; Masood, A.; Anjum, N.A.; Khan, N.A. Hydrogen peroxide potentiates defense system in presence of sulfur to protect chloroplast damage and photosynthesis of wheat under drought stress. *Physiol. Plant.* **2020**, *172*, 922–934. [CrossRef]
41. Zhao, W.; Liu, L.; Shen, Q.; Yang, J.; Han, X.; Tian, F.; Wu, J. Effects of Water Stress on Photosynthesis, Yield, and Water Use Efficiency in Winter Wheat. *Water* **2020**, *12*, 2127. [CrossRef]
42. Shahzad, M.W.; Ghani, H.; Ayyub, M.; Ali, Q.; Ahmad, H.M.; Faisal, M.; Ali, A.; Qasim, M.U. Performance of some Wheat Cultivars against APHID and Its Damage on Yield and Photosynthesis. *J. Glob. Innov. Agric. Soc. Sci.* **2019**, 105–109. [CrossRef]
43. Li, Y.; Li, H.; Li, Y.; Zhang, S. Improving water-use efficiency by decreasing stomatal conductance and transpiration rate to maintain higher ear photosynthetic rate in drought-resistant wheat. *Crop. J.* **2017**, *5*, 231–239. [CrossRef]
44. Banerjee, K.; Krishnan, P.; Das, B. Thermal imaging and multivariate techniques for characterizing and screening wheat genotypes under water stress condition. *Ecol. Indic.* **2020**, *119*, 106829. [CrossRef]
45. Francesconi, S.; Balestra, G.M. The modulation of stomatal conductance and photosynthetic parameters is involved in Fusarium head blight resistance in wheat. *PLoS ONE* **2020**, *15*, e0235482. [CrossRef]

46. Ahmed, S.S.; Liu, D.; Simon, J.-C. Impact of water-deficit stress on tritrophic interactions in a wheat-aphid-parasitoid system. *PLoS ONE* **2017**, *12*, e0186599. [CrossRef]
47. Pimenta, A.M.; Scafi, S.H.F.; Pasquini, C.; Raimundo Jr, I.M.; Rohwedder, J.J.; Montenegro, M.d.C.B.; Araújo, A.N. Determination of hydrogen peroxide by near infrared spectroscopy. *J. Infrared Spectrosc.* **2003**, *11*, 49–53. [CrossRef]
48. Fuentes, S.; Tongson, E.; Chen, J.; Viejo, C.G. A Digital Approach to Evaluate the Effect of Berry Cell Death on Pinot Noir Wines' Quality Traits and Sensory Profiles Using Non-Destructive Near-Infrared Spectroscopy. *Beverages* **2020**, *6*, 39. [CrossRef]
49. De Bei, R.; Fuentes, S.; Wirthensohn, M.; Cozzolino, D.; Tyerman, S. Feasibility study on the use of Near Infrared spectroscopy to measure water status of almond trees. *Acta Hortic.* **2018**, 79–84. [CrossRef]
50. Terhoeven-Urselmans, T.; Bruns, C.; Schmilewski, G.; Ludwig, B. Quality assessment of growing media with near-infrared spectroscopy: Chemical characteristics and plant assays. *Eur. J. Hortic. Sci.* **2008**, *73*, 28.
51. Burns, D.A.; Ciurczak, E.W. *Handbook of Near-Infrared Analysis*; CRC Press: Boca Raton, FL, USA, 2007.
52. Osborne, B.G.; Fearn, T.; Hindle, P.H. *Practical NIR Spectroscopy with Applications in Food and Beverage Analysis*; Longman scientific and technical: London, UK, 1993.
53. Cooper, W.R.; Dillwith, J.W.; Puterka, G.J. Salivary Proteins of Russian Wheat Aphid (Hemiptera: Aphididae). *Environ. Entomol.* **2010**, *39*, 223–231. [CrossRef] [PubMed]
54. Elzinga, D.A.; De Vos, M.; Jander, G. Suppression of Plant Defenses by a Myzus persicae (Green Peach Aphid) Salivary Effector Protein. *Mol. Plant-Microbe Interact.* **2014**, *27*, 747–756. [CrossRef] [PubMed]
55. Bulak, P.; Proc, K.; Pawłowska, M.; Kasprzycka, A.; Berus, W.; Bieganowski, A. Biogas generation from insects breeding post production wastes. *J. Clean. Prod.* **2019**, *244*, 118777. [CrossRef]
56. Hansen, A.; Moran, N.A. Aphid genome expression reveals host-symbiont cooperation in the production of amino acids. *Proc. Natl. Acad. Sci. USA* **2011**, *108*, 2849–2854. [CrossRef]
57. Chou, S.; Chen, J.M.; Yu, H.; Chen, B.; Zhang, X.; Croft, H.; Khalid, S.; Li, M.; Shi, Q. Canopy-Level Photochemical Reflectance Index from Hyperspectral Remote Sensing and Leaf-Level Non-Photochemical Quenching as Early Indicators of Water Stress in Maize. *Remote. Sens.* **2017**, *9*, 794. [CrossRef]
58. Paz, V.S.; Mikkelsen, T.N.; Johnson, M.; Mo, X.; Morillas, L.; Liu, S.; Shen, L.; Garcia, M. Hyperspectral and thermal sensing of stomatal conductance and photosynthesis under water stress for a C3 (soybean) and a C4 (maize) crop. In Proceedings of the EGU General Assembly Conference Abstracts, Vienna, Austria, 7–12 April 2019.
59. Rossi, L.; Bagheri, M.; Zhang, W.; Chen, Z.; Burken, J.G.; Ma, X. Using artificial neural network to investigate physiological changes and cerium oxide nanoparticles and cadmium uptake by Brassica napus plants. *Environ. Pollut.* **2018**, *246*, 381–389. [CrossRef] [PubMed]
60. Park, S.; Ryu, D.; Fuentes, S.; Chung, H.; Hernández-Montes, E.; O'Connell, M. Adaptive Estimation of Crop Water Stress in Nectarine and Peach Orchards Using High-Resolution Imagery from an Unmanned Aerial Vehicle (UAV). *Remote. Sens.* **2017**, *9*, 828. [CrossRef]
61. Fuentes, S.; De Bei, R.; Pech, J.; Tyerman, S. Computational water stress indices obtained from thermal image analysis of grapevine canopies. *Irrig. Sci.* **2012**, *30*, 523–536. [CrossRef]
62. Fuentes, S.; Tongson, E.J.; De Bei, R.; Viejo, C.G.; Ristic, R.; Tyerman, S.; Wilkinson, K. Non-Invasive Tools to Detect Smoke Contamination in Grapevine Canopies, Berries and Wine: A Remote Sensing and Machine Learning Modeling Approach. *Sensors* **2019**, *19*, 3335. [CrossRef]
63. Yuan, W.-L.; Xu, B.; Ran, G.-C.; Chen, H.-P.; Zhao, P.-Y.; Huang, Q.-L. Application of imidacloprid controlled-release granules to enhance the utilization rate and control wheat aphid on winter wheat. *J. Integr. Agric.* **2020**, *19*, 3045–3053. [CrossRef]
64. Zhang, H.; Garratt, M.P.; Bailey, A.; Potts, S.G.; Breeze, T. Economic valuation of natural pest control of the summer grain aphid in wheat in South East England. *Ecosyst. Serv.* **2018**, *30*, 149–157. [CrossRef]
65. Liu, Y.; Liu, J.; Zhou, H.; Chen, J. Enhancement of Natural Control Function for Aphids by Intercropping and Infochemical Releasers in Wheat Ecosystem. *Integr. Biol. Control* **2020**, 85–116. [CrossRef]
66. Pearis, F. *Development of Integrated Pest Management Approaches for Russian Wheat Aphid in Colorado USA*; GRDC Update Paper: Barton, ACT, Australia, 2017.
67. Singh, B.; Jasrotia, P. Impact of integrated pest management (IPM) module on major insect-pests of wheat and their natural enemies in North-western plains of India. *J. Cereal Res.* **2020**, *12*. [CrossRef] [PubMed]
68. Dixon, A.; Kindlmann, P. Population dynamics of aphids. In *Insect Populations in Theory and in Practice*; Springer: Berlin/Heidelberg, Germany, 1998; pp. 207–230.
69. Ahmad, T.; Hassan, M.W.; Jamil, M.; Iqbal, J. Population Dynamics of Aphids (Hemiptera: Aphididae) on Wheat Varieties (Triticum aestivum L.) as Affected by Abiotic Conditions in Bahawalpur, Pakistan. *Pak. J. Zool.* **2016**, *48*, 1039–1044.
70. Brabec, M.; Honěk, A.; Pekár, S.; Martinkova, Z. Population Dynamics of Aphids on Cereals: Digging in the Time-Series Data to Reveal Population Regulation Caused by Temperature. *PLoS ONE* **2014**, *9*, e106228. [CrossRef] [PubMed]
71. Xuesong, S.; Zi, L.; Lei, S.; Jiao, W.; Yang, Z. Aphid Identification and Counting Based on Smartphone and Machine Vision. *J. Sens.* **2017**, *2017*, 1–7. [CrossRef]

72. Lins, E.A.; Rodriguez, J.P.M.; Scoloski, S.I.; Pivato, J.; Lima, M.B.; Fernandes, J.M.C.; Pereira, P.R.V.D.S.; Lau, D.; Rieder, R. A method for counting and classifying aphids using computer vision. *Comput. Electron. Agric.* **2020**, *169*, 105200. [CrossRef]
73. Chen, P.; Li, W.; Yao, S.; Ma, C.; Zhang, J.; Wang, B.; Zheng, C.; Xie, C.; Liang, D. Recognition and counting of wheat mites in wheat fields by a three-step deep learning method. *Neurocomputing* **2021**, *437*, 21–30. [CrossRef]

MDPI
St. Alban-Anlage 66
4052 Basel
Switzerland
Tel. +41 61 683 77 34
Fax +41 61 302 89 18
www.mdpi.com

Sensors Editorial Office
E-mail: sensors@mdpi.com
www.mdpi.com/journal/sensors

www.ingramcontent.com/pod-product-compliance
Lightning Source LLC
LaVergne TN
LVHW070729100526
838202LV00013B/1201

9 7 8 3 0 3 6 5 2 9 0 4 2